Blueberry Culture

This volume is one of a number, publication of which is scheduled to celebrate the two hundredth anniversary of the Rutgers charter. The publishers believe it typifies the academic virtues of scholarship, historical continuity, and the development of fresh and stimulating insights into the human experience and meaning.

BLUEBERRY CULTURE

by
PAUL ECK, *Editor*
Rutgers University, New Brunswick, New Jersey

and

NORMAN F. CHILDERS, *Co-editor*
Rutgers University, New Brunswick, New Jersey

with

DENNIS A. ABDALLA
JOHN S. BAILEY
WALTER E. BALLINGER
W. THOMAS BRIGHTWELL
GEORGE M. DARROW
WALTER J. KENDER
CHARLES M. MAINLAND

PHILIP E. MARUCCI
JAMES N. MOORE
FREDERICK A. PERKINS
DONALD H. SCOTT
VLADIMIR G. SHUTAK
WARREN C. STILES
ALLAN W. STRETCH

EUGENE H. VARNEY

Rutgers University Press
New Brunswick   London

Copyright © 1966 by Rutgers, The State University
All rights reserved
Manufactured in the United States of America

Fourth printing, 1989

Library of Congress Cataloging—in—Publication Data

Main entry under title:
Blueberry Culture
Bibliography: p
Index
1. Blueberries. I. Eck, Paul. II. Title. III. Childers, Norman F.
LOC: 66–18880
ISBN 0–8135–0535–6

British Cataloging—in—Publication information available

To
GRADUATE STUDENTS, TEACHERS, RESEARCH WORKERS,
AND GROWERS
Who Are Contributing to Our Knowledge
in Blueberry Science

Preface

Blueberries have always been prized fruit in North America, but until the introduction of fine hybrid varieties and improvements in processing and shipping, only those who lived near stands of the wild bushes were able to enjoy the fresh fruit. Today blueberries are grown in great quantities for the commercial market, and one variety or another is available throughout the summer season in most parts of the United States. Growing blueberries is already a 20 million dollar industry, and large increases in acreage and production can be expected in the near future.

Many people are responsible for the phenomenal development of the blueberry industry, but the credit goes first to Dr. F. V. Coville and other horticultural scientists in the United States Department of Agriculture and federal and state agricultural experiment stations who have collaborated over half a century to develop plants far superior to the wild ones and to determine the horticultural practices that will produce the finest fruit. In this task they have had expert cooperation from growers in several parts of the country.

This book, too, is a cooperative venture, and it brings together all the scientific and cultural information that has been accumulated about the breeding, growing, harvesting, and marketing of the three types of blueberries that are grown commercially in North America: the highbush, the lowbush, and the rabbiteye. It is a book for professional horticulturists, commercial growers, and home gardeners, wherever climate and soil are suitable for growing one of these types. The editors and contributors hope very much that the book will prove valuable to European as well as American horticulturists, for blueberries of the same types are native to Great Britain and Continental Europe.

Blueberry Culture is based not only on a wealth of scientific literature but also on hitherto unrecorded observations of outstanding workers in the field. As members of agricultural experiment stations with active blueberry research projects, the contributors had access to highly specialized, up-to-date information for each growing area.

The editors wish to acknowledge the assistance of the following persons who reviewed portions of the manuscript: J. C. Cain, New York Agricultural Experiment Station, Geneva; C. A. Doehlert, Columbus, New Jersey; I. V. Hall, Canada Department of Agriculture, Kentville, Nova Scotia; Robert G. Hill, Ohio Agricultural Research and Development Center, Wooster; F. E. Scammell, grower, Toms River, New Jersey; R. A. Struchtemeyer, University of Maine, Orono; and W. E. Tomlinson, University of Massachusetts, East Wareham.

We are grateful to Mrs. Marion Savoca for her untiring efforts in typing and proofreading the manuscript and in helping to plan the illustrations.

PAUL ECK
NORMAN F. CHILDERS

New Brunswick, New Jersey
June, 1966

Contents

	Preface	vii
I	THE BLUEBERRY INDUSTRY Paul Eck and Norman F. Childers	3
II	BOTANY Paul Eck	14
III	BREEDING James N. Moore	45
IV	ENVIRONMENTAL RELATIONSHIPS Walter J. Kender and W. Thomas Brightwell	75
V	VARIETIES AND THEIR CHARACTERISTICS George M. Darrow and Donald H. Scott	94
VI	PROPAGATION AND PLANTING Charles M. Mainland	111
VII	SOIL MANAGEMENT, NUTRITION, AND FERTILIZER PRACTICES Walter E. Ballinger	132
VIII	PLANT AND FRUIT DEVELOPMENT Vladimir G. Shutak and Philip E. Marucci	179
IX	INSECTS AND THEIR CONTROL Philip E. Marucci	199
X	DISEASES AND THEIR CONTROL Eugene H. Varney and Allan W. Stretch	236

Contents

XI	HARVESTING, PROCESSING, AND STORAGE Warren C. Stiles and Dennis A. Abdalla	280
XII	ECONOMICS AND MARKETING Frederick A. Perkins	302
XIII	BLUEBERRIES IN THE HOME GARDEN Warren C. Stiles and John S. Bailey	320
	Bibliography	329
	Notes on the Contributors	355
	Index	359

List of Tables

	page
TABLE I. North American blueberry species, chromosome classification, common name, and proposed equivalent name.	16–18
TABLE II. Comparison of morphological characteristics of various species of *Vaccinium*.	25–27
TABLE III. Ancestry of the new highbush blueberry varieties.	46
TABLE IV. Some successful interspecific blueberry hybridizations made under controlled conditions.	47
TABLE V. *Vaccinium* species possessing desirable characters of potential value in a blueberry breeding program.	48
TABLE VI. Blueberry varieties originated by the United States Department of Agriculture cooperative breeding program and the estimated percentage of the 1963 acreage planted to each.	65
TABLE VII. Cooperating agencies and individuals of the United States Department of Agriculture blueberry breeding program, 1963.	66
TABLE VIII. Seedlings distributed from the United States Department of Agriculture Research Center, Beltsville, Maryland, to cooperators between 1946 and 1962.	67
TABLE IX. Segregation for fruit color in F_2 progenies from highbush × lowbush crosses.	71
TABLE X. Average monthly precipitation in major blueberry producing regions of the United States.	76
TABLE XI. Average frost dates and average length of growing season for major blueberry producing regions of the United States.	77

List of Tables

	page
TABLE XII. Relative cold resistance of highbush blueberry varieties.	80
TABLE XIII. Percentage of highbush blueberry buds injured at 21 F.	82
TABLE XIV. Percentage of crop picked each week during the ripening season for several blueberry varieties in eastern North Carolina, southern New Jersey, and Michigan.	99
TABLE XV. Ratings of highbush blueberries for some berry characteristics.	100
TABLE XVI. Ratings of rabbiteye blueberries for bush size, season, and some berry characteristics.	109
TABLE XVII. Per cent survival of hardwood cuttings of three varieties of blueberries as affected by position of cut and wood type.	113
TABLE XVIII. Mineral composition of blueberry leaves.	154
TABLE XIX. Composition of 100 grams of raw, canned, and frozen blueberries.	157
TABLE XX. Changing soil pH with sulfur or aluminum sulfate.	159
TABLE XXI. Fertilizer recommendations for commercial blueberry production.	164–165
TABLE XXII. Cross-pollination of Coville variety.	184
TABLE XXIII. Blueberry pest control.	200–201
TABLE XXIV. Importance of blueberry insects in the various growing areas.	203
TABLE XXV. Insect larvae that may be found in blueberry fruit at harvest time.	209
TABLE XXVI. Leafhopper survey of blueberry fields, 1947 to 1948.	214
TABLE XXVII. Comparison of harvesting results using various combinations of equipment in a commercial planting of highbush blueberries in Michigan.	297
TABLE XXVIII. Value of United States fruit, nut, and berry production for 1949, 1954, and 1959.	302

List of Tables

	page
TABLE XXIX. Value of United States blueberry production, by states, 1949, 1954, and 1959.	303
TABLE XXX. Major blueberry producing states showing acreage, production, and yield per acre, 1949 to 1959.	304
TABLE XXXI. Production, acreage, and value of production in major blueberry producing states, 1960, 1962, 1964.	305
TABLE XXXII. Weight loss of Croatan blueberries packaged in wood-veneer pint containers and moisture condensation on various film caps during storage.	311
TABLE XXXIII. Monthly shipments of fresh blueberries to 41 United States cities, by state of origin, 1963 season.	316
TABLE XXXIV. Fresh blueberry shipments to 41 United States and five Canadian cities, by state of origin, 1963 season.	317
TABLE XXXV. Wholesale blueberry prices in leading terminal markets for New Jersey stock for selected periods, 1963 season.	318

List of Figures

	page
FIGURE 1. The late Elizabeth C. White of Whitesbog, New Jersey, pioneer and cooperator with Dr. F. V. Coville in the development of the cultivated highbush blueberry for over two decades.	4
FIGURE 2. A typical highbush blueberry field in New Jersey.	5
FIGURE 3. Distribution of the major blueberry species harvested in the United States.	6
FIGURE 4. The lowbush blueberry *Vaccinium angustifolium* in the Blueberry Barrens of Maine.	7
FIGURE 5. The rabbiteye blueberry *Vaccinium ashei*, native to the southeastern United States.	8
FIGURE 6. Branches of the evergreen blueberry *Vaccinium ovatum*, harvested in the Pacific Northwest for ornamental purposes.	9
FIGURE 7. Blueberry production for the major blueberry producing states in the United States, by year and state, 1939 to 1965.	10
FIGURE 8. Fruit from the variety Rubel, the best selection from the wild, and Berkeley, a variety improved by hybridization.	11
FIGURE 9. Highbush blueberry acreage harvested in the United States, by year and state, 1939 to 1965.	12
FIGURE 10. Fruited branches of *Vaccinium myrtilloides* (*V. canadense*), a commercially important lowbush species native to Canada and Maine, and *V. amoenum*, an ancestor of the rabbiteye blueberry of northern Florida and South Carolina, west to Louisiana.	20

FIGURE 11. Dr. George M. Darrow, retired United States Department of Agriculture horticulturist, inspecting *Vaccinium tenel-*

List of Figures

	page
lum, an ancient diploid species from which the hexoploids *V. amoenum* and *V. ashei* were in large part derived. *V. myrsinites*, a tetraploid evergreen lowbush of the deep South, derived from the hybridization of *V. tenellum* and *V. darrowi*.	23
FIGURE 12. Typical raceme type of inflorescence of the cultivated highbush blueberry.	28
FIGURE 13. Important stages in the development of fruit and vegetative buds in *Vaccinium angustifolium*.	30
FIGURE 14. Lowbush blueberry rhizome.	33
FIGURE 15. The corolla of *Vaccinium corymbosum*.	35
FIGURE 16. The filament and anther of *Vaccinium corymbosum*.	36
FIGURE 17. The style of *Vaccinium corymbosum*.	37
FIGURE 18. The blueberry fruit.	39
FIGURE 19. The leaf of *Vaccinium corymbosum*.	40–41
FIGURE 20. The stem of *Vaccinium corymbosum*.	43
FIGURE 21. Dr. George M. Darrow inspecting seedlings that resulted from crossing a lowbush from Maine with a highbush. Another hybrid, produced by crossing *Vaccinium darrowi* with *V. tenellum*.	
FIGURE 22. Cluster of blueberry flowers with two flowers emasculated and ready for pollination.	54
FIGURE 23. Blueberry fruit rated for size.	55
FIGURE 24. Blueberry fruit rated for color and for scar.	56
FIGURE 25. A hybrid highbush selection, GM-37, that has produced berries more than an inch in diameter.	59
FIGURE 26. Rubel, the best blueberry selection from the wild in New Jersey.	60
FIGURE 27. Brooks, the best selection from the wild in New Hampshire.	61
FIGURE 28. Fruiting branches of US-4 and US-37.	62
FIGURE 29. Fruit clusters of rabbiteye and highbush blueberry selections.	64

List of Figures xvii

page

FIGURE 30. Typical soil of the New Jersey commercial blueberry growing region. 87

FIGURE 31. The profile of a virgin podzol soil compared to a podzol that has been in commercial lowbush blueberry production for at least fifty years. 88

FIGURE 32. Blueberry plants lifted out of the ground by freezing of the soil, in a location that should have been drained before the plants were set. Mounding in a blueberry planting in eastern North Carolina to keep the plant roots out of the water. 91

FIGURE 33. Berkeley, a Big Seven variety of the highbush blueberry noted for its large, firm, light blue fruit. 98

FIGURE 34. Bluecrop, a Big Seven variety of the highbush blueberry noted for its capacity to yield consistently good crops, and Stanley, an early introduction used extensively as a parent in the development of new varieties. 102

FIGURE 35. Cabot, one of the first highbush blueberry varieties to be introduced. 103

FIGURE 36. Branches of outstanding highbush blueberry varieties: Coville, a Big Seven variety with excellent fruit quality, the latest ripening of the recently introduced varieties. Earliblue, a Big Seven variety with excellent fruit quality, the earliest ripening of the new varieties. Herbert, a Big Seven variety with excellent dessert quality, recommended for home gardens. Ivanhoe, a variety with some canker resistance, developed for the industry in North Carolina. 105

FIGURE 37. Jersey, a productive and vigorous variety, grown extensively in Michigan. 106

FIGURE 38. Selection of hardwood cuttings for propagation. 112

FIGURE 39. Low propagating frame with glass sash and burlap shade in place. Movable trays, which sit in the top of the propagating frame, provide a convenient way of handling cuttings prior to planting in the nursery. 116

FIGURE 40. Preparation of a commercial propagating bed in New Jersey. 119

FIGURE 41. Softwood propagation with mist. 124

FIGURE 42. Handling rooted cuttings. 127

xviii List of Figures

page

FIGURE 43. A two-year-old blueberry plant grown from a rhizome cutting. 129

FIGURE 44. The exposed root system of a mature blueberry bush. 133

FIGURE 45. A tractor-drawn spring-tooth harrow and specially designed discs for cultivating blueberries. 135

FIGURE 46. An automatic grapehoe attachment designed for blueberries. 136

FIGURE 47. The bush and root system of a mulched plant compared with those of a plant that was cultivated. 141

FIGURE 48. Pumping water from main drainage ditch to portable sprinkler-irrigation pipe lines. Control gate and irrigation ditch used in sub-irrigation of blueberries. 145

FIGURE 49. Typical irrigation pond on a blueberry farm. 147

FIGURE 50. Iron chlorosis on highbush blueberries. 150

FIGURE 51. A tractor-drawn mechanical fertilizer distributor in eastern North Carolina. 166

FIGURE 52. Leaf deficiency symptoms of the rabbiteye blueberry when grown in sand culture. 176–177

FIGURE 53. Terminal shoots showing arrangement of blueberry flower and leaf buds. 180

FIGURE 54. Cluster of flowers of *Vaccinium angustifolium*, Washington County, Maine. 182

FIGURE 55. Growth curves for three Earliblue berries that ultimately attained a different size. 188

FIGURE 56. Pruning recommendations for highbush blueberries. 192

FIGURE 57. Burning equipment for lowbush blueberry fields: a Woolery oil burner and an LP-gas burner. 197

FIGURE 58. Larva, pupa, and moth stages of the cherry fruitworm, *Grapholitha packardi*. 212

FIGURE 59. The life history of the cherry fruitworm on blueberries in New Jersey in relation to specific stages of fruit development in the Jersey and Weymouth varieties. 213

List of Figures

	page
FIGURE 60. Average percentage emergence of *Rhagoletis pomonella* at various dates for the years 1952 to 1955 and probable emergence pattern under average conditions in New Jersey.	216
FIGURE 61. The blueberry fruit fly *Rhagoletis pomonella*, the most important enemy of cultivated and wild blueberries.	217
FIGURE 62. The kidney-shaped blueberry stem gall *Hemadas nubilipennis*.	224
FIGURE 63. Blueberry branches and berries blighted by mummy berry. Apothecia, or mummy cups of *Monilinia vaccinii-corymbosi*.	238
FIGURE 64. Typical stem cankers and fusicoccum cankers on the highbush blueberry.	242
FIGURE 65. Spots caused by *Septoria* on leaves and a twig of the cultivated blueberry variety June.	252
FIGURE 66. Witches'-broom, a rust fungus on blueberry bushes.	254
FIGURE 67. Leaf spotting on the lowbush blueberry.	259
FIGURE 68. Typical symptoms of stunt disease on the blueberry.	263
FIGURE 69. Typical red ringspot symptoms on the leaves and stems of the blueberry.	266
FIGURE 70. Stunted growth and dieback on a branch of the Pemberton blueberry infected with necrotic ringspot virus.	269
FIGURE 71. Blueberry leaves infected with mosaic virus.	271
FIGURE 72. Blueberry leaf malformations caused by the shoestring virus.	272
FIGURE 73. Root galls, branch galls, crown galls, and bud-proliferating galls on the blueberry.	275
FIGURE 74. Cuttings from an area in a cutting bed heavily infested with stubby root nematodes compared with cuttings from a nematode-free area.	279
FIGURE 75. Harvesting lowbush blueberries in Maine.	282–283
FIGURE 76. Vacuum mechanical harvesting unit for lowbush blueberries.	284
FIGURE 77. Tunnel of an IQF (Individually Quick-Frozen) freezer in Maine.	285

List of Figures

	page
FIGURE 78. Freezing lowbush blueberries by the flo-freeze method.	286–287
FIGURE 79. Containers used for packaging frozen blueberries.	289
FIGURE 80. A typical organization of labor for blueberry harvesting.	292
FIGURE 81. Harvesting highbush blueberries in New Jersey.	294–295
FIGURE 82. Harvesting highbush blueberries at the plantation of Elizabeth C. White at Whitesbog, New Jersey, in the 1930's.	298
FIGURE 83. A hand-held mechanical harvesting aid for highbush blueberries and an experimental highbush mechanical harvester.	299
FIGURE 84. Field-run Burlington blueberries prepared in different ways for storage in field boxes containing 12 pints each.	300
FIGURE 85. Four types of pint containers used in packaging highbush blueberries.	309
FIGURE 86. A 12-pint retail flat packed for market, showing condensation on various types of capping film.	310
FIGURE 87. A modern automatic packaging facility for filling pint containers with blueberries.	313
FIGURE 88. Flats of blueberries ready for shipment to market.	314
FIGURE 89. A home garden blueberry planting in a permanent enclosure.	327
FIGURE 90. Fish net covering on a blueberry planting.	327

Blueberry Culture

CHAPTER 1

The Blueberry Industry

by Paul Eck and Norman F. Childers, Department of Horticulture and Forestry, Rutgers University, New Brunswick, N.J.

Over the past thirty years growing blueberries has become a major fruit industry in the United States. Several species of this delicious berry are native to North America, and the wild fruit was greatly liked by Indians and early settlers alike; however, the industry of today owes its remarkable development to the skill and imagination of breeders and growers who began domesticating and crossing fine wild specimens of the highbush type in the early years of the twentieth century. With the introduction of named cultivated varieties, bred for excellence of flavor, size, light blue color, hardiness, and other desirable characteristics, the highbush blueberry became an attractive crop for commercial growers. Moreover, because of its preference for an acid soil and plenty of moisture many acres of land previously considered worthless for agriculture now yield profitable crops of blueberries.

Wild and semiwild berries are also harvested in several sections of the United States, but most fresh blueberries on the market in the North are the product of scientific breeding. They ship well, and like the wild berry require no hulling or peeling or pitting. This admirable fruit, wild or cultivated, may also be canned, frozen, or freeze-dried without appreciable loss of flavor, so that blueberry muffins, blueberry pies, blueberry cheesecakes, and dozens of other blueberry desserts and breakfast pastries can be enjoyed the year-round. Free of objectionable seeds and acids, the fruit is a favorite of young and old.

For all the great increase in blueberry production in recent years, many people in the United States have never tasted a fresh blueberry. Potential markets in the Midwest, the Far West, and the Southwest—at a considerable distance from the growing areas—have scarcely been tapped, and the export market is unknown. Meantime, however, more and more acres are being planted to blueberries, and each year scientific breeders in the

Figure 1. The late Elizabeth C. White inspecting a picker's flat in a highbush blueberry field at Whitesbog, New Jersey. Miss White was a pioneer and cooperator with Dr. F. V. Coville in the development of the cultivated highbush blueberry for over two decades. (*Newark News,* Newark, N.J.)

United States Department of Agriculture and several state experiment stations make perceptible progress in developing new hybrids suited to particular climates and disease hazards and destined to extend the fresh market season, both early and late.

The true blueberries belong to the genus *Vaccinium,* an ancient genus that became well established in North America following the glaciations of the Pleistocene period. The early settlers had many names for them because they closely resembled the European berries that they knew as whortleberries, whinberries, trackleberries, hurtleberries, or blaeberries. Through the years the blueberry has often been confused with the huckleberry (genus *Gaylussacia*), but the two are readily distinguishable by their seeds. Blueberries have many small soft seeds, whereas huckleberries have ten large bony seeds.

The blueberry industry in the United States is based on the harvest of three distinct types within the genus *Vaccinium*. In the northeastern United States and parts of Canada, the native species are lowbush types that thrive in the wild wherever the terrain is suitable, though to be profitable they require intensive care from the grower. The two most common species are *V. angustifolium* Aiton, formerly *V. lamarckii* Camp, and the Canadian blueberry, *V. myrtilloides* Michaux. A third species, *V. brittonii* Porter, which bears dark blue or black fruit, is often found growing in association with the lowbush blueberry. These lowbush species form dense and extensive colonies, usually 6 to 18 inches high. The fruit is gathered with hand rakes resembling small cranberry scoops during the months of July, August, and September, and the bulk of the crop is destined for processing plants. At present more than 100,000 acres of the lowbush species are under cultivation in the United States. Two-thirds of this acreage is harvested annually, and one-third is burned over each year.

In New Jersey, Michigan, North Carolina, and Washington, and to some extent also in Indiana, Ohio, Pennsylvania, New York, and Massachusetts, growers raise highbush types, 3 feet high or more, developed mainly from *V. australe* Small and *V. corymbosum* Linnaeus. The highbush blueberry is planted in clean-cultivated rows, and the large fruits are mostly hand picked for the fresh market in the months of June, July, and August. Almost 20,000 acres are planted to these superior varieties.

The rabbiteye blueberry, *V. ashei* Reade, native to North Carolina, South Carolina, Georgia, northwestern Florida, Alabama, Mississippi, and

Figure 2. A typical highbush blueberry field in New Jersey. The commercial highbush blueberry is planted in rows and is generally clean cultivated. (John Keller, Scotch Plains, N.J.)

Louisiana, is the tallest species commercially grown. The bushes attain heights of 30 feet and thrive on upland soils that are unsuitable to highbush species. More than 3,500 acres of this species are in cultivation.

Other blueberry species that are harvested commercially include the dryland blueberry, *V. pallidum*, of northeastern Alabama, northwestern Georgia, West Virginia, and western and northwestern Arkansas; the evergreen blueberry, *V. ovatum* Pursh, harvested extensively in northern California, along the coast of Oregon and Washington, and in the Puget Sound area; and the mountain blueberry, *V. membranaceum*, harvested primarily in the vicinity of Crater Lake, Mount Hood, Mount Adams, and Mount Rainier.

Neither the dryland blueberry nor the evergreen blueberry nor the mountain blueberry is cultivated; all three are simply harvested from the wild. The annual value of the dryland crop nevertheless averages $300,000, and the other two each average $200,000. The dryland blueberry grows only a little higher than the Maine lowbush, and the fruit has a good flavor. The evergreen blueberry, which sometimes reaches a height of 20 feet, is an attractive ornamental shrub and the sale of branches to florists brings in a good deal more than the berry harvest. The

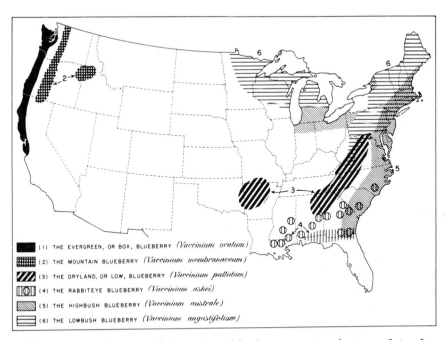

Figure 3. Distribution of the major blueberry species harvested in the United States. (USDA, Beltsville, Md.)

Figure 4. The lowbush blueberry *Vaccinium angustifolium* in the Blueberry Barrens of Maine. (USDA, Beltsville, Md.)

shiny black berries of this species are used mainly in pies. The mountain blueberry, 3 to 5 feet high, bears a pear-shaped black or maroon fruit, borne singly rather than in clusters. It is rather tart but may be eaten fresh as well as cooked.

The lowbush blueberry was the first to be cultivated commercially. Perhaps the first attempts at cultivation were made by the Indians, for they practiced burning as a pollarding technique on the blueberry barrens. They apparently knew from experience that the bushes do not bear when overgrown with brush and weeds. Today this same technique of promoting the growth of new shoots from underground stems is practiced but with huge self-propelled oil and gas burners. Within the last few decades, moreover, important advances have been made in the cultivation of this wild crop. The employment of fertilizers, insecticides, herbicides, and fungicides has made the farming of this crop an exacting science. The success of the new techniques is such that the lowbush harvest is larger than that of any other commercially grown species. However, since most of the fruit is canned or frozen, the value of the crop has been less than that for the highbush blueberry. The lowbush crop in

Figure 5. The rabbiteye blueberry *Vaccinium ashei*, native to the southeastern United States, is the most vigorous of all the blueberry species. (USDA, Beltsville, Md.)

Figure 6. Branches of the evergreen blueberry *Vaccinium ovatum* are harvested in the Pacific Northwest for ornamental purposes. The fruit is black with fair flavor. (USDA, Beltsville, Md.)

Maine alone is valued at $3 million annually. The total lowbush crop produced in the United States has been estimated at over $5 million.

The development of the highbush blueberry industry was made possible by the successes of the late Dr. F. V. Coville of the United States Department of Agriculture in domesticating and improving wild highbush species. Having determined that the highbush blueberry requires an acidic soil environment to survive, Dr. Coville applied plant breeding techniques of interspecific hybridization, plant selection, and plant evaluation to originate varieties that were superior to his original selections from the wild. In this work he had the valuable help of Elizabeth C.

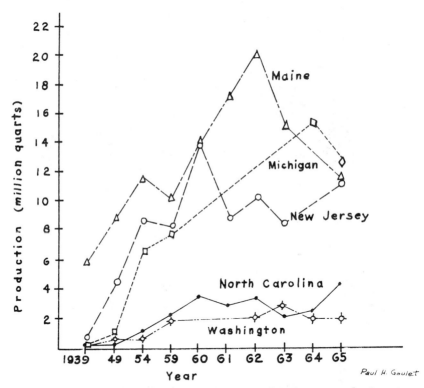

Figure 7. Blueberry production for the major blueberry producing states in the United States, by year and state, 1939 to 1965. (Prepared by J. C. Torio.) (Sources: United States Department of Commerce, *Census of Agriculture*, 1939, 1949, 1954, 1959; New Jersey State Department of Agriculture, Crop Reporting Service; Michigan Blueberry Growers Association; North Carolina State Department of Agriculture, Crop Reporting Service; Tru-Blu Cooperative; Washington State Department of Agriculture, Crop Reporting Service.)

White, a commercial cranberry grower in Whitesbog, New Jersey, who induced her pickers to search for exceptionally fine bushes in the wilds of the New Jersey Pine Barrens. The best of these were moved to Whitesbog for further study, and new plants were propagated from cuttings. When the young plants reached bearing age, cross pollinizing was done and the seed was used to produce hybrids. Over a period of thirty-five years Coville produced about 100,000 seedlings from which 15 improved varieties were named and introduced by the time of his death in

Figure 8. Fruit from the variety Rubel, the best selection from the wild, and Berkeley, a variety improved by hybridization (*left and right*). (USDA, Beltsville, Md.)

1937. Ever since, the industry has continued to benefit from Coville's pioneering efforts, for by 1959 the number of new varieties directly attributed to his original efforts had risen to 30. Concomitantly, land devoted to highbush blueberry production rose steadily from less than 200 acres in 1930 to nearly 20,000 acres in 1965 with an annual crop valued at $18 million. In the decade and a half between 1950 and 1965, the acreage quadrupled. No other temperate climate fruit crop has increased at this rate.

The new rabbiteye varieties recently introduced are greatly superior in flavor and palatability to the original selections from the wild. Improvement of this species did not begin until the 1940's and therefore has not progressed as far as with the highbush blueberry, but both the United States Department of Agriculture and state researchers in Georgia, North Carolina, and Florida are engaged in very active programs to develop commercially acceptable varieties of this species.

The blueberry industry in the United States is still in its infancy, for the production of all three types of cultivated berries is bound to expand phenomenally in the decades to come.

The lowbush industry is about to introduce mechanical harvesting, and improved methods of freezing and drying the berries have already had an important effect on this industry. Further gains can be expected in the future from an organized effort at plant improvement through plant breeding and the development of methods of propagation that can be utilized to upgrade indigenous plantings.

The highbush industry also stands to benefit from mechanization of the harvesting and processing operations. Spurred on by the promise of new varieties of superior fruit quality, this industry will continue to expand, making use of land that had been unproductive for centuries. It is probable that, within the next few decades, the greatest expansion in blueberry production will be in the highbush industry, as national and worldwide demand for its fruit increases. The limits of the industry's expansion

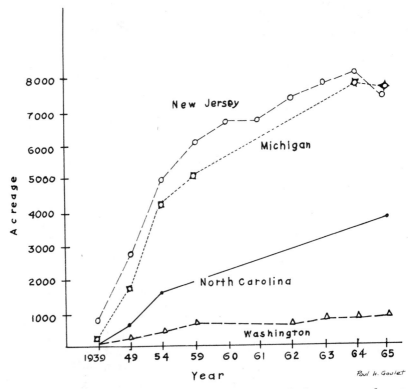

Figure 9. Highbush blueberry acreage harvested, by year and state, 1939 to 1965. (Prepared by J. C. Torio.) (Sources: see Figure 7.)

will be determined by the availability of sites whose soil and climate are favorable to highbush culture. On a small scale, in the home garden, the soil environment can usually be changed sufficiently by mulching and composting to meet requirements of the highbush, but there is no economical way to do this in large commercial plantings.

Breeding experts believe that the rabbiteye offers the greatest potential for future expansion of the blueberry industry in the United States. This species is capable of adjusting to upland soils and is a heavy bearer. It also possesses such desirable characteristics as heat and drought resistance and requires only a short winter rest period. However, the fruit of unselected bushes is usually small and black, gritty in texture, and lacking in flavor, and ripens over a long period. The few superior varieties of the rabbiteye that have been introduced already are indicative of what is in store for this part of the blueberry industry.

CHAPTER 2

Botany

by **Paul Eck**, *Department of Horticulture and Forestry, Rutgers University, New Brunswick, New Jersey*

TAXONOMY

The blueberry is a member of the *Ericaceae*, heath family, and has been further classified into the subfamily *Vaccinioideae* (Gray and Fernald, 1950). Two agriculturally important genera of *Vaccinioideae* are *Vaccinium*, which contains the blueberry, and *Gaylussacia*, of which the huckleberry is a member. Species of *Vaccinium* and *Gaylussacia* are often confused by the layman but may be readily distinguished by examination of the ovary. In *Vaccinium* the ovary has four or five cells, or sometimes eight or ten if there are false partitions, and contains many small seeds. In *Gaylussacia* the ovary is ten-celled, with one ovule in each cell, and the mature fruit contains only ten seeds. These seeds are much larger than the seeds in the blueberry.

The blueberry is further classified into the subgenus *Cyanococcus*, separating it from the subgenus *Batodendron*, which includes the farkleberry, sparkleberry, deerberry, and southern gooseberry. The bilberry and whortleberry are assigned to the subgenus *Euvaccinium* and the cranberry to the subgenus *Oxycoccus* (Gray and Fernald, 1950; Bailey, 1919).

Within the genus *Vaccinium* and the subgenus *Cyanococcus* are a large number of species of blueberries indigenous to North America. Species differentiation and subsequent nomenclature have presented a taxonomic problem. Notable efforts to dispel the confusion include the work of Camp (1940, 1942a, 1942b, 1945). In addition to outlining the major morphological differences in the species, Camp made extensive use of existing knowledge of chromosome numbers to classify many of the important North American blueberries. Changes in the nomenclature of certain species have necessarily resulted.

It is generally believed that the basic genome in the genus *Vaccinium* is 12 (Longley, 1927; Newcomber, 1941). Included in the diploid populations (2n = 24) are the species *Vaccinium myrtilloides* Michaux, *V. vacillans* Torrey, *V. pallidum* Aiton, *V. darrowi* Camp, *V. atrococcum* Heller, *V. caesariense* Mack, *V. elliottii* Chapman, and *V. tenellum* Aiton. These diploids are low-growing types except for *V. caesariense*, *V. atrococcum*, and *V. elliottii*, which are the highbush diploids. There are no fundamental sterility barriers between the members of *Vaccinium* as long as they have the same number of chromosomes. The result has been the formation of a large number of polyploids, mostly tetraploids, of taxonomic importance.

These tetraploids may have resulted from crosses between members of the same species, resulting in autotetraploids, or between members of different species, in which case they are classified as allotetraploids. Important tetraploids listed by Camp (1945) are *Vaccinium lamarckii* Camp, now identified as *V. angustifolium* Aiton, *V. brittonii* Porter, *V. corymbosum* Lamarck, *V. hirsutum* Buckley, *V. simulatum* Small, *V. alto-montanum* Ashe, *V. myrsinites* Lamarck, *V. australe* Small, and *V. virgatum* Aiton. Although not definitely established, it is believed that *V. marianum* Watson and *V. fuscatum* are also tetraploid.

A third important group of species is represented by the hexaploid (2n = 72) population of which *V. ashei* Reade, *V. constablaei* Gray, and *V. amoenum* Aiton are members.

The important North American species of *Vaccinium* and their chromosome classification are listed in Table I. Other species names appearing in the literature, as well as the common and a proposed name, are included (Camp, 1945).

DISTRIBUTION AND ECOLOGY

Canada to Northern Florida

The northernmost species of blueberry of commercial importance is *V. myrtilloides* Michaux (*V. canadense*) (Camp, 1945; Eaton and Hall, 1961). It is found from southeastern British Columbia east to Labrador and south to New York, Pennsylvania, Indiana, and West Virginia. The species may be found in boggy areas, uplands, and mountain meadows. Commercially, it is the most important species of the lowbush blueberry in Canada (Fig. 10). *V. myrtilloides* is less tolerant of burning than *V. angustifolium* but is more tolerant of shade and is therefore more common in second-growth forests.

V. angustifolium is the most important commercially of the wild species native to the northeastern United States and parts of Canada (Chandler and Hyland, 1941). It is common to open, rocky uplands and dry, sandy

Table I. North American Blueberry Species, Chromosome Classification, Common Name, and Proposed Equivalent Name

Vaccinium Species	Chromosome Classification	Other Names in Literature	Common Name (c.n.) Proposed Equivalent (p.e.)
V. alto-montanum Ashe	tetraploid	Cyanococcus liparis Small; C. subcordatus Small; C. tallopusae Coville; V. liparum (Small) Uphof; V. subcordatum (Small) Uphof; V. tallopusae (Coville ex Small) Uphof	c.n. huckleberry p.e. mountain dryland blueberry
V. amoenum Aiton	hexaploid		c.n. tall huckleberry p.e. large cluster blueberry
V. angustifolium Aiton	tetraploid	V. fissum Schrank; V. pennsylvanicum var. angustifolium Gray; V. pennsylvanicum var. alpinum Wood; V. angustifolium var. laevifolia House V. lamarckii Camp	c.n. lowbush blueberry, sugarberry p.e. low sugar blueberry
V. arkansanum Ashe	probably tetraploid		c.n. black highbush blueberry p.e. Arkansas highbush blueberry
V. ashei Reade	hexaploid: a complex allopolyploid	V. virgatum Aiton	c.n. rabbiteye blueberry p.e. Arkansas highbush blueberry
V. atrococcum Heller	diploid	V. corymbosum var. atrocarpum Gray; V. corymbosum var. atrococcum Gray; Cyanococcus atrococcus Small	c.n. black highbush blueberry p.e. pubescent-leafed blueberry

V. australe Small	tetraploid; an autopolyploid derivative of *V. caesariense*		c.n. highbush blueberry p.e. Southeastern highbush blueberry
V. brittonii Porter	tetraploid	*V. nigrum* Britton	c.n. huckleberry p.e. Britton's blueberry
V. caesariense Mackenzie	diploid		c.n. highbush blusberry p.e. New Jersey blueberry
V. constablaei Gray	hexaploid; an allopolyploid derived from combinations between *V. simulatum* and *V. alto-montanum*	*V. pallidum* of various authors; not *V. pallidum* Aiton	c.n. blueberry p.e. Constable's blueberry
V. corymbosum Linnaeus [a]	tetraploid: an allotetraploid showing characteristics of *V. arkansanum*, *V. simulatum*, *V. australe*, and *V. marianum* at its southern boundaries and *V. lamarckii*-like or *V. brittonii*-like at its northern distribution	*V. albiflorum* Hooker; *V. corymbosum* var. *glabrum* Gray; *V. corymbosum* var. *amoenum* Gray; *V. corymbosum* var. *pallidum* Gray; *V. corymbosum* var. *fuscatum* Gray; *V. vicinium* Bicknell; *Cyanococcus corymbosum* Rydberg	c.n. highbush blueberry p.e. northern highbush blueberry
V. darrowi Camp	diploid	*V. myrsinites* var. *glaucum* Gray	c.n. none p.e. Darrow's evergreen blueberry
V. elliottii Chapman	diploid	*Cyanococcus elliottii* Small	c.n. Mayberry p.e. Elliott's blueberry
V. fuscatum	probably tetraploid; appears to be an allopolyploid derived from the diploid hybrids of *atrococcum* and *darrowi*	*V. formosum* Andrews; *V. fuscatum* var. *pullum* Ashe; *V. holophyllum* (Small) Uphof; *Cyanococcus holophyllus* Small; *C. fuscatus* Small	c.n. none p.e. thick-leafed blueberry

Table I. North American Blueberry Species, Chromosome Classification, Common Name, and Proposed Equivalent Name (Continued)

Vaccinium Species	Chromosome Classification	Other Names in Literature	Common Name (c.n.) Proposed Equivalent (p.e.)
V. hirsutum Buckley	tetraploid	Cyanococcus hirsutus Small	c.n. hairy huckleberry p.e. hairy-fruited blueberry
V. lamarckii Camp	tetraploid	V. pennsylvanicum Lamarck V. angustifolium Aiton	c.n. upland blueberry, huckleberry p.e. Lamarck's sugar blueberry
V. marianum Watson	probably tetraploid; an allopolyploid derived from the tetraploid V. caesariense and the diploid V. atrococcum		c.n. highbush blueberry p.e. Maryland highbush blueberry
V. membranaceum Douglas	tetraploid	V. macrophyllum	c.n. mountain blueberry, broadleafed huckleberry p.e. mountain blueberry
V. myrsinites Lamarck	tetraploid; an allopolyploidic species derived out of the hybrids between tenellum and darrowi	V. nitidum Andrews; V. nitidum var. decumbens Sims; Cyanococcus myrsinites Small	c.n. huckleberry p.e. Florida evergreen blueberry
V. myrtilloides Michaux	diploid	V. myrtillus Linnaeus V. canadense Kalm; V. pennsylvanicum var. myrtilloides Fern; Cyanococcus canadense Rydberg	c.n. sourtop or velvet leaf blueberry p.e. Canadian blueberry
V. ovatum Pursh	probably tetraploid		c.n. evergreen or coast huckleberry p.e. box blueberry
V. pallidum Aiton	diploid	V. viride Ashe	c.n. huckleberry p.e. upland low blueberry

V. simulatum Small	tetraploid; an autopolyploid derivative of *V. pallidum*	*Cyanococcus simulatus* Small	c.n. highbush blueberry p.e. upland highbush blueberry
V. tenellum Aiton	diploid	*V. galezans* Michaux; *V. virgatum* var. *tenellum* Gray; *Cyanococcus tenellus* Small	c.n. low huckleberry p.e. small cluster blueberry
V. vacillans Torrey	diploid	*V. vacillans* var. *crinitum* Fern; *V. vacillans* var. *columbianum*; *V. vacillans* var. *columbianum* f. *molbifolium* Ashe; *V. torreyanum* Camp; *Cyanococcus vacillans* Rydberg	c.n. low huckleberry p.e. dwarf dryland blueberry
V. virgatum Aiton	tetraploid; an autopolyploid of *V. tenellum*	*V. ashei* Reade; *V. amoenum* Aiton	c.n. huckleberry p.e. medium cluster blueberry

Source: Camp, 1945; Darrow, 1957.

[a] A misnomer, applied by Linnaeus because this was the only species then known to him with this type of inflorescence. The correct botanical term for this type of inflorescence is racemose.

Figure 10. Left. A heavily fruited branch of *Vaccinium myrtilloides* (*V. canadense*), a commercially important lowbush species native to Canada and Maine. *Right.* *V. amoenum*, an ancestor of the rabbiteye blueberry, is the widely distributed hexaploid blueberry of northern Florida and South Carolina west to Louisiana. (Walter J. Kender, University of Maine, Orono; USDA, Beltsville, Md.)

areas, sometimes referred to as the blueberry barrens, is harvested extensively along the coast of Maine, and provides a major product for the canning industry. *V. angustifolium* also frequently migrates to abandoned hayfields. It is favored by periodic burning, a pollarding technique that is an essential part of the culture of this commercial species.

V. lamarckii Camp and *V. brittonii* Porter grow in the same areas as *V. angustifolium* and under similar ecological conditions, but they do not range as high in altitude or as far north as *V. angustifolium*. More recently *V. lamarckii* has been considered the same species as *V. angustifolium* (Aalders and Hall, 1961). The greater height of these tetraploids, which may leave them with less snow cover, makes them more susceptible to injury by desiccation and browsing animals during the winter.

The diploid *V. caesariense* Mackenzie is found in the coastal plain swamps and inland boggy areas from southern Maine to northern Florida. This highbush hybridizes freely with *V. atrococcum*, a diploid highbush blueberry found from western Tennessee and Arkansas south to the Gulf of Mexico, east to northern Florida, and north to central New York and Maine. Like *V. caesariense*, it is typically found in the coastal swamps and along rivers. Where populations of *V. caesariense* and *V. atrococcum* have merged, there occurs an allopolyploid given the name *V. marianum* (Camp, 1945).

Throughout southeastern Alabama and northern Florida and northward to New Jersey is found *V. australe*, the autopolyploid derivative of *V. caesariense*. It is indigenous to coastal marshes and swamps and inland boggy areas. *V. australe* is the species that has contributed most of the excellent qualities of the modern, commercially grown highbush blueberries. Many of the original commercially grown highbush blueberries were selections of *V. australe* taken from the wild.

From southern Ontario to Missouri, Tennessee, and Georgia, in dry, open woods, along rocky ledges, or in abandoned farm lands, may be found *V. vacillans*, a low-growing diploid species. This species hybridizes readily with *V. pallidum* on the Ozark plateau, and with *V. caesariense* on the coastal plain.

A common highbush species prevalent from Michigan east to Nova Scotia and south to the glacial boundary is *V. corymbosum*. This extremely complex hybrid may take on characteristics of *V. lamarckii* or *V. brittonii* in the South. It is common in swamps, stream margins, and moist, sandy areas, and along hillside seepages.

V. simulatum is distributed throughout northern Alabama and Georgia and northward in the mountains to Kentucky and Virginia. It is found on open mountain slopes and meadows between 1,000 and 2,500 feet above sea level. It is believed to be an autopolyploid derivative of *V. pallidum*. Within the same geographic area and often found in conjunction with *V. simulatum* is *V. alto-montanum*. Soil conditions common in open woods and rocky uplands favor this species. Its proximity to *V. simulatum* has enabled them to hybridize freely.

Although not as widely distributed as *V. vacillans*, *V. pallidum* is another dry upland species found mainly in the mountains and rugged country between 2,500 and 3,500 feet in elevation. Below 2,500 feet, *V. pallidum* often grades into one of the phases of *V. vacillans*. *V. pallidum* is found from northern Georgia and Alabama northward to West Virginia, Pennsylvania, and New York and also in the Ozark Mountains of Arkansas.

V. hirsutum Buckley, the hairy-fruited blueberry, is common to the high elevations throughout North Carolina and Tennessee. It is indigenous to dry ridges and mountain meadows, usually between 2,000 and 5,000 feet high.

Located on the mountain tops at elevations above 3,500 feet throughout western North Carolina and eastern Tennessee is a highly variable hexaploid population described by Gray as *V. constablaei*. Its morphological characteristics are derived from combinations of *V. simulatum* and *V. alto-montanum*.

Common to the Coastal Plain from southern Virginia to northern Florida is *V. tenellum* Aiton. This lowbush species is found mostly in open forests and meadows and is favored by periodic burning, which eliminates competition from tall grass and other shrubs (Fig. 11). Another species, *V. virgatum* Aiton, has similar ecological requirements and is apparently an autopolyploid of *V. tenellum*. It differs from *V. tenellum* in its ability to persist in competition with taller grass.

V. elliottii Chapman is a diploid species found in open, flat woods and river flood plains but rarely in swampy areas. Its distribution is from southeastern Virginia south to Florida and west to Louisiana and Arkansas. Also from northern Florida west to Texas and Arkansas, along sandy lake and stream margins or in swamps, can be found the highbush blueberry, *V. arkansanum* Ashe.

V. ashei Reade, the rabbiteye blueberry, is a plant of various habitats extending throughout a range of Georgia, southern Alabama, and northern Florida (Darrow, Wilcox, and Beckwith, 1944). It is found along the larger rivers such as the Yellow River in northwest Florida and southeast Alabama, the Suwannee River in northern Florida, and the Satilla River in southeastern Georgia (Brightwell, 1962). This hexaploid species is indigenous to stream banks and lake margins where it is protected from forest fires. It thrives on dry uplands and in abandoned agricultural areas when transplanted. The rabbiteye blueberry adapts better to open-field culture than do the highbush species of the area because of its resistance to drought and its tolerance of high temperatures. Other factors favoring its cultivation throughout this area are its requirement for a very short rest period and its tolerance of a relatively wide soil pH range.

Another hexaploid occurring from South Carolina to northern Florida and west to Texas and Arkansas is *V. amoenum* Aiton (Fig. 10). This species is common to the uplands and open woods.

Southern Georgia, Alabama, Louisiana, and Florida

V. darrowi Camp, named in honor of G. M. Darrow, is a lowbush species adapted to full sun exposure and the dry, sandy soils of southern Louisiana, Georgia, and Florida. *V. myrsinites* is an allopolyploidic species derived from hybridization between *V. tenellum* and *V. darrowi* (Fig. 11). Its ecological requirements are much the same as those of *V. darrowi*, and it is distributed throughout the same area.

Figure 11. Above. Dr. George M. Darrow, retired United States Department of Agriculture horticulturist, inspecting *Vaccinium tenellum,* an ancient diploid species from which the hexaploids *V. amoenum* and *V. ashei* were in large part derived. *Below. V. myrsinites,* a tetraploid evergreen lowbush of the deep South derived from the hybridization of *V. tenellum* and *V. darrowi.* (USDA, Beltsville, Md.)

Another species common to this geographic region is *V. fuscatum* Aiton. The habitat of this probable tetraploid, however, is sandy, flat woods and bottom lands and the edges of streams and lakes.

Pacific Northwest

The evergreen blueberry, *V. ovatum* Pursh, is native along the Pacific coast from central California to British Columbia (Crowley, 1933; Figs. 3 and 6). It thrives in the mild climate along the Pacific coast and around Puget Sound, reaching a height of 20 feet in the open woods.

MORPHOLOGY

Gross Morphology

Representatives of the North American blueberries range in size from a shrub less than a foot in height to huge bushes attaining a height of 30 feet (Table II). A convenient grouping of species is offered by the three categories lowbush, half-high, and highbush (Camp, 1945). The lowbush category consists of species which do not exceed a meter in height. Included within this category are the species *V. myrtilloides*, *V. angustifolium*, *V. lamarckii*, *V. vacillans*, *V. brittonii*, *V. pallidum*, and *V. alto-montanum*. Species ranging in height from 1.5 to 7 m are classified as highbush. Located both north and south of the glacial boundary, this group includes the species *V. elliottii*, *V. corymbosum*, *V. simulatum*, *V. constablaei*, *V. marianum*, *V. caesariense*, *V. australe*, *V. fuscatum*, and *V. arkansanum*.

The intermediate half-high group consists of a series of hybrids especially common north of the glacial boundary. Representative of these half-high species are the hybrids *V. corymbosum* × *V. lamarckii*, *V. atrococcum* × *V. vacillans*, *V. corymbosum* × *V. brittonii*, *V. atrococcum* × *V. myrtilloides*, and *V. caesariense* × *V. vacillans*.

The lowbush species most frequently form extensive colonies. Typical of these are *V. darrowi* in the south and *V. angustifolium* in the north. In contrast, the larger species represented by the half-high and highbush plants most frequently form several single stems and tend to form crowns. If disturbed, these species may sucker and form limited colonies.

Inflorescence

The most common type of inflorescence in blueberries is the raceme (Fig. 12), a simple inflorescence of pediceled flowers upon a common elongated axis (Robbins, 1931). The inflorescences are sometimes terminal, but in most species they are axillary. In the mountain blueberry,

Table II. Comparison of Morphological Characteristics of Various Species of *Vaccinium*

Vaccinium Species	Growth Habit and Height	Leaf Characteristics [a]	Flower (Corolla)	Fruit
V. amoenum Aiton	dense clumps, 1.5–2.5 m	g., dark green, pubescent, obovate, serrate	cylindro-urceolate	black, 8–10 mm, poor flavor
V. angustifolium Aiton	dense colonies, 5–20 cm	n.g., glabrous, elliptic, serrate	cylindraceous, white	bright blue, 5–7 mm, excellent flavor
V. arkansanum Ashe	crown-forming, 2–4 m	n.g, green, pubescent, elliptic or ovate, entire	cylindro-urceolate, greenish-white	
V. ashei Reade	crown-forming, 1.5–6 m	dark green to glaucous, scattered glands, densely pubescent to glabrous, elliptic to ovate, serrate to entire	variable, usually broadly urceolate, pale to bright pink	black to dull or glaucous, 8–18 mm, insipid, poor quality
V. atrococcum Heller	crown-forming, 1.5–2.5 m	n.g., deep green, pubescent, elliptic, entire	ovate to cylindro-urceolate, greenish-white	dull black, 5–8 mm, fair flavor
V. australe Small	several stems, 2–4 m	n.g., glaucous, glabrous, elliptic-ovate, entire	cylindraceous, white	blue, 7–12 mm, excellent flavor
V. brittonii Porter	compact colonies, 15–35 cm	n.g., glaucous, glabrous, elliptic, serrate	cylindro-campanulate, white	dark blue or black, 8–12 mm, excellent flavor
V. caesariense Mackenzie	crown-forming, 1.5–2 m	n.g., glaucous, glabrous, elliptic, entire	cylindraceous, white	blue, 5–7 mm, good flavor
V. constablaei Gray	crown-forming, 1–5 m	n.g., glaucous, glabrous, elliptic, serrate	cylindro-campanulate, white	glaucous, 7–12 mm, excellent flavor

Table II. Comparison of Morphological Characteristics of Various Species of *Vaccinium* (Continued)

Vaccinium Species	Growth Habit and Height	Leaf Characteristics [a]	Flower (Corolla)	Fruit
V. *corymbosum* Linnaeus	crown-forming, 1–3 m	n.g., glaucous, glabrous, elliptic or ovate, entire or serrate	cylindraceous or urceolate, white-pink tinged	blue to black, 5–10 mm, good or excellent flavor
V. *darrowi* Camp	extensive colonies, 15–40 cm	n.g., coriaceous, glaucous, glabrous, elliptic-spatulate, entire	urceolate, pink-red	blue, 4–6 mm, fair flavor
V. *elliottii* Chapman	crown-forming, 2–4 m	n.g., green, glabrous or pubescent, elliptic, serrate	narrowly urceolate, pink	dark, 5–8 mm, poor flavor
V. *fuscatum*	crown-forming, 1.5–3 m	n.g., coriaceous, pubescent, elliptic	pink to red	dark, 6–10 mm, poor flavor
V. *hirsutum* Buckley	dense colonies, 40–70 cm	n.g., pubescent, ovate-elliptic, entire	cylindro-urceolate, glandular, pubescent	black, 6–10 mm, pubescent
V. *lamarckii* Camp	small colonies, 15–40 cm	n.g., glabrous, broad elliptic, serrate	broadly cylindraceous, white	bright blue, 4–8 mm, excellent flavor
V. *marianum* Watson	crown-forming, 2–3 m	n.g., glaucous, glabrous to pubescent, elliptic, entire	cylindraceous, dull white	dull black, 6–10 mm, fair flavor
V. *membranaceum* Douglas	erect shrub with 4-angled branchlets and exfoliating bark	ovate, acute or pointed, thin bright green, sharply serrulate	subglobose-urceolate, pinkish	purplish to black, slightly glaucous, good flavor
V. *alto-montanum* Ashe	dense colonies, 50 cm–1 m	n.g., glabrous, elliptic, entire	broadly urceolate, greenish-white	dull, 7–10 mm, glaucous, excellent flavor

Species	Habit/Size	Leaves	Flower	Fruit
V. myrsinites Lamarck	extensive colonies, 25 cm–1 m	g., coriaceous, subglaucous, obovate-elliptic, entire	cylindro-urceolate, white-pink	dark to subglaucous, 5–7 mm, fair quality
V. myrtilloides Michaux	dense colonies, 20–40 cm	n.g., pubescent, elliptic, entire	cylindro-campanulate, white	frosty blue, 4–7 mm, good flavor
V. ovatum Pursh	extensive colonies, 1–5 m	ovate, pointed, bright green, serrate	cylindro-urceolate, white-pink	black, fair flavor
V. pallidum Aiton	large colonies, 30–80 cm	n.g., subcoriaceous, subglaucous, glabrous, elliptic, serrate	cylindraceous, greenish-white	dark blue or black, 5–7 mm, glaucous, fair quality
V. simulatum Small	crown-forming, 1.5–3 m	n.g., pale green, glabrous, ovate-elliptic, serrate	cylindro-campanulate, greenish white	dark blue, 6–10 mm, good flavor
V. tenellum Aiton	extensive colonies, 15–40 cm	g., green, pubescent, spatulate, sharply serrate	narrowly urceolate, pink to red	shiny black, 5–8 mm, poor flavor and texture
V. vacillans Torrey	colonies, 15–40 cm	n.g., subglaucous, glabrous to pubescent, elliptic or ovate, entire or serrate	cylindro-urceolate to narrowly campanulate, white	glaucous to dull, 5–8 mm, fair flavor
V. virgatum Aiton	extensive colonies, 30–50 cm	g., green, pubescent, spatulate or elliptic, serrate	cylindro-urceolate, pink-tinged	shiny black, 6–10 mm, poor flavor

Source: Camp, 1945; Darrow, 1957; Gray and Fernald, 1950.

[a] n.g. = nonglandular; g = glandular.

Figure 12. Typical raceme type of inflorescence of the cultivated highbush blueberry. (D. L. Craig, North Carolina State University at Raleigh.)

Vaccinium membranaceum, the flowers are borne singly or in pairs in the leaf axils.

The growth cycle of a flowering branch of *V. angustifolium* has been described by Bell (1950, 1953). The pertinent morphological characteristics described are applicable to the highbush blueberry species. Before the start of the growing season the distal portion of a branch is characterized by a large, almost spherical terminal flower bud (Fig. 13). On occasion there may be a second flower bud penultimate to the terminal bud. Below the terminal flower bud are located four to six narrow and pointed vegetative buds. At the base of the branch is a withered twig, the flowering branch of the preceding season.

After growth resumes in the spring, the flower buds burst open and the plants are in full bloom by the third or fourth week. As the flowers emerge, the vegetative buds grow into vegetative branches and attain their approximate full length by the time the flowers are in full bloom. The vegetative branch of *V. angustifolium* averages 3 inches in length and produces a varying number of bright green leaves. The axillary bud primordia are not conspicuous at this time.

The first indication of the determinate growth habit of the blueberry is the appearance of a small black speck at the tip of the vegetative branch. This speck is the dried and withered remains of the last leaf of the vegetative branch. As this leaf withers the axillary bud primordia enlarge. At the same time the corolla withers and falls off, leaving the current season's fruit (ovaries) small and green but somewhat swollen.

Noticeable differentiation of the axillary buds along the vegetative branch occurs early in June. By mid-June the withered tissue at the tip of the branch has usually dropped off. The axillary primordium in the axil of the leaf enlarges and takes the form of a flower bud. The flower bud which will produce the next season's bloom enlarges rapidly and eventually occupies the terminal position formally held by the withered apex. A secondary flower bud may be produced in the axil of the next lower leaf. The flower buds enlarge as the season progresses and assume the typical size and almost spherical shape. The vegetative buds also enlarge and assume their characteristic long and pointed shape. As the flower and vegetative buds enlarge, the current season's fruit matures. When the fruit has fallen off, the branch that bore the fruit cluster withers and may be seen as a slender, brittle twig at the base of the distal vegetative branch.

Flower

The corolla commonly is more or less closed, is inverted, and may be shaped like a globe, bell, urn, or tubular. The petals of the corolla are united, have four or five lobes, and are commonly white or pink. The

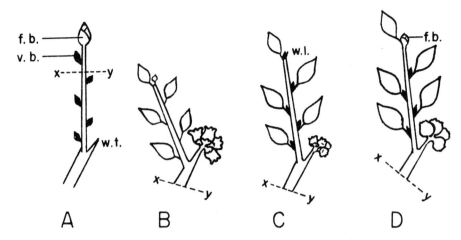

Figure 13. Above. Important stages in the development of fruit and vegetative buds in *Vaccinium angustifolium*. A shows the state in the dormant season. In B, C, and D, the growth line is distal to line *x-y* in A; B shows the development in early May, C in early June, and D in early August. *f.b.* = flower bud, *v.b.* = vegetative bud, *w.t.* = withered twig, and *w.l.* = withered leaf. (From Bell, 1950.) *Below.* The flower parts in *Vaccinium*. (From Robbins, 1931.)

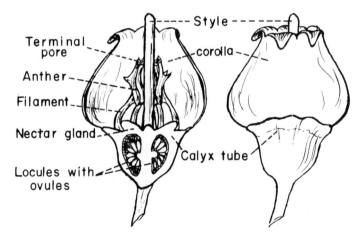

calyx forms a tube which is divided into four or five lobes (Fig. 13). The calyx is adnate to the ovary and remains with the fruit until maturity.

The ovary in the blueberry is united to the calyx tube and is consequently classified as an inferior ovary. It usually has four or five cells, or may have eight or ten by virtue of false partitions, with one to several ovules in each locule. The pistil consists of a filiform style topped by a small stigma, unmodified at the tip.

There are eight to ten stamens, usually twice as many as there are corolla lobes. The stamens are inserted at the base of the corolla and hang in a close circle about the style. The stamens are shorter than the style, and the pollen presentation area usually lies within the mouth of the corolla, an arrangement which is secured mainly by the varying length of the antherine awns. The stamen consists of an anther and a branch filament laced into a tight tube by the interweaving of their marginal hairs. The upper half of the anther is composed of two tubular structures, or awns, which provide for the escape of the pollen. Dehiscence is effected by means of pores which arise at the ends of the awns. Pollination is generally effected by an insect.

Fruit

The blueberry fruit is a many-seeded berry which matures in two to three months after flowering. Most blueberries are blue-black in color, although albino forms are known to occur. After reaching its permanent color the berry changes little in size, but for several days it continues to improve in sweetness and flavor.

The blueberry differs from the huckleberry in its seed count and seed size. An average blueberry may contain as many as 65 seeds, as in the lowbush V. angustifolium (Bell, 1957). Rarely does the number of seeds exceed 10 in the huckleberry. The seeds of hexaploid varieties are larger than those of diploid and tetraploid varieties (Darrow, 1941). Very large berries of tetraploid varieties may average 33 plump seeds per fruit, whereas very small berries may average only two plump seeds per fruit. The small seed of the blueberry does not interfere with the fruit's palatability.

Leaf

The leaves of the blueberry are simple and alternately arranged on the stem. The majority of the blueberry species, such as V. angustifolium, V. corymbosum, and V. caesariense, bear deciduous leaves. The evergreen species are represented by V. darrowi, V. ovatum, V. myrsinites, V. fuscatum, and V. ashei, which are native to the warmer localities.

The leaves vary in size from 0.7 to 3.5 cm in length for the lowbush species and up to 8 cm in length for the highbush species. The diploid species of both the lowbush and highbush groups generally have smaller leaves than the tetraploid species (Darrow, Camp, Fisher, et al., 1942). The most common leaf shape is elliptic, as characterized by V. *myrtilloides* and V. *angustifolium*. The leaves may be broadly spatulate to oblanceolate, as in V. *ashei*, or ovate, as in V. *corymbosum*.

The presence of a pubescence on the underside of the leaf is common to many species of the blueberry. These fine hairs are typical of V. *corymbosum*. In V. *hirsutum* the pubescence also covers the flower and fruit. Other species, as typified by V. *australe*, are glabrous. The lowbush species V. *angustifolium* is rarely found with pubescent leaves.

Other leaf characteristics commonly used to identify the different blueberry species are the presence or absence of glandlike structures on the underside of the leaf and whether the leaf margin is serrate or entire (Bailey and French, 1946; Clark and Gilbert, 1942; Clark, 1941). Species bearing the minute stalked glands include V. *myrsinites*, V. *fuscatum*, V. *ashei*, V. *tenellum*, V. *virgatum*, and V. *amoenum*. Important nonglandular species are V. *angustifolium*, V. *corymbosum*, and V. *australe*. Species characterized by entire margins include V. *australe*, V. *caesariense*, V. *myrtilloides*, V. *darrowi*, V. *ashei*, and others listed in Table II. Species having very sharp leaf serrations include V. *angustifolium*, V. *amoenum*, V. *virgatum*, V. *tenellum*, V. *simulatum*, and V. *brittonii*. The leaf margin of V. *corymbosum* may range from sharply serrate to entire, depending on its location. Other species with serrated margins are given in Table II.

Root

The highbush blueberry root system is devoid of root hairs (Coville, 1910). Because of this, the absorptive capacity of blueberry roots is small when compared with root systems with hairs. An interesting comparison made by Coville showed that a section of a blueberry rootlet with no root hairs presents about one-tenth of the absorptive surface of an equal area of a wheat rootlet bearing root hairs. Further observation revealed that the blueberry rootlet grows 1 mm per day under favorable conditions, whereas the wheat rootlet often grows twenty times as fast. The young blueberry rootlets are exceedingly slender before they branch, varying from 50 to 75 μ in diameter.

The rootlets of vigorous blueberry plants are inhabited by an endotrophic mycorrhiza (Coville, 1910). This fungus appears as a threadlike, transparent or pale brown network on the exterior of the epidermal cells (Doak, 1928). The presence of this fungus growth does not appear to be detrimental to the blueberry plant. It is suggested that the mycorrhizal fungus of the swamp blueberry transforms the nonavailable nitrogen of peaty soils into a form of nitrogen available to the blueberry plant.

In the lowbush blueberry, exemplified by V. *angustifolium* and V. *myrtilloides*, the major portion of the root system is adventitious, originating from extensive rhizome activity (Fig. 14; Mahlstede and

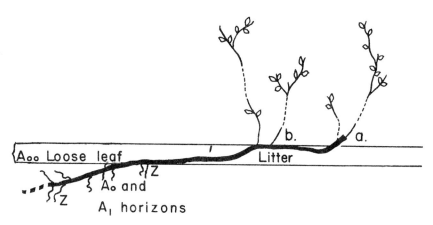

Figure 14. Lowbush blueberry rhizome, showing stems originating terminally at point *a.* and mid-rhizome at *b.* and roots z in A_0 and A_1 horizons. (From Trevett, 1956.)

Watson, 1952). It has been estimated that approximately 85 per cent of the stem tissue of the lowbush blueberry is in the form of underground rhizomes, the remainder consisting of short aerial shoots arising from thickened areas in the rhizomes. The rhizomes are generally sympodial in shape and measure 3 to 6 mm in diameter. They branch frequently, forming a close interlacing network 6 to 25 mm below the soil surface. The young rhizomes are pink with brownish scale leaves, whereas the older rhizomes are dark brown and suberized.

The bulk of the lowbush blueberry's root system is located in the surface organic mat and in the A_1 horizon (Trevett, 1956). Recent observations by Hall (1957) suggest that an extensive tap root in the root system of V. *angustifolium* also plays an important role in water and nutrient uptake because of its deeper penetration into the soil profile.

ANATOMY

Flower

In the blueberry flower, the united sepals that make up the calyx consist of an upper and lower epidermis, with mesophyll tissue in between, and a few highly branched veins (Wasscher, 1947). The upper epidermis in *V. corymbosum* consists of a single layer of cells irregularly aligned. The lower epidermis is composed of smaller cells with stomates present. The mesophyll is characterized by thin-walled parenchyma cells forming a layer 100 to 125 mμ thick. This mesophyll tissue may contain many chloroplasts and many large intercellular spaces. The conducting tissue is made up of phloem consisting of narrow elements and of xylem composed of narrow spiral vessels and slightly larger sieve vessels.

The corolla is made up of united petals which contain a main vein and a number of smaller branched veins (Fig. 15). The upper epidermis consists of irregular cells with straight walls. Stomates may be present on the upper half of the petal. The lower epidermis is similar to the upper epidermis, but the cuticle is not as thick and the number of stomates is greater. The mesophyll consists of elongated cells two or three layers thick. Large intercellular spaces are common. The main vein has two layers of parenchyma on the upper side, one layer on the lower side, phloem elements 3 to 7 mμ in width and spiral tracheids 5 mμ in width.

In *V. corymbosum* the stamen consists of a tapelike hairy filament and an elongated anther which is inserted on the filament at an angle (Fig. 16). The upper epidermis consists of elongated cells covered with a thin cuticle on the outer wall. Hairs, which may attain a length of 500 mμ, are inserted in the epidermal cell. On the outer surface of the anther's upper epidermal cells there are long, pointed papillae. The papillae are situated in the middle of the cell and may reach a size of 15 by 8 mμ.

Common characteristics of stamens in the blueberry are the frequent development of appendages (awns), a strong tendency toward terminal dehiscence of the anther, the absence in the wall of the anther of a well-marked fibrous layer, and the inversion of the anther during its growth so that the apical pores of the anther are really basal (Matthews and Knox, 1926). The anther is commonly of the adnate type. The paired awns associated with the anther arise from the top of the filament or from the connective. Awns may be present in some species of *Vaccinium* and absent in others.

Dehiscence in *Vaccinium* occurs after the thin-walled tissue on the inner face begins to disintegrate in the region where the opening will develop. This initiates the concavity, which gradually deepens to form a pore. The destruction of the tissue proceeds inward and downward, forming tubular structures that connect with the pollen sacs below. The pollen grains are formed in tetrads. The nucleus in each pollen grain divides to form the vegetative and generative cells about 15 days after the formation of this tetrad. When mature, the pollen drops down from the two anther sacs through two anther tubes and exits at the terminal pores.

In *Vaccinium* the pistil consists of an inferior ovary, style, and stigma. The ovary of V. *corymbosum* is divided into 10 parts by five partitions (Fig. 13) of which the last are formed because the ovary

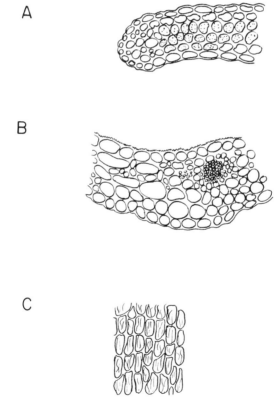

Figure 15. The corolla of *Vaccinium corymbosum.* A. Cross section of the edge of a sepal. B. Cross section of a flower petal. C. Upper view of the epidermis of a petal. (From Wasscher, 1947.)

wall possesses an outgrowth up to the middle of the space. The placenta is pointed in the upper corner of the locules against the central pillar. The ovary wall gradually merges with the sepals. At the top, the flat ovary carries a ring-formed disc. The style is straight and carries a budlike stigma.

The ovary wall is generally 300 to 500 mμ thick and consists of an outer and inner epidermis with mesophyll interlying. In *V. corymbosum*

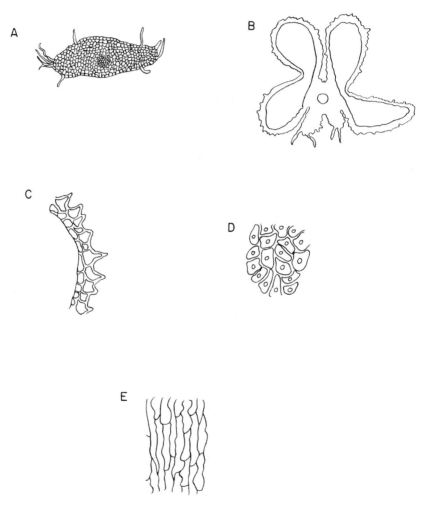

Figure 16. The filament and anther of *Vaccinium corymbosum*. A. Cross section of a filament, 170✕. B. Cross section of a ripe anther, 70✕. C. Cross section of an anther wall, 170✕. D. Upper view of the anther wall, 170✕. E. Wall of the anther tube, 170✕. (From Wasscher, 1947.)

there are 10 large conducting vessels and some small conducting vessels (Bell and Giffin, 1957). Half of the large conducting vessels are situated in front of the two inner walls; the remainder are in between.

In the central pillar there are five conducting vessels that join at the style after sending side branches to the placenta. The central pillar consists of seven layers of cylindrical cells, the length axis in a crosswise direction. The conducting tissue is similar to the wall but is differently situated in relation to the inner wall. The short, bell-shaped placenta, located in the inner and upper part of the locule, consists of parenchyma tissue, in which there is conducting tissue covered by epidermal tissue.

The straight, cylindrical style consists of epidermis, mesophyll, and conducting tissue, and an inner epidermis that borders in cross-section to four or five branched style canals (Fig. 17). The mesophyll consists of regular hexagonal parenchyma cells 40 to 50 mμ in diameter. The cell size increases toward median direction. The cells between the branches of the style canal are only 15 mμ in diameter. The conducting tissues are similar to the ovary wall and are found in front of the flat sides of the style canal.

The stigma consists of small isodiametric parenchyma cells with strong walls. The contents of the cells are darker than the walls and the epidermal cells are thick-walled.

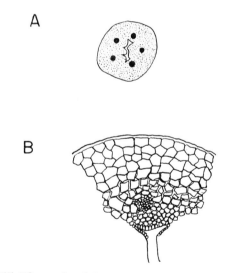

Figure 17. The style of *Vaccinium corymbosum*. A. Cross section of the style, 35×. B. Cross section of the style, 170×. (From Wasscher, 1947.)

Fruit

A detailed study of the anatomy of the blueberry fruit has been made by Yarbrough and Morrow (1947). Examinations were made of *V. ashei, V. alto-montanum, V. constablaei, V. tenellum,* and *V. australe* (horticultural varieties Rubel and Dixi).

The outer epidermis of the blueberry fruit consists of a single layer of cells devoid of stomates. In the epidermis and in the several layers of hypodermis lying beneath there occurs a dark purple pigment which gives the fruit its characteristic color. In the varieties Rubel and Dixi, this pigment is present in the entire fleshy pericarp.

In the plants studied, the endocarp consists of a single layer of tightly fitted stone cells (Fig. 18) and constitutes a lining of the five compound locules. These stone cells are generally smaller and more elongated than the stone cells lying in the flesh of the mesocarp. Although the stone cells of the mesocarp are larger than those of the endocarp, they have the same fundamental structure of strongly lignified walls and many simple pits. The stone cells may occur in small groups or singly. In the Rubel and Dixi varieties the stone cells appear congregated in a zone around the locules.

Species with abundant stone cells exhibited the character of grittiness and were generally less palatable than those containing fewer stone cells. Rubel and Dixi contained the lowest number of stone cells of all the species investigated. The authors conclude that the degree of grittiness due to stone cells may serve as a useful characteristic for distinguishing varieties.

Bell (1957) found a relatively low average of viable seed per ripe berry. Of an average of 64 seeds per berry, approximately 78 per cent were imperfect. When examined by size, the larger berries had a much greater proportion of perfect seeds. Complete pollination increased the percentage of normally developing ovules. The largest and most perfect seeds in a berry were always clustered around the top of the central axis, with the imperfect seeds occupying the lower and basal portion of the loculi, suggesting that there had been an insufficient number of pollen tubes to fertilize all the ovules. Many embryos had died at some stage of development. Mature seeds were found to be axile linear. Imperfect seeds were either medium-sized and solid, with middle integumentary layers lignified, or small and collapsed, with the tissues inside the seed coat disintegrated.

Bell (1957) and Stevens (1919) give detailed accounts of the development of the endosperm in *V. angustifolium* and *V. corymbosum,* respectively.

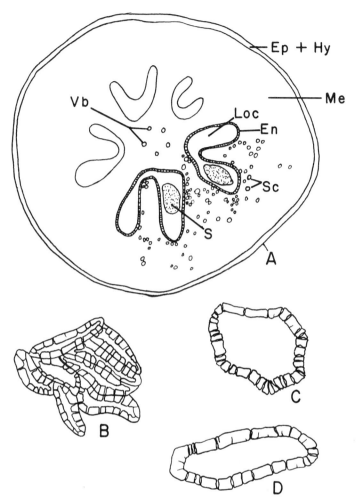

Figure 18. The blueberry fruit. A. Transverse section of the mature fruit of *Vaccinium ashei*. $Ep + Hy$ = epidermis and hypodermis; Me = mesocarp; En = endocarp; Loc = locule; S = seed; Sc = stone cells; and Vb = vascular bundles; 8×. B. A group of cells forming the endocarp, showing grouping and wall pits as seen in optical section, 215×. C. and D. Stone cells of the mesocarp as seen in optical section, 215×. (From Yarbrough and Morrow, 1947.)

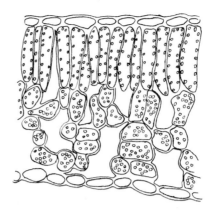

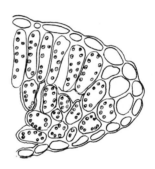

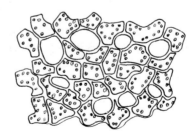

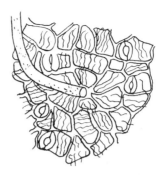

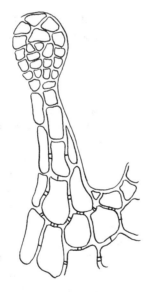

Figure 19. The leaf of *Vaccinium corymbosum*. A. Cross section of the leaf, 340×. B. Cross section of the leaf edge, 340×. C. Longitudinal section of the spongy parenchyma, 340×. D. Dorsal view of the lower epidermis showing the arrangement of epidermal cells and pubescence, 170×. E. Gland hair at the leaf edge, 340×. (From Wasscher, 1947.)

Leaf

The leaf anatomy of 11 species of *Vaccinium* has been studied by Wasscher (1947). Included in this study were a number of varieties of *V. corymbosum*. The leaf of *V. corymbosum* consists of an upper epiderm, composed of a single layer of irregularly rectangular cells, usually a palisade layer one cell in thickness, and a spongy parenchyma of varying thickness containing numerous air spaces (Fig. 19). Stomates occur primarily in the lower epidermis. They are medium in size and are encircled by two true guard cells. An average of 231 to 295 stomates per mm^2 were observed on the underside of a leaf. In most species the interveinal area of the upper epidermis has no hairs, but interveinal unicellular hairs occur on the lower epidermis in *V. corymbosum*.

The vascular bundle is composed of one layer of phloem and one layer of xylem, and it is surrounded by a completely or nearly closed sheath of fibers. Around the vascular bundle lies a layer of cortical parenchyma. The epidermis of the midrib of most species is composed of rectangular cells with straight walls. Club hairs occur frequently on the lower side of the midrib as in *V. hirsutum* and *V. corymbosum*.

In *V. corymbosum* vars. Harding and Rancocas, the margin of the leaves is toothed owing to the presence of large club hairs (Fig. 19). In most species the mesophyll of the margin is composed of parenchymatous tissue.

Stem

The blueberry stem has a thick cuticle which covers a single layer of epidermal cells (Fig. 20). In *V. corymbosum* the cuticle is interrupted by stomates formed by two thin-walled guard cells. The epidermal layer consists of closely packed polyhedral cells possessing blunt, saw-toothed apices on their exterior surfaces. The outer cortical region is composed of five to 12 rows of chlorophyll-containing cells. These cells form a solid tissue between the epidermis and the air canals. The air canals are so oriented that two ducts are adjacent to one another along the radial axis. These canals vary in thickness and are bounded by one and two rows of cortical parenchyma.

Centripetal to the ring of air canals is a continuous band of thick-walled pericyclic fibers with a narrow lumen. The adjacent phloem tissue consists of nearly fiber-shaped sieve tubes with obliquely suited septa. In the xylem no great difference exists between the spring and autumn wood. The xylem is composed of long, angular vessels surrounded by thick-walled fibers and is dissected radially by rows of thick-walled xylem parenchyma.

In *V. corymbosum* the rays may be either broad or uniseriate (Flint, 1918). The broad rays are composed of large, light-colored cells and smaller dark cells. The broad rays are compound in structure, resulting from the fusion of small rays or the transformation of fibers into parenchyma cells. The pith is composed of parenchyma with thin cellulose walls.

Root

Adventitious roots of hardwood cuttings arise from within at the base of the cutting (Mahlstede and Watson, 1952). The root initial originates in the cambium and phloem region, often immediately adjacent to a xylem ray. The cells which are to become a root initial are characterized by a large, centrally located nucleus and small vacuole. After numerous anticlinal and periclinal cell divisions, the root primordium pushes toward the periphery of the stem and may be projected

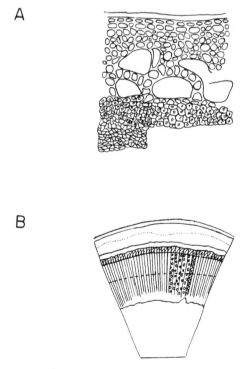

Figure 20. The stem of *Vaccinium corymbosum*. A. Cross section of a year-old branch, showing the phloem ring, 170×. B. Cross section of a two-year-old stem. The stippled area = cork; black = strengthening tissue; rays = phloem; blocked = xylem; 35×. (From Wasscher, 1947.)

through the callus or through the outer stele, cortex, and epidermis. Failure of the primordium to emerge may be caused by inhibition by the lignified pericyclic fibers or by the epidermis of closely packed cells and the thick cuticle.

The young blueberry rootlets vary from 50 to 75 μ in diameter and may contain three to five rows of epidermal cells. The epidermal cells which are devoid of fungus hyphae are extremely transparent, and the protoplasmic membrane lining the cell is visible only where it is thickened to envelop the nucleus. The remainder of the cell is filled with a colorless cell sap. The anatomy of an older root is similar to that of the stem.

In the lowbush blueberry the rhizome differs from a typical root by virtue of the scales borne near the distal end. Anatomically the rhizome also resembles the stem. Another way of distinguishing the rhizome from the lowbush tap root, however, is to compare the number of growth layers in the cross section of the proximal end of the rhizome and tap root (Hall, 1957). The older tap root would reveal the greater number of growth layers.

CHAPTER 3

Breeding

by *James N. Moore, Department of Horticulture and Forestry, University of Arkansas, Fayetteville*

PRINCIPLES, TECHNIQUES, AND OBJECTIVES

Interspecific Hybridization

Interspecific hybridization represents the foundation on which today's blueberry breeding programs have been built. Coville (1937) undertook interspecific hybridization at the very beginning of his blueberry breeding experiments. The first varieties to be introduced, Cabot, Katharine, and Pioneer, were first generation hybrids of *V. corymbosum* and *V. australe* (Table III). *V. lamarckii* (*V. angustifolium*), the tetraploid lowbush blueberry, was also used by Coville in crosses with other species. The variety June is one-fourth *V. lamarckii*, Weymouth is one-eighth *V. lamarckii*, and Earliblue is one-sixteenth *V. lamarckii*.

There are eight diploid (2n = 24), eight tetraploid (2n = 48), and three hexaploid (2n = 72) cluster-fruited blueberry species indigenous to eastern North America (Aalders and Hall, 1961; Darrow, 1949, 1960; Darrow, Camp, Fisher, *et al.*, 1944). Coville (1927) reported that all diploids were interfertile, as were all tetraploids, and that *V. ashei*, a hexaploid species, would cross with the tetraploid highbush. A study of hybrids between many species showed that nearly all homoploid species were fully interfertile, whereas heteroploid crosses yielded seedlings with both fertile and sterile pollen and ovules (Darrow and Camp, 1945).

Since there are few if any sterility barriers between species with the same chromosome number, a great many crosses have been made to combine the various desirable characteristics of the many species. Some of the interspecific hybridizations that have been made are listed in Table IV. In addition, many of the hybrids from these crosses have been crossed with species other than those from which they were derived.

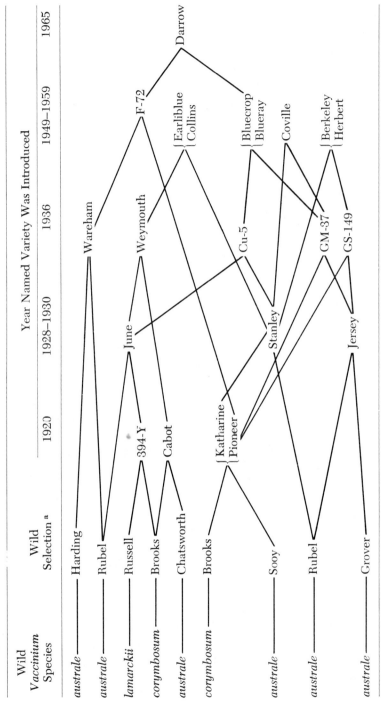

Table III. Ancestry of the New Highbush Blueberry Varieties.

Source: Darrow, 1960.
[a] Rubel, Chatsworth, Sooy, Harding, and Grover were selected in New Jersey, and Russell and Brooks in New Hampshire.

Table IV. Some Successful Interspecific Blueberry Hybridizations Made under Controlled Conditions

Crosses Between Diploids

V. tenellum × V. darrowi
V. pallidum × V. myrtilloides
V. myrtilloides × V. angustifolium [a]
V. pallidum × V. atrococcum
V. angustifolium × V. atrococcum
V. atrococcum × V. angustifolium

Crosses Between Tetraploids

V. australe × V. corymbosum
V. australe × V. simulatum
V. australe × V. alto-montanum
V. australe × V. myrsinites
V. corymbosum × V. lamarckii
V. corymbosum × V. brittonii
V. lamarckii × V. hirsutum
V. lamarckii × V. myrsinites
V. australe × V. lamarckii
V. australe × V. hirsutum
V. australe × V. arctostaphyles
V. brittonii × V. myrsinites
V. corymbosum × V. myrsinites
V. lamarckii × V. alto-montanum
V. lamarckii × V. brittonii

Crosses Between Hexaploids

V. ashei × V. amoenum
V. ashei × V. constablaei

Heteroploid Crosses

V. ashei (6X) × V. tenellum (2X)
V. ashei (6X) × V. myrsinites (4X)
V. ashei (6X) × V. atrococcum (2X)
V. australe (4X) × V. amoenum (6X)
V. australe (4X) × V. darrowi (2X)
V. simulatum (4X) × V. ashei (6X)
V. alto-montanum (4X) × V. ashei (6X)
V. myrsinites (4X) × V. constablaei (6X)
V. ashei (6X) × V. vacillans (2X)
V. ashei (6X) × V. darrowi (2X)
V. ashei (6X) × V. corymbosum (4X)
V. australe (4X) × V. ashei (6X)
V. corymbosum (4X) × V. constablaei (6X)
V. angustifolium (2X) × V. alto-montanum (4X)
V. lamarckii (4X) × V. ashei (6X)

Source: Darrow and Camp, 1945.
[a] A probable misnomer, since V. angustifolium has been definitely established as tetraploid (Aalders and Hall, 1961).

To this list of successful interspecific crosses may be added a large number of naturally occurring species hybrids observed by Darrow and Camp (1945). These authors noted the ease with which hybrids could be produced between homoploid species and the value of this breeding approach in combining the numerous divergent characters of the various species to produce desired forms for the selection of new horticultural varieties.

Records left by Coville indicate that he obtained viable hybrids between the deerberry, Vaccinium stamineum Linnaeus and V. myrtilloides Michaux, both diploids. In addition, he recorded a successful cross between V. melanocarpum Mohr and V. myrtilloides.

Recently Rousi (1963) in Finland reported successful hybridizations between *Vaccinium uliginosum* Linnaeus, a bilberry, and the cultivated highbush varieties Rancocas and Pemberton. Such crosses might be of value in utilizing the genes for extreme cold resistance of *V. uliginosum* in breeding programs for areas where low winter temperatures are a limiting factor in blueberry production.

There are many species in the three chromosome groups that have characters that would be desirable in cultivated varieties, such as *V. darrowi* and *V. myrsinites* for low chill requirement in Florida and *V. lamarckii* for earliness and winter hardiness in Maine. Other species and the characters of value that they possess are listed in Table V.

Table V. *Vaccinium* Species Possessing Desirable Characters of Potential Value in a Blueberry Breeding Program

Species	Desirable Characters
DIPLOID ($2n = 24$)	
V. *atrococcum* Heller	very early ripening; uniform ripening; adaptation to upland soils
V. *darrowi* Camp	low chilling; heat and drought resistance; light blue fruit color
V. *elliottii* Chapman	adaptation to less acid soils
V. *myrtilloides* Michaux	cold hardiness; light blue color
V. *myrtillus* Linnaeus	extreme cold hardiness
V. *tenellum* Aiton	low chilling; heat and drought resistance
V. *vacillans* Torrey	drought resistance
V. *uliginosum* Linnaeus	cold hardiness; resistance to stem diseases
TETRAPLOID ($2n = 48$)	
V. *alto-montanum* Ashe	drought resistance
V. *angustifolium* Aiton	early ripening; cold hardiness
V. *australe* Small	resistance to fungus diseases
V. *brittonii* Porter	cold hardiness
V. *corymbosum* Linnaeus	cold hardiness; early ripening
V. *lamarckii* Camp	cold hardiness; early ripening
V. *membranaceum* Douglas	extreme drought resistance
V. *myrsinites* Lamarck	heat and drought resistance
HEXAPLOID ($2n = 72$)	
V. *ashei* Reade	small fruit scar, bush vigor, heat and drought tolerance, tolerance to fungus diseases, tolerance to upland soils
V. *constablaei* Gray	long rest period, early ripening, cold hardiness

Source: Moore, 1965.

Many of these species have been used in crosses, and some show promise as parents in transferring desired characters to cultivated forms (Fig. 21). Diploid species have been utilized by first crossing with hexaploid species and subsequently using the synthetic tetraploid seedlings in crosses with cultivated tetraploid varieties (Darrow, Dermen, and Scott, 1949; Darrow, Scott, and Dermen, 1954). Examples of this type of breeding are the crosses of (*V. tenellum* Aiton × *V. ashei*) × highbush and (*V. darrowi* × *V. ashei*) × highbush.

Darrow, Morrow, and Scott (1952) reported on the evaluation of several interspecific blueberry crosses. In addition to highbush × lowbush crosses discussed later, they studied crosses of highbush × Florida evergreen (*V. myrsinites*), lowbush (*V. angustifolium*) × dryland lowbush (*V. alto-montanum* Ashe), dryland lowbush × highbush, rabbiteye (*V. ashei*) × *V. constablaei*, and rabbiteye × highbush.

Plants of highbush × Florida evergreen were unusually vigorous, had black to very dark blue fruit, and ripened very late. This cross appeared to the authors to be very promising because of the drought resistance of the evergreen parent and the increased hardiness of the hybrids.

The lowbush × dryland lowbush plants were not promising when grown in Maryland.

The dryland lowbush × highbush plants were low-growing and spread freely. Although the fruit of some was rather large, the color was dark.

Rabbiteye × highbush crosses were very vigorous and hardier than the rabbiteye. Though not self-fruitful, they set good crops when pollinated by either highbush or rabbiteye. Some produced large-sized fruit of high flavor.

Seedlings of rabbiteye × *V. constablaei* were hardy in Maryland when rabbiteye varieties were not. Many of the seedlings had light blue fruit.

Brightwell, Darrow, and Woodard (1949) also studied seedlings from crosses of the rabbiteye and *V. constablaei*. They found that the hybrids generally lacked vigor and bore smaller fruit than the rabbiteye and that only 28 per cent had fruit of acceptable color. The hybrids exhibited greater resistance to leaf diseases and generally had small scars and good quality. The most desirable qualities of the seedlings from this cross were their early maturity and small, dry scars. The authors considered this approach to be very promising in obtaining varieties for the South that ripen earlier and over a shorter period of time than varieties of *V. ashei*.

Interspecific hybridization is among the most promising breeding approaches for the future. Since all existing highbush varieties can be traced back to only a few wild selections, the genetic base of blueberry breeding needs to be widened by introducing new genes. As the industry continues to expand into new and sometimes marginal areas, new problems are sure to arise. For many of these problems the only answer will prob-

Figure 21. Above. Dr. Darrow inspecting seedlings that resulted from crossing a lowbush from Maine with a highbush. *Below.* Another hybrid produced by crossing *Vaccinium darrowi* with *V. tenellum.* (USDA, Beltsville, Md.)

ably be to introduce new genes into the breeding program. Species hybridization appears to be the easiest approach to introducing new genes into the cultivated blueberry, although the gene pool may be increased by utilizing intergeneric crosses, colchiploidy, induced mutations, and further exploration of the wilds in the search for new characters.

Inheritance

The improvement of the blueberry by means of plant breeding is based on the fundamentals of inheritance. Inheritance makes it possible, by choosing selected parents having desirable characteristics, to produce progeny that are commercially superior to their parents.

The inheritance of certain characters in the blueberry was first reported by Darrow, Clark, and Morrow (1939). Some 3,000 seedlings from controlled crosses were rated for size, scar type, color, and season of ripening. Their results indicated that certain blueberry varieties transmitted desirable characters to seedling progenies with a higher frequency than other varieties. Some characters and the varieties which transmitted them are:

Large size: Stanley, GM-37, FI-66, GS-149, and probably Wareham.
Small size: Russell and Allen (varieties of *V. lamarckii*) and Cabot.

Good scars: Russell and Allen.
Fair scars: Stanley, GS-149, and GM-37.
Bad scars: Grover.

Light blue color: Stanley.
Dark blue to black color: Russell, Allen, and Grover.

Early ripening: Russell, Allen, June, and Cabot.
Late ripening: Jersey, selections of *V. ashei*.

Although the authors did not attempt any genetic interpretation of their results, the data indicate that the characters are probably inherited quantitatively.

Johnston (1942a) studied the transmission of bush size and growth habit, productivity, berry size, color, firmness, flavor, ease of picking, and season of ripening to seedling progenies by the varieties Adams, Pioneer, Rubel, Harding, and Cabot. High transmission of characters to offspring of these varieties is as follows:

Upright growth habit: Rubel and Adams.

Large bush size: Pioneer and Rubel.

High productivity: all varieties used.

Large berry size: results variable, Pioneer best.

Light blue color: Rubel and Cabot.

Firm berry: Rubel.

Sweet flavor: Pioneer and Harding.

Ease of picking: all varieties used.

Early ripening: Adams.

Late ripening: Rubel.

Johnston also evaluated the parents used in the crosses by the number of promising selections made from each cross. The cross of Pioneer × Rubel gave the greatest number of promising selections by far.

The Ashworth blueberry, a *V. corymbosum* selection from the wild, has been prepotent in transmitting hardiness, earliness, and precocity to its seedlings (Darrow, Whitton, and Scott, 1960; Moore, 1965). However, it has the disadvantage of transmitting susceptibility to witches'-broom rust.

Sharpe and Darrow (1959) found that both *V. myrsinites* and *V. darrowi* were good parents in transmitting low winter cold requirement in the Florida breeding program. *V. myrsinites*, however, also transmitted black, unattractive fruit to its progenies, whereas some *V. darrowi* seedlings had attractive blue fruits. Consequently the breeding program in Florida at present is based on the use of *V. darrowi* and its seedlings.

Moore, Bowen, and Scott (1962) reported that susceptibility to powdery mildew appeared to be transmitted to blueberry progenies as a partially dominant character. The varieties Berkeley, Earliblue, and Ivanhoe were resistant and could be expected to transmit resistance to a large per cent of their progenies, whereas Collins, Stanley, Rubel, Blueray, Burlington, Herbert, Jersey, and Atlantic were very susceptible. These authors also found that the response of very young seedlings in the greenhouse to powdery mildew was closely correlated with their response later in the field and that many susceptible seedlings could be eliminated at an early age in the greenhouse.

Brightwell, Woodard, Darrow, *et al.* (1955) found that the rabbiteye varieties Myers and Walker transmitted small fruit size when used as parents, whereas Black Giant and Ethel transmitted large size to their progenies. Fruit color in the rabbiteye appeared to be governed by multiple factors, with dark color partially dominant. An evaluation of interspecific crosses indicated that different genes for color exist in the different species of blueberries. Crosses of *V. myrsinites* × *V. australe* indicated that the small, compact plant and small leaf size of V. *myrsinites* are dominant characters.

The highbush varieties evaluated for transmission of desirable characters by Darrow, Clark, and Morrow (1939) and by Johnston (1942) and the rabbiteye varieties evaluated by Brightwell, Woodard, Darrow, *et al.* (1955) have largely been superseded in breeding programs by more recently introduced varieties. Information is needed on the value of these newer varieties in transmitting certain characters to their progenies.

Techniques of Breeding

The techniques of blueberry breeding are similar to those used in other fruit crops. Crossing may be performed in a greenhouse or in the field, but the greenhouse has many advantages. In a greenhouse, plants may be brought into flower during the winter as soon as they have gone through an adequate chill period. At Beltsville, plants forced after February 1 flower normally. Since blueberry varieties and species vary considerably in the time they require to blossom after forcing, the plants to be used as pollen parents are forced several days earlier than the seed parents.

Cut branches may also be forced for pollen. Branches should be cut so that some two-year-old wood is attached at the basal end. Branches are kept in a jar of water under mist until the bud scales of the blossom buds separate. They are then moved to a greenhouse bench. Because conducting vessels at the cut ends of branches tend to become plugged after a few days, it is suggested that small sections be removed from time to time. Snipping tips off flower buds on cut branches insures the shedding of pollen by weak buds that otherwise might not open.

Emasculation is performed by removing the corolla and the ring of stamens with a small forceps just before anthesis (Fig. 22). Pollination is effected by touching the stigmatic surface of the pistil with a glass slide or thumbnail on which pollen has been collected. The fingertip may also be used to apply pollen to the stigma and gives more complete coverage than a camel's-hair brush (Merrill, 1936). Pollination may be performed immediately or up to four days after emasculation in the greenhouse (Moore, 1964). The optimum time for pollination is 24 to 48 hours after emasculation (Merrill, 1936). In heteroploid species crosses, naphthalene acetamide in lanolin may be applied to tissue torn by removal of the corolla during emasculation to increase the berry set (Darrow, 1956b).

The berries are harvested when fully colored, that is, 45 to 60 days after pollination for the highbush varieties. The seeds are removed by shredding the berries in a food blender for about 30 seconds and pouring off the pulp and skins (Morrow, Darrow, and Scott, 1954). Sound seeds settle to the bottom and abortive seeds and pulp float. The seeds are air-dried, placed in a container, and stored in an ordinary refrigerator. Seeds of the blueberry stored in this manner have remained viable up to 12 years (Darrow and Scott, 1954).

Seeds are planted on shredded sphagnum moss and covered very lightly with sphagnum. Stratification does not seem to be necessary provided the seeds have been removed from the fruit for five to six months (Scott and Ink, 1955). Seedlings begin to emerge about a month after seeding and continue to emerge over a long period. When the seedlings are a few inches in height, they are transplanted to small peat pots in a mixture of 2 parts peat moss, 1 part sand, and 1 part composted soil. They are normally grown the first summer in this manner and trans-

Figure 22. Cluster of blueberry flowers with two flowers emasculated and ready for pollination. (USDA, Beltsville, Md.)

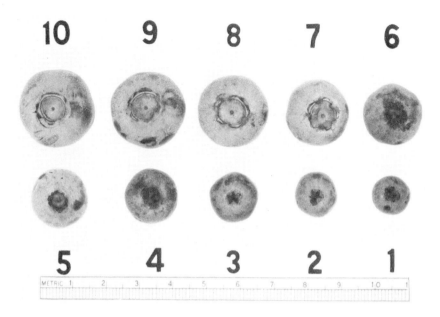

Figure 23. Blueberry fruit rated for size on scale of 1 to 10. (USDA, Beltsville, Md.)

planted to nursery rows after they become dormant. Following a year in the nursery they are planted in the field to fruit.

Seedlings and selections are evaluated by scoring individual characters on a scale of 1 to 10, with 10 representing the most desirable manifestation of the character and 6 the lowest commercially acceptable score (Morrow, Darrow, and Rigney, 1949). Figures 23 and 24 show representative samples for the 1-to-10 scoring system for size, color, and scar. With experience, a breeder may become quite adept at determining by observation slight variations in the various characters.

Berry color, however, can be determined by comparison with a standard color chart. In one study the color of the most glaucous berries had a value of about 6.5 and was within the chroma range of 2 to 3 on the Munsell Purple-Blue Chart (Morrow, Darrow, and Rigney, 1949).

Berry size can be measured with a metal gauge containing holes of various sizes. Berries are sized by finding the hole that the berry will just pass through. Another method is to weigh a given number of berries picked at random from each bush. The method of determining berry size most used by breeders is to count the number of berries required to fill a half-pint measuring cup.

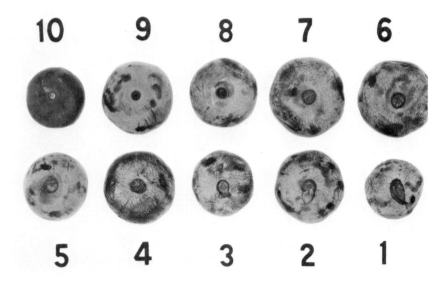

Figure 24. Above. Blueberry fruit rated for color on scale of 1 to 10. *Below.* Blueberry fruit rated for scar on scale of 1 to 10. (USDA, Beltsville, Md.)

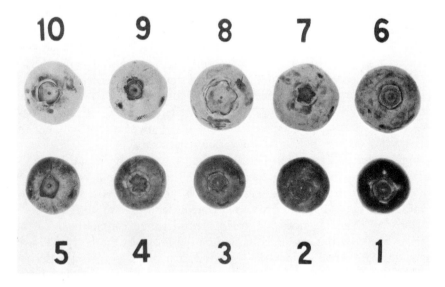

Breeding 57

Objectives

The early objectives in blueberry breeding included large berry size, light blue color, small scar, firmness of fruit, good dessert quality, and productivity. Each of these objectives has been realized, but all are not yet combined to perfection in any single variety. Attempts are currently being made to combine these desirable characters and to extend the harvest period in a series of varieties that ripen at about five-day intervals. With the expansion of blueberry acreage, new objectives for blueberry breeding have been introduced. Breeding objectives for the various areas where active breeding programs are in progress have been listed by Scott and Moore (1961) as follows:

A) Major objectives of work in the North:
  1) Greater winter hardiness.
  2) Highly productive plants with firm, large, flavorful berries.
  3) Varieties maturing earlier and later.
  4) Varieties resistant to mummy berry.
  5) Berries with a small, dry scar.
  6) Species hybridization for drought resistance and wide adaptation.
  7) Inheritance and polyploid studies.
  8) Adaptation to mechanical harvesting.
B) Major objectives of work in North Carolina:
  1) Varieties resistant to canker and mites.
  2) Early varieties with light blue, large, firm, flavorful fruit.
  3) Very vigorous, productive midseason varieties.
C) Major objectives of work in Georgia and Florida for the deep South:
  1) Large, flavorful, very productive rabbiteye.
  2) Early ripening tetraploid highbush with a short chill requirement.
  3) Species hybridization to obtain heat-resistant and drought-resistant tetraploids.

Breeding for Specific Characters

Desirable fruit characters to be considered in selecting blueberries are large size, light blue color, small, dry scar, firmness, and good dessert quality. Desirable plant characters include a vigorous, upright bush, dependable production, loose clusters, heat and drought resistance, and resistance to diseases. Other important characters are time of fruit maturity, resistance to premature fruit abscission at maturity, and ease of propagation. Most, and probably all, of these characters are inherited quantitatively.

Fruit Size. Large berry size is a desirable and important character in blueberries. Large berries not only make a more appealing fresh market pack but also are easier to harvest than small ones. The large size of the present newer varieties has come about through an increase in size with each generation of seedlings (Fig. 25). Since all of the current Big Seven varieties were originated through three generations of seedlings from only six wild selections, it is apparent that dramatic increases in size have been achieved (Figs. 26 and 27). Some of the newer varieties are over five times as large as their wild progenitors. Further size increases of this magnitude in future seedlings are doubtful, but there is no reason to believe that the ultimate size has been attained. Parents that transmit large fruit size to a high percentage of their progenies are Berkeley, Herbert, US 11-93 ([Jersey × Pioneer] × [Stanley × June]), and E-30 (Berkeley × Earliblue).

Color. The intensity of the bloom, a waxy covering on the skin of the berry, determines the external color of the berry. A very light blue is highly desirable, especially for blueberries to be marketed as fresh fruit. Berries of light color appear fresh even after several days on the market counter. The layer of wax perhaps helps to prevent loss of moisture and shriveling. The varieties Berkeley and Bluecrop, which are among the best-colored blueberries at present, transmit this character well in breeding.

Scar. Large, wet scars (point of attachment of fruit to pedicel) admit fungi that cause post-harvest decay. Loss of moisture through the scar may also result in shriveling. Burlington and Bluecrop transmit small scars to many of their progenies. The best scars are found in varieties of the rabbiteye blueberry, *V. ashei*. With new developments, it should be possible soon to utilize these genes for good scar in highbush breeding.

Firmness. Firmness of berry is essential to good handling and shipping quality. As acreage increases and long-distance shipping becomes more commonplace, firmness will take on greater importance. All the newer highbush varieties, except Herbert, are firm and contribute genes for firmness to their progenies.

Dessert Quality. The best flavors in seedlings have been obtained by the use of Herbert and F-72 (an unnamed selection from a Wareham × Pioneer cross) in breeding. Coville and Bluecrop are quite acid but have produced some flavorful offspring, especially in crosses with low-acid berries.

Productivity. All blueberry varieties have the capacity to produce full crops of fruit under ideal conditions. Only a few, however, have proved to be good producers year after year and under various con-

Figure 25. A hybrid selection, GM-37, that has produced berries more than an inch in diameter. (USDA, Beltsville, Md.)

Figure 26. Rubel, the best blueberry selection from the wild in New Jersey. (USDA, Beltsville, Md.)

Figure 27. Brooks, the best selection from the wild in New Hampshire. (USDA, Beltsville, Md.)

ditions. Consistent production is the aspect of productivity needed in all varieties. The outstanding variety in this respect is Bluecrop. Blueray and Herbert are also dependable producers. Seedlings of these varieties have proved to be better than average for this character. Heavy and consistent cropping has also been obtained in progenies of Michigan-19H (a half-high selection from highbush × lowbush). Crosses of Michigan-1, a lowbush selection, with highbush have given both good production and earliness (Fig. 28). Perhaps the most productive of all blueberries is the rabbiteye, *V. ashei* (Fig. 29). It has not been possible to transfer this character to the highbush, however.

Early Ripening. Since most blueberry varieties bloom at nearly the same time, the time of ripening depends largely on the period re-

Figure 28. Above. Fruiting branch of US-4, a selection from a cross of Dixi (highbush) × Michigan-1 (lowbush). *Below.* US-37 (*V. tenellum* × Callaway) × Earliblue. (USDA, Beltsville, Md.)

quired from bloom to maturity. This may be as short as 45 days for some early-maturing highbush varieties and as long as 120 days for certain rabbiteye varieties. In general, seedlings tend to be intermediate in time of ripening when crosses are made between parents ripening at different intervals. Weymouth and Earliblue have been used most as sources of earliness in intervarietal crosses. Certain selected *Vaccinium angustifolium* clones, particularly Michigan-1, have been prepotent in transmitting earliness. *Vaccinium atrococcum* Heller, a highbush diploid species, matures its fruit in 35 days after bloom and, if made into a tetraploid, might be a valuable addition to the gene pool for earliness.

Late Ripening. Coville, Herbert, and Burlington are late-maturing varieties and tend to transmit this character in breeding. US-1, a selection from Dixi × (Jersey × Pioneer), has been the best source of late-maturing seedlings. Certain selections from US-1 crosses appear to have commercial possibilities to extend the harvest season 10 to 14 days. Recently in Massachusetts, a wild *V. corymbosum* selection that ripens much later than any other known highbush was made, and it has been used in crosses.

With new material and techniques, it appears that the genes for late maturity of *V. ashei* varieties may be transferred to the highbush. Hybrids of the rabbiteye and highbush blueberry mature their fruit from August 15 to frost in Maryland. This should make possible the development of blueberry varieties that will ripen over the entire summer.

Other Characters. Some other characters which breeders are presently seeking but for which known sources are scarce or absent are resistance to viruses, mummy berry, anthracnose, mildew, bud mites and witches'-broom rust, adaptation to soils of high pH, and uniform ripening or resistance to fruit abscission until all berries are ripe for mechanical harvesting.

HIGHBUSH BLUEBERRY

History of Improvement

Dr. F. V. Coville of the Bureau of Plant Industry, United States Department of Agriculture, began selecting blueberries from the wild for breeding purposes in 1908. His first selection was Brooks (*V. corymbosum*), in New Hampshire. In 1909 a lowbush, Russell (*V. angustifolium*), was also selected in New Hampshire.

The first recorded attempts at blueberry breeding were made in 1909 and 1910, when Coville attempted to self-pollinate the Brooks selection. The flowers set very poorly, and no seedlings resulted from these efforts. In 1911 cross-pollinations were made between Brooks and Russell, and

Figure 29. Fruit clusters of rabbiteye and highbush blueberry selections (*above and below*). (USDA, Beltsville, Md.)

Table VI. Blueberry Varieties Originated by the United States Department of Agriculture Cooperative Breeding Program and the Estimated Percentage of the 1963 Acreage Planted to Each

Variety	Parentage	Year Introduced	Estimated Per Cent of Acreage [a]
Pioneer	Brooks × Sooy	1920	0
Cabot	Brooks × Chatsworth	1920	0
Katharine	Brooks × Sooy	1920	0
Greenfield	Brooks × Russell	1926	0
Rancocas	(Brooks × Russell) × Rubel	1926	2
Jersey	Rubel × Grover	1928	28
Concord	Brooks × Rubel	1928	0
Stanley	Katharine × Rubel	1930	1
June	(Brooks × Russell) × Rubel	1930	1
Scammell	(Brooks × Chatsworth) × Rubel	1931	1
Redskin	Brooks × Russell (F_2)	1932	0
Catawba	Brooks × Russell (F_2)	1932	0
Wareham	Rubel × Harding	1936	0
Weymouth	June × Cabot	1936	12
Dixi	(Jersey × Pioneer) × Stanley	1936	1
Atlantic	Jersey × Pioneer	1939	0
Burlington	Rubel × Pioneer	1939	1
Pemberton	Katharine × Rubel	1939	1
Berkeley	Stanley × (Jersey × Pioneer)	1949	5
Coville	(Jersey × Pioneer) × Stanley	1949	8
Wolcott	Weymouth × (Stanley × Crabbe 4)	1950	9
Murphy	Weymouth × (Stanley × Crabbe 4)	1950	2
Angola	Weymouth × (Stanley × Crabbe 4)	1951	1
Ivanhoe	(Rancocas × Carter) × Stanley	1951	0
Bluecrop	(Jersey × Pioneer) × (Stanley × June)	1952	12
Earliblue	Stanley × Weymouth	1952	5
Herbert	Stanley × (Jersey × Pioneer)	1952	0
Croatan	Weymouth × (Stanley × Crabbe 4)	1954	1
Blueray	(Jersey × Pioneer) × (Stanley × June)	1955	4
Collins	Stanley × Weymouth	1959	0
Morrow	Angola × Adams	1964	0
Darrow	F-72 × Bluecrop	1965	0

Source: Moore, 1965.

[a] Approximately 5 per cent of the estimated acreage planted to Rubel, a variety selected from the wild.

some of the resulting hybrids were sib-crossed in 1913. About 3,000 first and second generation hybrids were grown to maturity in the field. Thus, interspecific hybridization was the very foundation of the blueberry breeding program.

In 1911, after reading of Coville's interest in the blueberry, Miss Elizabeth White of Whitesbog, New Jersey, perceived the significance of blueberry improvement and offered her assistance in the project. Miss White not only furnished land on which thousands of seedlings were fruited but also was instrumental in enlisting the aid of pickers of wild blueberries in selecting native New Jersey material for use in the breeding program. Varieties selected from the wild by Miss White and Dr. Coville were Adams, Chatsworth, Dunfee, Grover, Harding, Rubel, Sam, and Sooy.

Intercrosses between a number of these wild selections resulted in the introduction of the first improved varieties, Pioneer, Cabot, and

Table VII. Cooperating Agencies and Individuals of the United States Department of Agriculture Blueberry Breeding Program, 1963

Seedling Growers	Selection Test Cooperators

Georgia Coastal Plain Experiment Station
  W. T. Brightwell
	Michigan State University
	  Stanley Johnston

North Carolina State College
  G. J. Galletta
	Oregon State University
	  Harry Lagerstedt

Rutgers University
  L. F. Hough
	University of Massachusetts
	  J. S. Bailey

University of Florida
  Ralph Sharpe
	University of New Hampshire
	  E. M. Meader

University of Maine
  Russell Bailey
	Washington State University
	  C. D. Schwartze

Arthur Elliott,
  Otter Lake, Mich.
	H. B. Scammell,
	  Toms River, N.J.

Carlton Shaw,
  Middleboro, Mass.
	I. C. Haut,
	  Allenwood, Pa.

Gale Harrison,
  Ivanhoe, N.C.
	William Darrow,
	  Putney, Vt.

George Spayd, Currie, N.C.	
J. H. Alexander, Middleboro, Mass.	
S. A. Galletta, Hammonton, N.J.	
Sayre Rose, S. Glastonbury, Conn.	

Source: Moore, 1965.

Katharine, in 1920. These varieties served as the basis for the establishment of a new fruit industry. At the time of Dr. Coville's death in 1937 a total of 68,000 seedlings from controlled crosses had been fruited, and 15 varieties had been introduced. Since 1937 an additional 15 highbush varieties have been developed from seeds and seedlings left by Coville (Darrow, 1956a; Darrow and Clark, 1940; Darrow, Scott, and Galletta, 1952; Darrow, Scott, and Gilbert, 1949; Darrow, Scott, and Hough, 1956; Darrow, Woodward, and Morrow, 1944; Scott, Moore, Knight, et al., 1960). These varieties, their parentage, and estimated 1963 acreage are given in Table VI.

In 1937 Dr. George M. Darrow of the United States Department of Agriculture, Beltsville, Maryland, assumed leadership of the blueberry breeding program. One of his first efforts was to initiate a cooperative program of seedling and selection testing with various state experiment stations and private growers. This greatly expanded program facilitated the development of varieties adapted to widely differing conditions. By 1965 experiment stations and private growers in 13 states were cooperating with the department's blueberry breeders in developing varieties adapted to their areas. These agencies and individuals are listed in Table VII. The distribution of Beltsville seedlings to cooperating seedling growers from 1946 to 1962 is given in Table VIII.

Table VIII. Seedlings Distributed from the United States Department of Agriculture Research Center, Beltsville, Maryland, to Cooperators Between 1946 and 1962

Year	N.J.	Mich.	Mass.	Me.	Wash.	Conn.	N.C.	Ga.	Other
1946	0	0	900	0	1,400	700	0	3,300	80
1947	0	450	1,800	0	1,500	500	5,200	1,000	0
1948	0	800	5,200	0	350	600	4,800	2,700	670
1949	3,480	1,100	4,900	0	800	400	1,200	4,000	0
1950	1,400	1,200	2,750	0	0	0	550	600	0
1951	975	0	2,800	0	0	0	250	300	0
1952	3,880	700	0	0	0	1,575	1,000	125	0
1953	860	1,960	0	640	0	500	675	0	2,500
1954	1,300	1,450	0	2,100	0	0	100	370	0
1955	5,000	2,150	0	2,580	0	0	350	0	600
1956	3,300	0	0	2,400	0	0	0	6,400	660
1957	10,600	1,600	750	5,500	0	0	0	1,100	0
1958	5,140	0	0	2,580	0	0	0	14,000	0
1959	9,945	300	1,200	6,750	0	3,150	0	0	4,143
1960	5,375	500	1,150	4,497	0	0	0	0	0
1961	4,181	2,305	1,275	3,991	0	0	0	0	997
1962	4,464	528	0	2,938	0	0	0	0	0
Total	59,900	15,043	22,725	33,976	4,050	7,425	14,125	33,895	9,650
Grand Total	200,789								

Source: Scott and Moore, 1961.

Progress and Present Status

Maine. A cooperative breeding program was initiated in 1954 between the United States Department of Agriculture and the Maine Agricultural Experiment Station with the primary objective of originating highbush blueberry varieties that would withstand the low winter temperatures of that state and ripen during the short, cool growing season. Experimental plantings indicated that no highbush variety was sufficiently hardy for the area. Breeding for hardiness is being done through the use of hardy native selections from Maine and New York and hardy hybrids of highbush × lowbush (*V. angustifolium*). At present, hardy selections are being outcrossed to the better highbush varieties and new hardy native selections are being tested for their breeding value in transmitting hardiness.

Ashworth, a wild *V. corymbosum* selection made in northern New York, is one such hardy selection from the wild and is being used in breeding as a source of hardiness in Maine. Several thousand seedlings now growing have Ashworth as one parent. None of the seedlings fruited to date has shown commercial possibilities, but outcrosses of hardy selections to highbush varieties are expected to combine hardiness with superior horticultural characters in future seedling families (Scott and Moore, 1961). Ashworth also transmits precocious bearing to its progenies, an important characteristic for the short growing season and cool summer temperature conditions of Maine.

New Jersey, Michigan, and New England. New Jersey, Michigan, and southern New England were the first commercial highbush blueberry growing areas in the United States. The original wild selections used by Coville in his breeding work were found in New England and New Jersey, and the majority of seedlings from the USDA breeding program have been fruited and evaluated in these areas. Approximately 750 preliminary selections have been made from several thousand seedlings grown since 1946. Thirty-five of these have been designated as outstanding and are being propagated for extensive testing (Scott and Moore, 1961).

Over much of this region the newer large-fruited varieties have largely replaced the older varieties (Darrow, 1956a; Table VI). Some of these varieties, however, do not produce dependably year after year. The consistent heavy yields of the hardy varieties Bluecrop and Blueray in New Jersey as compared with the other new varieties have demonstrated the importance of winter hardiness and blossom hardiness. Darrow (F-72 × Bluecrop), released in 1965, G-80 (US-1 × Berkeley), and G-90 (US-1 × Coville) have been outstanding in producing consistently large crops.

There exists an opportunity to extend the harvest season by originating varieties that mature both earlier and later than those presently available. G-80 ripen five days later and G-90 10 days later than any existing variety. Early-ripening selections made at Otter Lake, Michigan, are very promising and are being tested as possible replacements for early varieties that do not produce consistently.

Mass inoculation of small seedlings with mummy berry disease resulted in severe infection in all seedlings, including those of which both parents appear resistant; a new approach is necessary (Scott and Moore, 1961).

North Carolina. The limiting factor in the blueberry variety picture in North Carolina is cane canker, caused by *Botryosphaeria corticis*. All major northern highbush varieties are more or less susceptible to the fungus and over a period of years are usually killed (Demaree and Morrow, 1951).

When the problem became apparent, some selected North Carolina wild clones were observed to be resistant and were hybridized with horticultural clones. The first outcross of the resistant F_1 hybrids to cultivated varieties resulted in four varieties, Wolcott, Murphy, Croatan, and Angola, now grown almost exclusively on North Carolina's 3,900 blueberry acres.

The current emphasis is on improving the horticultural characters of the present varieties by further breeding and on extending the harvest season with earlier and later varieties. The lack of a controlled screening technique to eliminate seedlings susceptible to cane canker at an early stage has handicapped the program. However, it has been possible to select seedlings that show superior tolerance in the field. Among the more promising of the canker-tolerant selections being evaluated for possible introduction are NC-683 (Coville × wild *V. australe* selection), a late-ripening selection with very good flavor; and NC-690 (NC-245 × wild *V. australe* selection), a very productive, large-fruited selection.

Florida. The cooperative program in Florida was begun in 1949 with the objective of developing blueberries with a low chill requirement (Sharpe, 1954). Most northern highbush varieties require 1,000 to 1,200 hours below 45 F to grow properly (Darrow, 1942). The rabbiteye blueberry requires around 500 hours below 45 F. Two *Vaccinium* species, *V. myrsinites* and *V. darrowi*, both native to Florida, are available for use as a source of the low chill character. Initial crosses were made between *V. myrsinites* and northern highbush varieties, both tetraploid with 48 somatic chromosomes, but seedlings from these crosses were not promising. Growth was badly delayed, indicating that they required a high chill period, and the fruits were generally black and unattractive.

The next series of crosses involved *V. darrowi*, a blue-fruited 24-chromosome species, and the rabbiteye blueberry, *V. ashei*, which has 72 chromosomes. From more than 7,500 pollinations only five hybrids were obtained. These are intermediate in morphological characters between the parents, and dormancy break is excellent suggesting a low chill requirement.

Since crosses of highbush and *V. darrowi* should theoretically produce sterile triploid hybrids, such crosses were not made initially. When such crosses were attempted, however, some fertile tetraploid seedlings resulted, presumably through the formation of some unreduced gametes by *V. darrowi*. These hybrids have been used recently in crosses with the northern highbush to increase fruit size. Some of the seedlings have produced light blue fruit up to 20 mm in diameter but not of commercial quality (Sharpe and Darrow, 1960). Further outcrossing and selection should yield adapted varieties with good horticultural characters.

Highbush × Lowbush Crosses

There has been considerable interest in hybridizing the highbush and lowbush species to obtain earliness and hardiness. Since one lowbush species, *V. lamarckii* (*V. angustifolium*), is tetraploid, crosses can be made readily and the resulting seedlings are fully fertile. Coville (1937) crossed a lowbush selection (Russell) to a highbush (Brooks) in his first cross-pollinations. The first named varieties resulting from this cross were Greenfield, Redskin, and Catawba. Five commercially important varieties eventually resulted from these crosses including the varieties June and Rancocas (one-fourth *V. lamarckii*), Weymouth (one-eighth *V. lamarckii*), and Earliblue and Collins (one-sixteenth *V. lamarckii*).

Johnston (1946) made a series of highbush × lowbush crosses with the hope of obtaining plants of intermediate size with large berries. He found that the lowbush growth habit was almost completely dominant in crosses between lowbush and highbush types. Ninety-seven per cent of the hybrids were of the lowbush type. He also found that 65 per cent of the hybrids produced berries of very dark color, indicating that dark blue fruit color is dominant over light blue color. None of the hybrids produced berries as large as the highbush parent.

Meader, Smith, and Yeager (1954) studied the inheritance of bush type and fruit color in hybrids of highbush and lowbush blueberries in New Hampshire. At 11 years of age the first generation plants were intermediate between parents in height; all were definitely taller than the lowbush parents. These results differ from those of Johnston (1946) reported above. There is agreement, however, on the predominance of dark fruit color in the hybrid progenies.

Breeding

In the second generation of highbush × lowbush crosses, Meader, Smith, and Yeager (1954) observed segregation for bush type. Of 954 F_2 seedlings, 246 were in the lowbush class (1 foot or less in height), 699 were of intermediate height (13 to 35 inches) and 9 plants were in the highbush class (3 feet or more).

Segregation for fruit color in the F_2 generation compared closely with the results obtained by Coville (1937) in an F_2 population from the cross of Russell × Brooks (Table IX).

Table IX. Segregation for Fruit Color in F_2 Progenies from Highbush × Lowbush Crosses

Fruit Color	Coville (1937) Percentage of Seedlings	Meader, Smith, and Yeager (1954) Percentage of Seedlings
Black	18.0	18.4
Dark blue	65.0	64.6
Light blue	15.0	12.8
Aluminum	0.5	0.3
Albino	1.5	0.1

Source: Coville, 1937; Meader, Smith, and Yeager, 1954.

Darrow, Morrow, and Scott (1952) reported on the characteristics of highbush × lowbush blueberry progenies grown in New Jersey, Maryland, and North Carolina. Most of the F_1 hybrid population ranged from 2.5 to 3.5 feet in height and were relatively uniform in height. All were very productive and most were early ripening but dark in color (rating 1 to 5 on a scale of 1 to 10 for color) and only fair in flavor.

The B_1 plants (half-high backcrossed to highbush) had dark to light blue fruit and a height ranging from half-high to 5.5 feet.

Plants of a small F_2 population were grown and found to be vigorous and no higher than the half-high F_1 plants, but the fruit color ranged from shiny black to light blue.

Other characters possessed by selected lowbush clones that would be of value in breeding highbush blueberries are drought tolerance, winter hardiness (Whitton, 1960), and early fruit maturity (Moore, 1965).

RABBITEYE BLUEBERRY

The rabbiteye blueberry (*V. ashei*) is commercially important to the southern United States because of its tolerance to heat and drought.

It thrives on upland soils and requires less chilling to satisfy the rest requirement than does the northern highbush. Darrow (1947) reported that the rabbiteye blueberry of the South had merit in its own area and contained important horticultural characters that might be useful in a breeding program.

The first commercial planting of the rabbiteye blueberry was probably made by M. A. Sapp in western Florida about 1893, using bushes transplanted from the wild (Darrow and Moore, 1962). Between 1920 and 1930 some 2,225 acres were planted in Florida with transplanted wild bushes, and similar plantings were made in other southern states. In general these bushes were not selected for good horticultural characteristics and were abandoned after improved varieties became available.

History of Improvement

A collection of wild plants from Florida and Georgia was planted in 1925 at the Georgia Coastal Plain Experiment Station, Tifton, Georgia (Brightwell, Woodard, Darrow, et al., 1955; Brightwell and Woodard, 1960). The vigor and productivity of these plants indicated possibilities for commercial use if the better characteristics could be combined through breeding. A cooperative breeding program between the Georgia Coastal Plain Experiment Station and the United States Department of Agriculture was initiated in 1940.

Rabbiteye blueberry breeding has also been conducted in North Carolina for several years (Darrow, Woodard, and Morrow, 1944), and a program was initiated at the University of Florida, Gainesville, Florida, in 1948 (Sharpe and Darrow, 1959).

Progress and Present Status

Considerable progress has been made in improving the color, size, and quality of the rabbiteye blueberry. Brightwell (1960) reported that only 25 per cent of the first seedlings grown in 1941 scored 5 or above on color on a rating scale of 1 to 10, while 97 per cent of the seedlings fruited in 1959 scored 5 or above. New varieties of rabbiteye are larger and less seedy than older ones (Scott and Moore, 1961).

Native selections that produce the largest-fruited seedlings have been Clara, Myers, Black Giant, and Ethel (Brightwell, 1960). In 1959 the number of berries required to fill a half-pint measuring cup ranged from 69 to 116 among various selections on test in Georgia, whereas native varieties normally range from 150 to 200 berries per cup.

In the Georgia breeding program about 25,000 seedlings from controlled crosses have been fruited and 116 selections have been made. Of these, 89 are *V. ashei* × *V. ashei*, 3 are *V. ashei* × *V. australe*, 4 are

V. *ashei* × V. *constablaei*, 7 are V. *australe* × V. *myrsinites*, and 13 are V. *australe* × V. *australe*. Five varieties have been introduced from the Georgia breeding program; these are Callaway and Coastal in 1950, Homebell and Tifblue in 1955, and Woodard in 1960.

Two new rabbiteye varieties, Garden Blue (NC-468) and Menditoo (NC-478), were introduced jointly with the North Carolina Station in the winter of 1958-59 (Scott and Moore, 1961).

The breeding work at Gainesville is aimed at the origination of varieties with a low chill requirement to adapt to Florida's climatic conditions, since abnormal bud break is a problem in many of the named varieties after mild winters. Eighteen selections were made in 1948 from old plantings of wild seedlings in the Gainesville area. Since 1952 more than 4,000 seedlings have been fruited at Gainesville, about one-half resulting from controlled crosses and the remainder originating from open-pollinated seeds from unnamed Georgia selections. Twenty-three selections have been saved for further evaluation. The fruit of all twenty-three selections ripen rather late in the season. A few plants grow normally even after the mildest winters.

Crosses of the rabbiteye blueberry with V. *constablaei* appear to be very promising in developing varieties that ripen much earlier for the South (Brightwell, Darrow, and Woodard, 1949).

Rabbiteye × Highbush Crosses

Both the rabbiteye (V. *ashei*) and the highbush (V. *corymbosum*) have characters that would be of value if transferred to the other species. Highbush varieties have a much shorter bloom-to-maturity period, and a cross with the rabbiteye might yield a variety for the South that ripens much earlier. The large fruit size of the highbush is also needed in rabbiteye varieties. The rabbiteye, on the other hand, could contribute genes to the highbush for much smaller fruit scars, greater vigor, late ripening, resistance to heat and drought, tolerance to certain fungus diseases, and tolerance to more varied soil types. However, crosses of these species (6X × 4X) result in pentaploids that are self-sterile (Coville, 1927; Darrow, Morrow, and Scott, 1952; Darrow, Scott, and Dermen, 1954), and irregular meiosis makes them useless for further breeding (Darrow, Dermen, and Scott, 1949; Darrow, Scott, and Dermen, 1954; Longley, 1927).

This problem is being approached in two ways. Synthetic tetraploids have been developed by crossing the rabbiteye with diploid species (Darrow, Dermen, and Scott, 1949; Darrow, Scott, and Dermen, 1954). These tetraploids have proved fully compatible with highbush species. This approach is expected to require several generations of breeding and selection to give practical results, since the "bridge" must be made through the use of diploid species which in general transmit many undesirable characters.

A more promising approach now appears to be available as a result of the recent development at Beltsville of a decaploid blueberry by colchicine treatment of a pentaploid seedling of rabbiteye × highbush (Moore, Scott, and Dermen, 1964). Crosses have already been made between this selection and the rabbiteye and the selection has been self-pollinated, but the character of the seedlings has not yet been determined.

CHAPTER 4

Environmental Relationships

by **Walter J. Kender**, *Department of Plant and Soil Sciences, University of Maine, Orono, and* **W. Thomas Brightwell**, *Department of Horticulture, University of Georgia, Tifton*

Although the blueberry is found in many different parts of the country, the commercial production of the blueberry is dependent on specific climatic and edaphic conditions. The lowbush blueberry (*Vaccinium angustifolium*), the most important commercial species, is harvested from Maine to Minnesota and southward in the Alleghenies to West Virginia. Commercial plantings of the highbush blueberry are found in eastern North Carolina, northward to southern Maine and Nova Scotia, west to northern Michigan, and in western Oregon, Washington and British Columbia. The rabbiteye blueberry was first grown commercially in Florida about 1893 (Camp, 1945) and now comprises some 3,500 acres of commercial blueberry production in Florida and Georgia.

CLIMATIC FACTORS

Climate of the Major Blueberry Growing Regions

New England. New England, lying in the middle latitudes, is subject to constant conflicts between cold, dry air masses from the subpolar regions to the northwest and the warmer, moisture-bearing marine air from the south. This condition brings about a succession of storms with intervening periods of fair weather. Precipitation is evenly distributed throughout the 12 months (Table X). The relative humidity is high, and heavy fogs are common during bloom and harvest (Chandler, 1943).

The average temperature during the growing season in the lowbush blueberry belt of the coastal counties of Maine is usually above 55 F and does not exceed 70 F. The length of the frost-free season (Table XI) determines to a large extent where the blueberry can be grown success-

Table X. Average Monthly Precipitation in Major Blueberry Producing Regions of the United States (Inches)

Region	Jan.	Feb.	Mar.	Apr.	May	June	July	Aug.	Sept.	Oct.	Nov.	Dec.	Annual
Eastern Maine (Ellsworth)	3.98	3.43	3.76	3.55	3.26	3.07	3.10	2.56	3.96	3.83	4.85	3.93	43.28
Massachusetts (Worcester)	3.81	2.98	4.22	3.96	3.84	3.78	3.58	4.08	3.89	3.54	4.16	3.53	45.37
Southwestern Michigan (South Haven)	1.98	1.76	2.12	2.98	3.52	3.43	2.60	3.00	3.04	3.11	2.65	2.06	32.25
Southern New Jersey (Hammonton)	3.57	3.25	4.26	3.60	3.92	3.95	4.56	5.64	3.77	3.69	3.74	3.73	47.68
Western Oregon (Newport)	10.01	8.38	8.38	4.16	2.93	2.39	0.81	0.92	2.31	6.12	8.80	11.02	66.23
Washington (Seattle Air Port)	5.73	4.24	3.79	2.40	1.73	1.58	0.81	0.95	2.05	4.02	5.35	6.29	38.94
Eastern North Carolina (Wilmington Air Port)	2.85	3.42	4.03	2.86	3.52	4.26	7.68	6.86	6.29	3.01	3.09	3.42	51.29
Southern Georgia (Tifton Exp. Sta.)	3.39	3.83	4.66	4.26	3.39	4.11	6.30	5.02	3.74	2.04	1.80	3.23	45.77
Northwestern Florida (Marianna Indian School)	3.78	4.34	5.70	5.02	4.30	4.82	7.67	6.47	4.81	2.08	3.27	4.07	56.33

Source: United States Department of Commerce, Weather Bureau, 1964.

Table XI. Average Frost Dates and Average Length of Growing Season for Major Blueberry Producing Regions of the United States

Region	Last Frost in Spring	First Frost in Fall	Length of Growing Season (Days)
Eastern Maine (Old Town)	May 8	Sept. 29	144
Massachusetts (Worcester)	May 7	Oct. 3	149
Southwestern Michigan	May 5	Oct. 16	164
Southern New Jersey	Apr. 15	Oct. 21	189
Western Oregon (New Port)	Mar. 1	Dec. 10	284
Washington (Puget Sound)	Feb. 23	Dec. 1	281
Eastern North Carolina	Mar. 8	Nov. 24	262
Southern Georgia	Feb. 25	Dec. 3	281
Northwestern Florida	Mar. 3	Nov. 24	266

Source: United States Department of Commerce, Weather Bureau, 1964.

fully. In the blueberry barrens of eastern Maine, the center of the lowbush industry, the length of the growing season is from 125 to 160 days (Chandler, 1943).

For best performance, the highbush blueberry requires a growing season of at least 160 days. It is commercially unwise to plant present highbush varieties in areas where the growing season is appreciably shorter. One reason for the poor performance of highbush blueberries in northern New England is the short growing season. The last killing frosts in spring occur usually between May 3 and 18 in eastern Maine, but in the cultivated parts of northern Maine, frosts may occur as late as June 10. In the interior of Massachusetts, destructive frosts may come as late as May 10 but usually 15 to 20 days earlier along the coast.

The first killing frosts in the fall normally occur sometime during the first half of September in the north and one or two weeks later in the central part. Thus, the average length of the growing season ranges from 100 to 125 days in northern Maine, through about 150 in central New England, to as many as 175 to 200 days near the coast of Massachusetts.

New Jersey. New Jersey has a varied topography, but in the southern interior, the blueberry growing center, the land is low and some of it swampy. The land is naturally covered with forests of scrub pine and scrub oak. The annual precipitation averages about 47 inches.

In the central and southern interior of New Jersey, the average date of the last killing frost in spring is April 15 and that of the first in fall October 21, giving an average growing season of 189 days.

Michigan. Much of the southern half of the Lower Peninsula of Michigan is level or gently rolling, though some parts reach elevations of 1,000 to 1,200 feet above sea level. Located in the heart of the Great Lakes region, Michigan is under the climatic influences of these large bodies of water. Lake Michigan seldom freezes over entirely, its temperature remaining above the freezing point in the coldest winter. The Canadian cold waves that sweep down from the northwest are tempered as they pass over the relatively warm lake. The eastern shore of Lake Michigan, therefore, does not experience the severe cold that prevails farther inland. Since large bodies of water are less responsive to temperature changes than land areas, the lake holds the winter cold longer in the spring and the summer heat longer in the fall. This stabilizing influence tends to retard the advance of spring along the shores, thus holding back the development of vegetation until the likelihood of frost is over. In the fall a reverse process slows the approach of cold weather until vegetation has matured and is safe from frost. Thus the growing season is twice as long on the peninsula's western shore as it is farther inland. This factor, together with well-adapted soils, sufficient rainfall during the growing season, and abundant sunshine in the summer, has been conducive to a successful blueberry industry in the southwestern counties bordering the eastern shore of the lake. Precipitation is well distributed throughout the year. Droughts occur occasionally but are never so severe and prolonged as in the states to the south and west.

In the Upper Peninsula of Michigan, the average date of the last killing frost in spring ranges from about May 20 along the shores of Lakes Superior and Michigan to about June 10 in the interior, and in the Lower Peninsula from May 1 to 10. The average date of the first killing frost ranges from about September 1 in the interior of the Upper Peninsula to October 1 to 10 along its shores and to October 20 in the southwestern Lower Peninsula. In the Upper Peninsula, the average growing season ranges from 80 to 90 days in the interior to 140 days along the lake shores, and in the Lower Peninsula the season lasts 90 to 100 days in the northern interior and 160 to 180 days in the southwest.

Eastern North Carolina. Blueberries are commercially important in the southeastern coastal plain, where the climate approaches the subtropical. Because of the modifying influence of the Gulf Stream, fluctuations in daily and seasonal temperatures are reduced and rarely go as low as zero. Precipitation is comparatively heavy along and near the coast, with a considerable part of the summer precipitation occurring as afternoon thunderstorms. The growing season averages 284 days.

Pacific Northwest. The Cascade Mountain range, 50 miles or more wide, extends southward from British Columbia through southern Oregon and divides the states of Washington and Oregon into western and eastern parts. Blueberries are grown commercially only west of the range. The climate is greatly affected by the Pacific Ocean and the mountain ranges. Air reaching the area from the west acquires considerable quantities of water vapor in passing over the ocean and has a cooling effect in summer and a warming influence in winter. The climate of western Washington and Oregon is much milder than that of any other section of the continent in the same latitude. Precipitation is infrequent in the summer months, the dry season reaching a peak in July and August, after which the rainy season comes on gradually. Precipitation is heaviest near the ocean and the southwestern slopes of the mountains. Irrigation is necessary for successful blueberry growing.

Over most of the area west of the Cascade Mountains in Washington and Oregon, the average period between the last killing frost in the spring and the first in fall is 150 days or more, reaching 280 days on the Pacific coast.

Georgia and Northern Florida. Rabbiteye blueberry plants have successfully fruited in the mountainous area of North Georgia, where there is an average of 160 frost-free days, and also near Gainesville, Florida, where the average is 266 days. Because of the low chill requirement the fruit crop is more subject to frost damage in colder areas, since plants tend to bloom before the danger of frost is past.

Temperature Influences

Ultimately the survival of the blueberry plant depends on the temperature extremes in its habitat. The effects of temperature are exceedingly complex. All plant processes are influenced to some extent by temperature, and rates of individual processes vary with change in temperature.

Cold Requirement. The lowest latitude for commercial blueberry production is determined by the amount of winter chilling required to fulfill the requirements for the blueberry's rest period. The rest period of the blueberry is that period during dormancy when the aboveground portion will not grow, even though temperature, moisture and other environmental conditions are favorable. If the rest period is not completely satisfied, poor and delayed opening of the buds results, with a consequent poor crop of fruit.

The cold requirement of the highbush blueberry is similar to that of the peach (Darrow, 1942). An accumulated minimum of 650 to 850 hours below 45 F (depending upon variety) is required to satisfy the blueberry's rest requirement. In general, the blueberry responds more favorably than the peach to an additional period of cold beyond 800 hours. The southern limit for highbush varieties is about 300 miles north of the Gulf of Mexico from Georgia to Louisiana. In California, highbush blueberry production is limited to a few small plantings in the central and northern coast regions where the climate is relatively cool.

In Florida, southern Georgia and southern Louisiana, highbush varieties do not get enough cold to satisfy the rest period and usually die without producing fruit. But the rabbiteye blueberry requires less than one-third the cold period of the highbush and is better adapted to these regions. Darrow (1942) found that for the rabbiteye variety Pecan, 200 hours below 45 F of chilling was not sufficient for any new shoot growth but that after 250 hours the plants appeared to have completed their rest period.

Hardiness. The northern limits for cultivated blueberries are determined not only by the length of the growing season but also by the extremes of winter temperatures and by prolonged severe cold periods that kill the roots if the soil is not protected by snow or other cover. The highbush blueberry is not hardy north of southern Maine and central Michigan and is generally winter-killed at temperatures lower than about -20 F. Temperatures below -20 to -25 may be expected to kill some of the shoots (Cain and Slate, 1953). These same authors reported that plant tops were killed to the ground or to snow level at temperatures of about -30. Certain varieties appear to be hardier than others (Table XII).

Table XII. Relative Cold Resistance of Highbush Blueberry Varieties

Decreasing Order of Hardiness		
NORTHERN VARIETIES		NORTH CAROLINA VARIETIES
1. Bluecrop	9. Weymouth	1. Scammell
2. Blueray	10. Atlantic	2. Murphy
3. Herbert	11. Collins	3. Wolcott
4. Jersey	12. Berkeley	4. Croatan
5. Burlington	13. Coville	5. Angola
6. Rubel	14. Pemberton	
7. Earliblue	15. Dixi	
8. Rancocas	16. Stanley	
	17. Concord	

Source: Darrow and Moore, 1962.

In sections where the blueberry is not generally hardy, local conditions of good air drainage or protection by deep snow may make blueberry growing possible. Temperatures are lower where air drainage is poor, and the difference may be sufficient to cause injury during periods of severe cold if the wood is not well hardened. Succulent growth late in the growing season is particularly vulnerable to winter killing.

The principal factor limiting highbush blueberry culture in northern areas such as Maine is cold injury. Because of the short growing season in this region present varieties of highbush are not able to harden sufficiently to sustain the low winter temperatures (frequently between -15 and -30 F). Although this injury is insignificant in some years, the heavy loss of crop following bad winters is sufficient to preclude commercial production.

The moderating influence of large bodies of water, however, may favor the success of highbush blueberries in areas ordinarily too cold for commercial production. For example, the highbush blueberry succeeds in the northwestern part of the Lower Peninsula of Michigan if the plantings are close to Lake Michigan whereas plantings a few miles inland do not (Johnston, 1951a). A similar situation exists in the Annapolis Valley of Nova Scotia, where the Bay of Fundy influences the climate.

It is suggested that the lowbush blueberry is found farther north than the highbush species, not because it is more hardy, but because it is covered by snow in the winter and is thus protected from cold injury (Eaton, 1949). Although the annual snowfall is more than 70 inches in eastern Maine, in many years snow does not stay on the land long enough to prevent the loss of some lowbush blueberry stems as a result of desiccation by the winter winds.

At Beltsville, Maryland, a temperature of -17 F in January, 1943, killed the fruit buds of *Vaccinium ashei* Reade and injured some young shoots on plants in a group of seedlings (Darrow, Woodard and Morrow, 1944). In Tennessee it has been reported that a drop to -18 before mid-December killed plants of the Homebell variety to the ground and damaged Tifblue plants slightly, but did no damage to plants of the Woodard variety. Many bearing apple trees were killed and many supposedly hardy shrubs were killed to the ground during this cold spell.

Frost Damage. The greatest losses from frost damage result from freezing of flower buds, flowers, or young fruits. The best time to determine frost damage is during the bloom period, when the injury will appear in the flower itself. If the pistil and ovaries are black several hours after the low temperatures occur, the damage is due to frost. Varietal difference in susceptibility of flower buds to frost injury (Table XIII) was observed by Bailey (1949).

Table XIII. Percentage of Highbush Blueberry Buds Injured at 21 F

Variety	Per Cent
Concord	10
Rubel	13
Rancocas	22
Cabot	28
Scammell	32
Stanley	42
Pioneer	44
Pemberton	52
Wareham	74

Source: Bailey, 1949.

Although blueberry blossoms are considered hardier than the blossoms of most fruits, temperatures below freezing can ruin fully opened blossoms, particularly if the cold follows a warm spell that has stimulated soft growth. Often the lack of a dramatic outright kill of the blossom, shown by rapid blackening of tissue, leads to the premature evaluation that frost damage has been light. Less visible damage, however, such as injury to the pistil or the ovules, may destroy the potential of the blossom to set fruit or of the fruit to reach a good size. Frosts early in the blossoming season are potentially less damaging than later ones, since unopened blossoms are innately more resistant to low temperatures than opened ones.

The frost in North Carolina on April 1, 1964, when temperatures plunged to near 20 F in many fields, is a good example of the resistance of unopened blossoms. This was widely publicized as one of the most disastrous blueberry frosts in the history of the state and was termed to be "nearly a complete wipe-out." Samples taken on the day after the frost were examined under the binocular microscope by officials of the Tru-Blu Blueberry Cooperative. These observations revealed that 62 per cent of the opened blossoms, 21 per cent of the closed blossoms, and 48 per cent of the total blossoms had been injured. On the basis of these examinations, the cooperative planned its 1964 North Carolina operation on an estimated 60 per cent of a normal crop, and this was close to the amount harvested.

In eastern Maine developing berries were severely injured by a hard frost on the fourth of July.

Occasional losses of commercial rabbiteye crops as a result of frost have been reported. The fruit crop was reduced at Ivanhoe, North Carolina, in 1943 and 1944 by late spring frosts (Darrow, 1944). A planting of about 125 acres in south-central Alabama has not sustained a crop loss due to frost in 22 years. Crops from a planting in south-central Georgia have been good in every year since 1944 except 1955, when a late spring freeze destroyed the entire crop.

Frost Types and Protection. The study of the frequency of frost types is of practical importance, since the efficiency of protective measures depends upon the weather conditions and particularly upon the vertical temperature profiles that characterize the different types of frosts.

Radiation frosts during fair, calm nights are spotty and shallow. All common protection methods work successfully. The temperature increases sharply from the ground upward (inversion), and air warmed by heaters does not rise higher than the layer that is the same temperature (convection lid). Continued heating will warm the entire mass below the convection lid. The sharper the inversion and the lower the convection lid, the shallower the layer and the smaller the volume of air to be heated by the fires. If ventilators are used, the propellers mounted on towers will readily mix the cold bottom air with warmer air from about 40 feet above the ground.

Advective frosts (freezes), on the other hand, are caused by large-scale transportation of cold air masses that bring widespread frosts without inversions. Neither heating nor ventilation provides efficient protection. Infrared radiators heated to approximately 1500 F, however, have been found to operate successfully against advective frosts.

The blueberry grower should know what type of frost he has to deal with in order to avoid wasted effort by the use of methods that do not work. Three-year records taken at Indian Mills, New Jersey, showed that 68 per cent of the frosts occurring during the growing season were radiation frosts (Janifer, 1952).

The selection of favorable planting sites is one of the best frost-protective measures. Pockets, with their regular nightly gravitational influx of cold air (air drainage), should be avoided. Blueberries do better at higher levels, along the warmer "thermal belts" above the bottoms. Areas completely surrounded by trees and brush, which tend to inhibit air circulation, are also more subject to frost than open sites. It is highly desirable, therefore, that the site chosen for planting have good air circulation. Such a condition is found in level, open country not surrounded by a dense growth of trees and shrubs, or on a gentle slope.

Summer Temperature. High summer temperatures and associated droughty conditions occasionally cause injury in blueberry plantings. It has been observed that a pocket may become excessively hot in summer because of poor air circulation, and young plants have been killed when temperatures in the sun reached 120 F. The blueberry's root system, which is not equipped with root hairs, cannot absorb enough water to replace transpiration losses during periods of extreme heat.

Although the adverse effects of high temperature are associated with drought, the temperature itself may be high enough to cause direct injury to the cell protoplasm. Leaf scorch has been observed especially in young blueberry plants growing under greenhouse conditions.

Fruit Maturation. Temperature has an important influence on the maturation of the blueberry fruit. The fruit of the highbush blueberry matures from 50 to 90 days after bloom, depending on the variety and the average temperature during the growing season (Bailey, 1947). But the lowbush blueberry, because it grows in a cooler climate, matures from 90 to 120 days after bloom.

In the Puget Sound region of Washington the crop ripens rapidly in the warm days of July and August, in contrast with the slower ripening in cooler districts, including the sea coast. Early fruit maturity in Washington is an important advantage since it allows the crop to be harvested before the end of August, when fall rains and lower temperatures adversely affect quality (Schwartze and Myhre, 1954b).

In Florida the period of maturation of the rabbiteye blueberry begins in late May or early June and lasts for 10 to 12 weeks, midseason being about the first or second week of July (Mowry and Camp, 1928).

Fruit Quality. Temperature affects the characteristics of the blueberry fruit as well as the plant. Blueberries are more highly flavored toward the northern limit of the blueberry growing areas, where the days are long and the nights are cool at the time of fruit ripening.

Light Influences

Photoperiod. Photoperiod studies indicate that the vegetative growth of blueberry plants is increased by long days and that short days are required for flower bud initiation (Perlmutter and Darrow, 1942; Hall and Ludwig, 1961; and Hall, Craig, and Aalders, 1963). Histological studies by Bell (1950), however, have shown that flower primordia were present on lowbush plants before the onset of short days. Bell's finding is supported by the fact that the lowbush blueberry may be found in latitudes where short day lengths do not occur during the growing season (Camp, 1945). Subsequent studies (Hall and Aalders, 1964) suggested that temperature may be an important factor in initiating flower buds under long days in the field.

Flower bud initiation has been induced in the lowbush blueberry in a greenhouse environment in the absence of a dormant period suggesting that cold treatment is not essential for growth and flowering (Hall and Ludwig, 1961; Hall, Craig, and Aalders, 1963). Under a 16-hour day the plants experienced only vegetative growth, but when they were given short day treatments, flower primordia were initiated. When the plants on short days were returned to long days, flowers were produced, indicating that the rest period is not an essential part of the flowering process.

Intensity. In lowbush blueberry fields, Hall (1958) observed that low light intensity results in shorter, thinner stems, fewer sprouts with flower buds, and a lower total of flower buds. Therefore shade produced by weedy shrubs may be an important factor in decreasing lowbush blueberry production.

When sunlight intensity was reduced 70 to 80 per cent with shade cloth, flower-bud formation and vegetative growth in highbush blueberry were greatly reduced (Shutak and Hindle, personal communication). During the following season many of the berries on these bushes dropped prematurely or failed to develop normally. When normal light intensity was reduced approximately 50 per cent, the effects were not so pronounced but were still quite obvious. These results emphasize the importance of good pruning practices to admit maximum light to all parts of the bush.

Although good light is essential for maximum growth and fruit production, it is not necessary for actual ripening and color development of the berry. Hindle (1955) found that berries picked while unripe turn blue in complete darkness. He also covered some blueberry clusters on bushes with light-tight aluminum foil and found that the berries ripened normally with no harm to the flavor.

Precipitation Influences

Normal rainfall usually keeps the soil adequately and uniformly moist in commercial plantings on native blueberry soils. As a general rule, 1 to 2 inches of rain per week (depending upon climate and soil) are required during the growing season. Irrigation may be necessary during unusual droughts to prevent drying out of the soil around the roots. Uniform and adequate soil moisture is of utmost importance in successful blueberry production, and the prospective grower should be cognizant of both a region's average annual precipitation and its seasonal distribution. A low monthly rainfall pattern during the growing season is usually a reliable indicator of the need for irrigation.

EDAPHIC FACTORS

Early attempts to grow the highbush blueberry on upland soils failed primarily because of a lack of information concerning the soil and moisture requirements of this crop. Low yields and poor plant vigor due to an unfavorable soil environment are evident in isolated fields today.

Soil Texture and Structure

Blueberry bushes can be found on nearly every type of soil but thrive on light, well-drained, acid soils. Open, porous soils conducive to good aeration, such as sandy loams with high organic matter content, make good blueberry soils (Fig. 30). A well aerated soil may be the result of light texture or of good structure in a heavier soil. Good soil structure in soils of medium or heavy texture may result from high organic matter content and good soil management. An excellent structure is often found under old sod. Soil structure can be improved by growing grass for several years, and the effects of poor structure can be offset to a considerable extent by the use of a mulch.

Coarse sand or sandy soils with a meager supply of organic matter are not desirable. Some commercial plantings may appear to be located on sand because rains have caused the organic matter to filter downward, leaving a thin layer of white sand on the soil surface while the organic matter content may actually be from 5 to 7 per cent.

Blueberry plants will grow on clay soils provided they are sufficiently acid and are well supplied with organic matter. But heavy clay soils generally have poor drainage and lack good structure and aeration, which may prevent the penetration of roots. Rabbiteye blueberry plants will grow better on the heavier upland soils than will the highbush varieties.

The best soils for highbush blueberries in New Jersey are the Leon and St. John sands with organic matter contents ranging from 3 to 15 per cent. These ground-water podzols are generally underlain with a hardpan within 3 or 4 feet of the surface (Beckwith and Coville, 1931). In Michigan the principal blueberry soil is the Saugatuck sand, a dark gray sand 2 to 10 inches thick with a subsoil of light gray, nearly white sand overlying a hard pan layer of rusty-brown or coffee-colored sand from 3 to 12 inches thick (Johnston, 1942b).

Trevett (1962) has characterized the lowbush blueberry soils of eastern Maine as typical podzol soils with an average organic matter content of 11.5 per cent (Fig. 31). The presence of a hardpan may vary from a few inches to a few feet below the surface. Two distinct bands comprise the organic layer. The topmost band, or A_{00} horizon, is composed of newly fallen leaves and twigs in a loose, relatively undecomposed condition. Beneath this litter is the A_0 horizon, a somewhat compacted band of organic matter in which decomposition has occurred.

The rabbiteye blueberry is grown in Florida on upland soils of sand and sandy loam textures underlain with a clay subsoil at 1 to 4 feet. The most satisfactory results have been attained from plantings made on moderately high, rolling lands with good drainage but underlain by a subsoil that prevents excessive loss of moisture during periods of drought

Figure 30. Typical soil of the New Jersey commercial blueberry growing region. Note the characteristic mixture of sand and peat. (Norman F. Childers, Rutgers University, New Brunswick, N.J.)

Figure 31. The profile of a virgin podzol soil compared to a podzol that has been in commercial lowbush blueberry production for at least fifty years. The soil at left has never been burned over, but the soil at right has been burned many times. (M. F. Trevett, University of Maine, Orono.)

(Mowry and Camp, 1928). Rabbiteye blueberries planted in 1945 on an eroded clay hillside near Clemson, South Carolina, produced good crops for many years. Growth and production are better, however, on low-lying and fairly well-drained soils. In southern Georgia the soils considered best for blueberries range from the well-drained Norfolk through the Goldsboro, Lynchburg, Irvington, and Kleg to the poorly drained Plummer and Raines. Peat and muck soils have certain limitations for blueberry culture. These soils are wet and cold in the spring and warm up slowly. They release an abundance of nitrogen in late summer, thus prolonging plant growth and delaying fall leaf drop and wood maturation. Plants are subject to damage by fall frost and winter freezing on these soils. Because peats and mucks are almost always in low-lying pockets, plantings in these soils may also suffer from spring frost damage (Schwartze and Myhre, 1954b).

Soil Organic Matter

Soil organic matter is of importance in the improvement of soil structure and in the retention of moisture and nutrients. The consequence of an inadequate supply of organic matter in sandy soils is readily observed in commercial fields of both highbush and lowbush during moderately dry seasons. In some areas young plantings fail before reaching maturity, and in others poor growth and reduced yields are evident. A number of studies have indicated that the growth and yield of blueberries are generally proportional to the amount of organic matter in the soil (Johnston, 1948, 1951b). The deeper the organic layer, the longer the stems of the lowbush blueberry (Trevett, 1962). This confirms observations that the more vigorous stems usually grow around decaying stumps and other places where organic matter abounds.

Doehlert (1937) has found that the downward spread of roots in the St. John sand is limited by the organic matter content of the soil. Observations on root distribution indicated that no roots were growing in the sand subsoil, which was void of organic matter. Roots growing below the peaty layer were in areas where a streak of peat, a decayed twig or another root penetrated the subsoil.

In addition to holding moisture, organic matter reduces the leaching of plant nutrients by retaining cations on its negatively charged exchange sites and appears to increase the availability of certain micronutrient elements, particularly iron. Good soil structure in moderately heavy soils is attributed largely to a relatively high organic matter content.

On sandy soils, elements such as potassium, calcium, and magnesium are retained in the soil on the exchange complex of organic matter. Nitrogen and phosphorus frequently are converted to organic forms or incorporated into organic complexes by the microorganisms and released later as available plant nutrients. Nitrogen also may be held on the exchange complex of organic matter in the ammonium form, which is absorbed readily by blueberry plants (Cain, 1952b).

The bulk of the underground portion of the lowbush blueberry plant is located in the surface organic mat and in the A_1 horizon of the soil because the organic layer provides both a path of least resistance and a favorable physical, chemical, and biological environment (Trevett, 1956). Frequent burning of lowbush blueberry fields has prevented the accumulation of fresh organic debris by destroying annual leaf litter, and occasional deep burns have destroyed a portion of the residual, decaying organic matter of the A_0 horizon. In fields that have been in commercial production for 25 to 75 years the surface soil consists of remnants of the A_0 horizon and usually an intact A_1 horizon.

Soil Aeration

Good soil aeration is important for optimum growth of the blueberry. Aeration depends on many factors, but predominantly on the soil's water content, structure, and texture. Many of the problems that confront a grower in the field may be aeration problems in disguise. Poor production from plants growing in waterlogged soils may be caused by the lack of aeration, and so may the unsuccessful culture of blueberries in some clay soils.

The atmosphere of the soil is not a continuation of the ordinary atmosphere. In the soil, for example, the carbon-dioxide content is normally 0.3 per cent, or 10 times the concentration found in our atmosphere. Under well-aerated conditions the oxygen content of the soil air will be approximately 20 per cent, the same as the atmosphere concentration. When aeration is poor, the oxygen content of the soil air will decrease markedly and the carbon dioxide will increase correspondingly. Most soil air occupies the larger pores that are not occupied by water. At least 50 per cent of the pore space of soils that have good structure and are well drained is in the form of large air-holding pores. The beneficial effects of mulches and of adding organic matter to the soil are partly the result of improved soil structure and moisture-holding capacity. The large pore space of the soil is increased as the soil structure is improved and this, in turn, improves drainage and aeration.

Soil Reaction

In their natural habitat blueberries occur predominantly on acid soils and so are called acid-loving plants (Emmett and Ashley, 1934). Coville (1910) was the first to establish that the blueberry requires an acid soil. The optimum growth of the cultivated highbush blueberry occurs on soils with a pH range of 4.3 to 4.8, but a soil pH range of 4.0 to 5.2 is usually satisfactory.

Soil Moisture

Blueberry plants are injured by dry soil conditions. Their reaction to drought is first noted by a reddening of the foliage. Extended droughts cause thin, weak shoot growth, decreased fruit set and early defoliation. Severe droughts during the growing season may kill shoot tips and severely injure or kill plants growing on marginal sand soils, particularly in slightly raised areas where the water table is lower.

On the other hand, poorly drained soils have an equally detrimental influence on blueberry plant growth. The highbush blueberry does not tolerate standing water during the growing season. Plants growing on wet soils may be heaved out of the soil by intermittent freezing and thawing during the winter. The water table must be at least 14 inches below the surface; if it is not, the land must be drained before blueberries are planted.

In sections of fields covered by water for several days after storms, plants are consistently smaller and less productive than plants in well-drained areas of these same fields (Doehlert, 1937). This effect of poor drainage on lowland soils has been a common experience for blueberry growers. To solve the problem, mounds or ridges can be built up around the plants by disking. The mounds become permanent as a result of the formation of a dense mat of roots at the soil surface (Fig. 32).

Figure 32. Above. Blueberry plants lifted out of the ground by freezing of the soil. This location is too wet and should have been drained before the plants were set. *Below.* Standing water after a shower in a blueberry planting in eastern North Carolina. Mounding kept the plant roots out of the water. (John S. Bailey, University of Massachusetts, Amherst; Walter E. Ballinger, North Carolina State University at Raleigh.)

In many commercial areas, ideal soil moisture conditions occur on a peat-sand or peat-loam soil underlain by a hardpan at a depth of 30 inches or more. The water table rests about 18 to 24 inches below the surface. If the water table is higher than this, adequate drainage and maintenance of an ideal water level can be accomplished by constructing drainage ditches around the field. The fields should drain rapidly during heavy summer rains to prevent injury from waterlogged soil. Adequate soil moisture during droughty periods can be maintained by subirrigation or by sprinkler systems using water from these ditches, as practiced in some fields in southern New Jersey.

SITE LOCATION AND LAND PREPARATION

For the prospective blueberry grower in search of a site, the natural vegetation is an indicator of soil type. The presence of wild blueberries and some related plants, such as huckleberries, azaleas, and laurel, suggests suitable blueberry soil. For upland soils, native blueberries and native spiraea or hardhack are considered favorable signs.

In New Jersey, a mixture of pine, red maple, and white cedar on a peat layer 4 to 8 inches thick generally indicates excellent blueberry soil (Beckwith, 1943b). Pure stands of white cedar indicate deeper peat that would not be entirely satisfactory for blueberries because of drainage and cultivation difficulties. A mixture of wild blueberry, leatherleaf (*Chamaedaphne calyculata* L.), and cranberry occurs where the peat layer is less than 4 inches thick. Blueberries may do well in such soil if proper moisture and drainage can be provided. The experienced grower, however, usually will not use virgin land with less than 4 to 6 inches of peat. Pine and oak forests indicate land that is too dry for blueberries. Some abandoned cranberry bogs in Massachusetts and New Jersey with a peat layer less than 6 inches thick have been used successfully for growing blueberries.

Peat and muck soils in Washington have certain limitations for blueberry culture (Schwartze and Myhre, 1954b). These soils are likely to be cold in the spring and warm up slowly. The danger of early fall frost damage and winter injury on these organic soils is greater than on mineral soils where plant growth stops earlier in the fall.

Lowbush blueberry plants are generally found growing in old pasture land, former hayfields or in cleared woodlots. When buying additional land for lowbush blueberries, growers in Maine and Canada prefer abandoned pasture land to areas that have recently been cleared of forest trees, since it is easier to manage and supports fewer plants that are serious competitors of the blueberry (Hall, 1959). After the land has been cleared, lowbush blueberry plants shoot up from rhizomes that were previously shaded out by the trees. Plants whose organs are far enough below ground level to survive burning grow readily in subsequent years.

The grower who acquires virgin land must clear it of all natural vegetation and fill or drain any wet areas so that the water table remains at least 14 inches below the surface. Drainage ditches can be installed to carry off water that would otherwise stand in the field. Low, wet areas in a field may also be drained by using tile. The main outlets should be large enough to remove the runoff from heavy rains within 12 hours.

After this preparation the soil can be improved by plowing and thorough harrowing. If the soil is sand with a peat cover, plowing should be deep enough to bring up some of the sand. The land should lie fallow for a year prior to planting. Occasional harrowing during the fallow year helps to control weeds and harmful grubs that may be harbored in the sod cover. Sand-peat soils often have "thin" spots where the peat deposit is very shallow. These spots can be improved by the addition of peat from some other location.

A grade of 2 to 3 inches in 100 feet has been recommended (Krohn and Mahn, 1962). This grade is usually started from the center of the field, producing a mound effect which provides adequate surface drainage for the entire field.

Basic methods of establishing lowbush blueberry fields have changed little from the time the American Indians, who understood that clearing forest land would encourage natural stands of wild blueberry plants. When a woodlot is cut and burned, the brush should not be piled on the field but rather on stone heaps or nonblueberry land, since excessive heat will destroy the existing blueberry rhizomes. Stands of evergreens can be destroyed by a single cutting. With deciduous species, however, repeated cutting or treatment with herbicides is necessary. Rotary brush cutters afford an excellent means of clearing brush land. The simplest situations, however, are those in which lowbush blueberry plants develop in abandoned hayfields without the complication of tree growth.

Very few fields cleared from woods are completely covered with blueberries, and any great increase in stand results not from the growth of new seedlings but from the initiation of stems from rhizomes present in the field before clearing (Hall, 1955). The length of time required to obtain a solid stand depends on the condition of the soil, the number of plants present, and the degree of competition from weeds. Repeated burning appears to encourage the growth and spread of lowbush blueberries and to retard weed development. If pruning or weed control is stopped, the area grows up to brush and woodland again.

Attempts to domesticate the lowbush blueberry by row cultivation similar to the highbush blueberry have failed because transplanted cuttings and seedlings develop rhizomes at an extremely retarded rate. Plants persist for many years, but generally no lateral spread occurs from the original transplant.

CHAPTER 5

Varieties and Their Characteristics

by George M. Darrow (retired) and Donald H. Scott, Small Fruit and Grape Investigations, Plant Industry Station, United States Department of Agriculture, Beltsville, Maryland

Blueberry variety improvement through interspecific hybridization and plant selection has been confined to the highbush and rabbiteye segments of the blueberry industry. No organized effort has been made to use similar plant breeding techniques to improve the lowbush blueberry, although selected clones from the wild are being tested.

HIGHBUSH BLUEBERRY

History of Varieties

The highbush blueberry varieties are largely derived from two wild highbush species, *V. australe* Small and *V. corymbosum* Linnaeus of the eastern United States. A few varieties are two to three generations from a cross of a highbush and a lowbush species (*V. corymbosum* × *V. angustifolium* Aiton). According to the classification of Camp (1945), Wareham and Rancocas belong to *V. corymbosum*, a species whose leaves have serrate edges and a pubescent undersurface. All other highbush varieties now propagated belong to *V. australe*, a species with nonpubescent, nonserrate leaves.

Rancocas and June were derived from a highbush × lowbush selection backcrossed to a highbush. Weymouth, Angola, Croatan, Murphy, and Wolcott are second to third backcrosses of the hybrid to the highbush. So far, there seems to be no character of value derived from the lowbush in the varieties just named. Keweenaw, introduced by the Michigan Agricultural Experiment Station (Johnston, 1951c), is a highbush × lowbush variety that may have two characters of value from the

Varieties and Their Characteristics

lowbush—adaptation to higher sites and a low-growing bush easily covered with snow for winter protection.

Most highbush blueberry varieties now grown in the commercial production areas resulted from crosses made by F. V. Coville of the United States Department of Agriculture (Coville, 1937; Darrow, 1947; Darrow, Demaree, and Tomlinson, 1951; Darrow, Scott, and Galletta, 1952; Darrow, Scott, and Hough, 1956; and Morrow and Darrow, 1950). With the aid of Miss Elizabeth C. White, Coville obtained selections from the wild, chiefly from New Hampshire and New Jersey, including the varieties Rubel, Sooy, Grover, Chatsworth, Brooks, Harding, Carter, and Russell, to provide a reservoir of genes from which to build the blueberry breeding program. This program, beginning with the introduction of three varieties in 1920, Pioneer, Cabot, and Katharine, served as the basis for the establishment of the cultivated highbush blueberry industry, which has grown with the introduction of 32 new varieties since that time (see Table VI).

Other early varieties originated and named by Dr. Coville included Greenfield, a half-high with small fruit; Redskin and Catawba, both half-highs with small reddish fruit. None of these is of value today, nor are several selections from the wild that were propagated—Sooy, Grover, Sam, Chatsworth, Harding, and Adams.

Highbush varieties other than those that have been associated with the United States Department of Agriculture breeding program have included Harbout (late) and Twearley (early), varieties that have been propagated by Tom Windom at Parsonburg, Maryland, and that may have interest because of their season and productiveness. Joseph Eberhardt, of Olympia, Washington, named three of his seedlings—Pacific (early), Olympia (midseason), and Washington (late) (Schwartze and Myhre, 1954b). Of these, Washington may be of some interest in the Pacific Northwest. E. W. Johnston of Lulu Island, British Columbia, named the Fraser, Shirley, Evelyn, and Charlotte, none of which seems of commercial value. Mrs. A. M. Drew (1953) of Gate, Washington, patented Gem, a Rancocas × June cross, of which all the berries could be picked at one time. Its fruit is about the size of the Rancocas berry, but not as large as the fruit of Earliblue and Collins, which can also be picked all at one time.

The Nova Scotia Experiment Station introduced Kenlate and Kengrape (Eaton, 1943) and Kenafter (Eaton, 1949). Kengrape may have value in Nova Scotia but hardly where the best highbush varieties can be grown.

Classification of Varieties

Leaf Characters. Two classifications of highbush blueberry varieties based on leaf and plant characters were those of Clark (1941) and

Bailey and French (1946). Clark and Gilbert (1942) based their system on serration of leaf margins, pubescence of undersurface, the leaf tip, leaf base angles and width-length ratios, as used for peaches, and gave a key for the varieties introduced up to 1940. Bailey and French followed with a revised key and a comparison of similar varieties. Their key can be used to separate the varieties in the trade up to 1945. No key that includes more recent varieties is available.

Bush Characters. Few reports on the relation of bush characters to yield have been published. Most information is observational. Occasionally, varietal characters of importance to the home gardener or in the development of a commercial industry are not initially recognized. Sometimes after the introduction of a new variety a character that makes the variety of less or greater value than was thought at first becomes evident. Thus in 1954, a year of severe drought, Bluecrop appeared to have a drought-resistant bush, and in that same year Ivanhoe buds appeared to be tender to low temperature.

Bush characters to be considered in judging the value of a variety are vigor, erectness, hardiness of bud, number of buds per branch, and disease resistance.

Vigorous bushes produce more berries and are easier to prune than nonvigorous ones. Jersey and Stanley were the standards of vigor until Pemberton was introduced. Pemberton is the most vigorous of the varieties grown in the North, and Wolcott, Angola, and Croatan of those grown in eastern North Carolina. The long, stout shoots of these varieties make them relatively easy to prune as compared with Rancocas, Rubel, Cabot, and Pioneer.

In general, bush types classed as erect-spreading are the most satisfactory. Varieties with spreading bushes are slower to come into full production and are more costly to prune than the more erect varieties.

Hardiness of the blueberry bud probably depends on several physiological responses: resistance to cold, length of rest period, rapidity of hardening with the onset of winter, and tenderizing following warmer intervals. There is little information available as to the minimum temperatures blueberry varieties will withstand at different times in the fall, winter, and spring. In Michigan, Jersey and Rubel are considered the hardiest of the older varieties. However, Burlington, Bluecrop, Earliblue, and Blueray may be as resistant to cold as Jersey and Rubel. Varieties that become dormant relatively early in the fall and require a long rest period should be among the hardiest and those with a short rest period the least hardy. Therefore, varieties selected in North Carolina, such as Wolcott, Murphy, Angola, and Croatan, are not considered adapted to New Jersey and northward. They apparently have a short rest period and seem to tenderize during warm spells more than most varieties selected farther north.

Concord and Burlington set fewer flower buds per branch than most varieties, and in most seasons their cluster size and bearing habit are considered ideal. Pioneer, Cabot, and Scammell set too many flower buds per branch, and if all buds are left, the berry clusters are long and the berries relatively small. Heading back such branches is a standard but expensive practice; varieties that produce them are costly to prune.

Resistance to at least one disease appears in most cultivated varieties of the blueberry, apparently because of its extremely heterozygous nature. Rancocas, for example, is relatively resistant to infection by stunt virus; Wolcott, Murphy, Croatan, and Angola to cane canker; Rancocas, Dixi, June, and Weymouth to mildew; June to double spot; and Cabot, Dixi, June, Rancocas, and Weymouth to leaf rust and mummy berry. It should be possible to obtain varieties highly resistant to these diseases through breeding.

Berry Characters. Berry characters that help to determine the value of a variety are size, color, firmness, resistance to cracking, type of scar, tendency of fruit to drop when ripe, keeping quality, aroma, flavor, and season of maturity. The steady increase in the acreage of the cultivated blueberry indicates that the berry qualities of present varieties are acceptable. However, to insure a ready market for future larger crops, planting improved varieties with the most acceptable qualities is necessary. The recent introduction of eight varieties with large berries should lessen picking costs and make the market products more attractive. The large size, beautiful light blue color, firmness, and resistance to cracking of Berkeley, one of these eight, are all notable qualities (Fig. 33). In contrast, the small size of Cabot, the dark color and softness of Weymouth, and the cracking of Cabot, Rancocas, and Dixi lessen their market value.

The type of scar (where the berry separates from the stem) largely governs the susceptibility to rots. Rots start very quickly after picking in varieties with large, moist scars. Among the commercially grown varieties, Pemberton has the poorest scar and hence is among the poorest varieties for keeping and shipping; Burlington and Bluecrop are among the best, for their scars are small and rather dry.

None of the commercially grown varieties drops berries to any great extent. But Coville, Earliblue, and Collins drop less than other varieties, so their berries can usually be left on the bushes until most of the crop is ripe.

Of the older varieties Stanley, Pioneer, and Wareham have the highest aroma and flavor. Of the newer ones Herbert and Darrow have high aroma and may have the best flavor of all varieties. The aroma and flavor of Blueray and Ivanhoe are nearly as good as those of Herbert. Coville has a distinctive high flavor and high acidity, making it especially desirable for pies. The sweet flavor of Berkeley is liked particularly in New England.

Season of Maturity. The importance of the season of maturity for a variety varies greatly in different blueberry growing regions. In North Carolina the varieties that ripen earliest are naturally of greatest cash value because the crop is shipped to the large markets before competition begins from the other production areas. In general the crop

Figure 33. Berkeley, a Big Seven variety of the highbush blueberry noted for its large, firm, light blue fruit. (A. F. Sozio, New York City.)

Varieties and Their Characteristics

Table XIV. Percentage of Crop Picked Each Week During the Ripening Season for Several Blueberry Varieties in Eastern North Carolina, Southern New Jersey, and Michigan

State and Variety	Week of Season													
	1 [a]	2	3	4	5	6	7	8	9	10	11	12	13	14
North Carolina														
Morrow	80	20	—	—	—	—	—	—	—	—	—	—	—	—
Angola	60	30	10	—	—	—	—	—	—	—	—	—	—	—
Wolcott	30	50	20	—	—	—	—	—	—	—	—	—	—	—
Weymouth	30	35	35	—	—	—	—	—	—	—	—	—	—	—
Croatan	20	50	30	—	—	—	—	—	—	—	—	—	—	—
Jersey	—	—	30	30	30	10	—	—	—	—	—	—	—	—
New Jersey														
Earliblue	—	—	—	—	40	60	—	—	—	—	—	—	—	—
Weymouth	—	—	—	—	50	35	15	—	—	—	—	—	—	—
Collins	—	—	—	—	—	50	50	—	—	—	—	—	—	—
Blueray	—	—	—	—	—	30	40	30	—	—	—	—	—	—
Bluecrop	—	—	—	—	—	20	40	30	10	—	—	—	—	—
Berkeley	—	—	—	—	—	—	20	30	30	20	—	—	—	—
Jersey	—	—	—	—	—	—	—	50	40	10	—	—	—	—
Herbert	—	—	—	—	—	—	—	20	40	30	10	—	—	—
Darrow	—	—	—	—	—	—	—	30	40	30	—	—	—	—
Coville	—	—	—	—	—	—	—	—	20	30	30	20	—	—
Michigan														
Earliblue	—	—	—	—	—	—	—	10	40	50	—	—	—	—
Stanley	—	—	—	—	—	—	—	—	20	30	40	10	—	—
Bluecrop	—	—	—	—	—	—	—	—	—	30	30	40	—	—
Berkeley	—	—	—	—	—	—	—	—	—	—	40	30	30	—
Herbert	—	—	—	—	—	—	—	—	—	—	40	30	20	10
Jersey	—	—	—	—	—	—	—	—	—	—	40	30	20	10

Source: Darrow, 1957.
[a] May 15–22 in North Carolina.

from that state supplies the market for about four weeks—May 15 to June 12—before harvest begins in Maryland and New Jersey (Table XIV). Most of the berries of the Angola, Croatan, Wolcott, and Murphy varieties in North Carolina mature within this four-week period. Smaller quantities of later varieties are in demand for the southern markets.

In New Jersey at least six varieties with different ripening dates are needed to supply the eastern markets during the eight-week harvest season.

In Michigan there is a good demand for early varieties, but the chief demand is for midseason to late varieties for the general market. Very late varieties, or those that mature in September, are not generally desirable because pickers are scarce.

Rating of Varieties

A rating system has been used to a considerable extent to express the relative value of the characters of the varieties (Darrow, Scott, and Gilbert, 1949; Morrow and Darrow, 1950). The scale from 10 for the best quality to 1 for the poorest has been employed. For season, 10 represents the earliest-ripening fruit and 1 the latest. A rating of less than 6 is considered unsatisfactory for commercial use. This makes it possible to compare the qualities of the different varieties maturing at the same season.

In Table XV the ratings of 29 highbush varieties for season, berry size, color, scar, and flavor are given. At a glance it can be seen that

Table XV. Ratings of Highbush Blueberries for Some Berry Characteristics [a]

Variety [b]	Season	Size	Color	Scar	Flavor	Remarks
Morrow	10	8	7	6	6	very early, canker-resistant
Angola	10	8	5	7	6	very fine bush, canker-resistant
Wolcott	9	8	6	9	6	very fine bush, canker-resistant
Croatan	9	8	7	8	6	very fine bush, canker-resistant
Weymouth	9	8	5	6	5	poor bush, productive, poorest flavor
Earliblue	9	9	8	7	7	fine bush, won't drop
Collins	8	9	8	7	8	fine bush, won't drop
Murphy	8	8	6	6	7	spreading bush, canker-resistant
Cabot	8	5	7	6	5	spreading bush, berries crack
June	8	5	6	5	6	bush usually weak
Rancocas	7	6	7	6	6	berries crack, resistant to stunt
Ivanhoe	7	9	7	9	10	buds not hardy, hard to propagate
Stanley	7	6	7	4	9	easy to prune, berry size runs down
Blueray	7	10	8	7	9	bush hardy, easy to propagate
Bluecrop	6	9	9	9	7	drought-resistant, hardy, fine color
Concord	6	7	7	6	8	fine cluster size
Pioneer	6	6	6	6	9	berries crack
Scammell	6	7	6	6	7	sets too large clusters
Berkeley	5	10	10	8	7	berries drop some, lightest blue
Atlantic	4	8	7	7	8	berries drop some
Pemberton	4	8	6	4	7	bush very vigorous, most productive
Rubel	4	6	7	7	6	bush hardy, hard to prune
Jersey	4	7	7	7	6	bush hardy, long picking season
Dixi	4	10	6	5	9	berries crack, run down in size
Herbert	3	10	7	7	10	berries tender, bush hardy
Darrow	3	10	8	8	9	bush erect, vigorous, berries do not drop
Wareham	2	6	6	7	10	berries tender, bad mildew
Burlington	2	7	7	10	6	bush hardy, berries store well
Coville	2	10	7	6	9	berries won't drop, fine processed

Source: Darrow and Moore, 1962.

[a] For season, 1 = latest, 10 = earliest; for color, 1 = dark, 10 = light; for other characters, 1 = poorest, 10 = best.

[b] No ratings available for Redskin, Katharine, Catawba, and Greenfield.

Morrow and Angola are the earliest and Coville among the latest; that Blueray, Berkeley, Dixi, Herbert, Darrow, and Coville are the largest and Cabot and June the smallest; that Berkeley and Bluecrop are the lightest blue in color and Angola and Weymouth the darkest; that Stanley and Pemberton have the poorest scars and Burlington and Bluecrop among the best; and that Blueray, Ivanhoe, Herbert, and Wareham have the best flavor and Cabot and Weymouth the poorest.

The ratings are not absolute, for all varieties mature earlier when heavily pruned than when they get little or no pruning, but some varieties respond more than others. The amount of pruning also greatly affects the berry size. Color is somewhat lighter in dry seasons than in wet ones, but in general is affected least by environmental conditions. In general, flavor is higher the farther north the variety is grown. In the cooler climate of the North, however, berry acidity may become too acid for the average taste.

Where exact measurements of a character are possible their use depends on the ease of obtaining them or on their importance. Thus berry size may be rated quickly on the basis of 1 to 10, or the actual number of berries per standard half-pint measuring cup may be obtained for one picking or for all pickings of one or more seasons (Chandler, 1941). Color may also be given a rating of 1 to 10 or may be measured with Munsell color standards.

Description of Varieties

Adams. Selection from wild. No longer propagated

Angola. Weymouth × (Stanley × Crabbe 4). Introduced 1951. Bush very vigorous, upright, productive; fruit cluster loose; berry large, dark, of about size and color of Weymouth, medium firm, slightly aromatic, medium dessert quality, scar above average, resistant to cracking; season earliest of all. Highly resistant to canker. May be replaced by Morrow.

Atlantic. Jersey × Pioneer. Introduced 1939. Bush vigorous, open-spreading, productive; fruit cluster loose; berry large, five-sided, medium blue, firm, slightly aromatic, medium dessert quality, scar above medium, resistant to cracking; season late, slightly earlier than that of Jersey, good for processing. Little planted now.

Berkeley. Stanley × (Jersey × Pioneer). Introduced 1949. Bush vigorous, open-spreading, productive; fruit cluster loose; berry very large, oblate, lightest blue, firm, slightly aromatic, medium dessert quality, scar above medium, resistant to cracking, drops somewhat; midseason, keeps well. Well liked for beauty of color, large size, firmness and productiveness.

Figure 34. Left. Bluecrop, a Big Seven variety, is noted for its capacity to yield consistently good crops. *Right.* Stanley, an early introduction, has been used extensively as a parent in the development of new highbush varieties. (USDA, Beltsville, Md.)

Bluecrop. (Jersey × Pioneer) × (Stanley × June). Introduced 1952. Bush vigorous, upright, very productive; fruit cluster loose; berry large, very light blue, firm, slightly aromatic, medium dessert quality, scar one of best, resistant to cracking, drops somewhat; early midseason; hardier than most others, drought-resistant. Well liked for consistent productiveness, hardiness, and light blue color (Fig. 34).

Blueray. (Jersey × Pioneer) × (Stanley × June). Introduced 1955. Bush vigorous, upright, spreading, productive; fruit clusters rather small and tight; berry very large, light blue, firm, aromatic, high dessert quality, scar medium, resistant to cracking; early midseason. Liked for its hardy bush, and very large high-flavored berries.

Burlington. Rubel × Pioneer. Introduced 1939. Bush vigorous, upright-spreading, productive; fruit cluster medium tight; berry medium size, medium blue, firm, slightly aromatic, medium dessert quality, scar best of all, resistant to cracking; season late, lasting about a week after Jersey; stores well. Liked for its fine scar and good keeping in cold storage, but berry much smaller than Coville of the same season.

Varieties and Their Characteristics 103

Cabot. Brooks × Chatsworth. Introduced 1920. Bush lacking in vigor, low-spreading, productive; fruit cluster long, tight; berry small, medium blue, not firm, slightly aromatic, poor dessert quality, scar below average, cracks badly, poor texture; season early and long; being discarded in all sections (Fig. 35).

Collins. Stanley × Weymouth. Introduced 1959. Very like Earliblue but ripens five days later.

Concord. Brooks × Rubel. Introduced 1928. Bush with medium vigor, upright-spreading, productive; fruit cluster small, tight; berry medium

Figure 35. Cabot, one of the first highbush blueberry varieties to be introduced, is now being replaced by better varieties. (USDA, Beltsville, Md.)

size, medium blue, firm, slightly aromatic, good dessert quality, scar poor; midseason. Berkeley of same season and Bluecrop slightly earlier are much better.

Coville. (Jersey × Pioneer) × Stanley. Introduced 1949. Bush vigorous, open-spreading, productive; fruit cluster loose; berry very large, medium blue, firm, highly aromatic, tart, very good dessert quality, scar medium size, resistant to cracking; season latest of all, lasting two weeks after Jersey; does not drop; excellent for processing. Well liked for its late season, large size, vigor and nondropping quality.

Croatan. Weymouth × (Stanley × Crabbe 4). Introduced 1954. Bush vigorous, upright, productive; fruit cluster medium, loose; berry large, dark blue, medium firm, slightly aromatic, sweet, medium dessert quality, scar above average, resistant to cracking; season early, with Weymouth. High resistance to bud mite and medium resistance to canker. Recommended for North Carolina.

Darrow. (Wareham × Pioneer) × Bluecrop. Introduced 1965. Bush vigorous, upright, consistently productive; fruit cluster medium; loose; berry large, light blue, firm, slightly acidic, high dessert quality at peak maturity, medium-sized scar, resistant to cracking; season late, ahead of Coville. Considered as a replacement for Coville.

Dixi. (Jersey × Pioneer) × Stanley. Introduced 1936. Bush vigorous, open-spreading, productive; fruit cluster medium, tight; berry very large, medium blue, firm, aromatic, high dessert quality, scar large, subject to cracking; season late, slightly after Jersey. Liked for its large size, high dessert quality and productiveness in Ohio and Pacific Coast states; too subject to cracking and too bad a scar elsewhere.

Earliblue. Stanley × Weymouth. Introduced 1952. Bush vigorous, upright-spreading, productive; fruit cluster medium loose; berry large, good blue, firm, good aroma, good dessert quality, scar medium, resistant to cracking; season early, with Weymouth; does not drop. Liked for its large size, light blue color, firmness, good dessert quality and nondropping character. Apparently hardier than most.

Herbert. Stanley × (Jersey × Pioneer). Introduced 1952. Bush vigorous, open-spreading, productive; fruit cluster medium loose; berry largest, medium blue, rather tender, aromatic, very high dessert quality, scar medium, resistant to cracking; season late, slightly after Jersey. Liked for its large size, high dessert quality and consistent productiveness (Fig. 36).

Ivanhoe. (Rancocas × Carter) × Stanley. Introduced 1951. Bush very vigorous, upright, productive in North Carolina and Maryland; fruit cluster medium tight; berry large, light blue, firm, aromatic, high dessert quality, scar above average, resistant to cracking; season second early.

Figure 36. Branches of outstanding highbush varieties. a. Coville, a Big Seven variety with excellent fruit quality, is the latest ripening of the recently introduced varieties. b. Earliblue, a Big Seven variety with excellent fruit quality, is the earliest ripening of the new highbush blueberries. c. Herbert, a Big Seven variety with excellent dessert quality, is recommended for home gardens. d. Ivanhoe, a variety with some canker resistance, was developed for the industry in North Carolina. (USDA, Beltsville, Md.)

Has some canker resistance. Liked for its large size, light blue color, and high dessert quality. Not fully hardy in New Jersey and Michigan.

Jersey. Rubel × Grover. Introduced 1928. Bush vigorous, erect, productive; fruit cluster loose; berry medium to large, medium blue, firm, lacking in aroma, fair dessert quality, scar medium, resistant to cracking; season late. Liked for its vigorous, hardy bush and open fruit cluster. Most grown of any variety, but being replaced in New Jersey by Bluecrop (Fig. 37).

June. (Brooks × Russell) × Rubel. Introduced 1930. Bush of below average vigor, erect; fruit cluster loose; berry small, dark blue, firm, slightly aromatic, low dessert quality, scar medium, season early.

Figure 37. Jersey, a productive and vigorous variety, is the most grown of any highbush in Michigan. Basket is quart size. (USDA, Beltsville, Md.)

Leaves subject to June spot. Being replaced by Earliblue, which is earlier, larger, and lighter blue.

Katharine. Brooks × Sooy. Introduced 1920. Not commercially grown because of bad tearing of skin and flesh when picked.

Morrow. Angola × Adams. Introduced 1964. Bush slow-growing, semi-upright, productive; fruit cluster large, loose; berry large, light blue, firm, mild, high dessert quality; scar large and moist; season early and short. Recommended as a replacement for Angola in North Carolina.

Murphy. Weymouth × (Stanley × Crabbe 4). Introduced 1950. Bush vigorous, spreading, productive; fruit cluster loose; berry large, dark blue, similar to Weymouth in color, firm, slightly aromatic, fair dessert quality, scar fair; season early, slightly later than Weymouth. Promising in North Carolina for its early season and canker resistance; its spreading bush takes longer to come into full bearing than erect-growing varieties.

Pemberton. Katharine × Rubel. Introduced 1939. Bush among the most vigorous, erect, very productive; fruit cluster very loose; berry large, dark blue, firm, slightly aromatic, medium dessert quality, scar poor; season late, slightly earlier than Jersey. Liked because of its vigor, productiveness, and ease of growing and pruning, but has a bad scar.

Pioneer. Brooks × Sooy. Introduced 1920. Bush of medium vigor, spreading; fruit cluster long and tight; berry small, dark blue, firm, aromatic, high dessert quality, scar medium to poor, cracks badly; midseason. Being replaced by Berkeley, whose berry is larger and much lighter blue.

Rancocas. (Brooks × Russell) × Rubel. Introduced 1926. Bush medium in vigor, erect, productive, small leaf with serrate margin; fruit is small, oblate, firm, good flavor, cracks badly after rain; clusters are tight. Very little is planted now.

Rubel. Selection from the wild in New Jersey. Bush vigorous, erect, productive; fruit cluster very loose; berry rather small, good blue, firm, slightly aromatic, fair dessert quality, scar medium; season late, with Jersey. Liked for its hardiness and productiveness but is being replaced by Coville, Herbert, and Pemberton.

Sam. Selection from wild. No longer propagated.

Scammell. (Brooks × Chatsworth) × Rubel. Introduced 1931. Bush erect, vigorous, productive; fruit cluster long and tight; berry medium size, dark blue, firm, slightly aromatic, good dessert quality, scar medium; midseason. Resistant to canker but very subject to stunt and being replaced by Berkeley.

Stanley. Katharine × Rubel. Introduced 1930. Bush erect, vigorous, but with few main stems; fruit cluster medium loose; berry medium size but often small in last pickings, medium blue, firm, very aromatic, high dessert quality, scar poor, resistant to cracking; season second early. Well liked for its vigorous bush, ease of pruning, and flavorful berries, but berries are often small and the scar is poor.

Wareham. Rubel × Harding. Introduced 1936. Though high in flavor, it is being discarded because of its dark color, bad cracking and rather small size.

Weymouth. June × Cabot. Introduced 1936. Bush small, open-spreading; fruit cluster medium loose; berry large, dark blue, lacking in aroma, poor dessert quality, scar medium, berries drop; season early, crop ripens quickly. Being replaced by Angola, Croatan, and Wolcott in North Carolina; northward it is being augmented by Earliblue.

Wolcott. Weymouth × (Stanley × Crabbe 4). Introduced 1950. Bush very vigorous, upright, productive; fruit cluster loose; berry large, round, dark blue, firmer than Weymouth, aromatic, medium dessert quality, scar small; season early and short. Replaced Weymouth in North Carolina because of its canker resistance, vigor, and productiveness.

RABBITEYE BLUEBERRY

History of Varieties

Most rabbiteye blueberry plantings consist of the native seedlings dug from the wild and transplanted into commercial fields (Darrow, 1947, and Brightwell and Darrow, 1950). However, most of these plantings are given little care and the berries are mostly black, small, and gritty-fleshed. The rabbiteye varieties that have been selected from the wild are derived from *V. ashei* Reade, a species native to northern Florida, southern Alabama, and southern Georgia.

The best of these early selections were Hagood, selected by the Reverend H. H. Hagood of Crestview, Florida; Black Giant, selected by W. B. Sapp, also of Crestview; Myers and Clara, selected by J. T. Bush of Valdosta, Georgia; and Walker and Ethel, selected by W. M. Walker of Brunswick, Georgia. Most of these and others were surveyed and evaluated from 1939 to 1941 by Jackson Batchelor, formerly of the United States Department of Agriculture. Many small plantings of these varieties have been made.

Breeding work was begun in 1940 by George M. Darrow of the United States Department of Agriculture, in cooperation with the Georgia Coastal Plain Experiment Station at Tifton, Georgia, and with the North Carolina State Experiment Station at Raleigh, North Carolina. In

1950 the first two varieties resulting from this work, Callaway and Coastal, both crosses of Myers × Black Giant, and in 1955 two others, Tifblue and Homebell, and in 1960, Woodard, were named and introduced by the Georgia Coastal Plain Experiment Station and the United States Department of Agriculture. In 1958 Garden Blue and Menditoo were introduced jointly by the North Carolina State Experiment Station and the United States Department of Agriculture.

Characteristics and Ratings of Varieties

Selections from west Florida grow to 15 feet in height and produce mostly dark berries; selections from southeast Georgia are mostly 6 to 8 feet high and produce light blue berries. The rabbiteye varieties are not fully hardy even as far south as Washington, D.C. They have a shorter cold requirement and are much better adapted to highland conditions than the highbush. The berries of many of the wild rabbiteye plants are small and seedy and not very juicy, but selected plants produce good-sized, juicy, high-flavored berries.

The rabbiteye blueberries mature four to six weeks later than the highbush varieties grown in the same area in North Carolina. The first picking of the earliest rabbiteye varieties in Georgia is made about the second week in June, about four weeks after the first picking of the highbush in eastern North Carolina. Because the North Carolina area is so much closer to the large northern markets, the rabbiteye from northern Florida and southern Georgia and Alabama cannot compete there. The rabbiteye must therefore be grown for the local and

Table XVI. Ratings of Rabbiteye Blueberries for Bush Size, Season, and Some Berry Characteristics [a]

Variety	Bush Size	Season	Berry Size	Berry Color	Flavor	Scar
Coastal	9	10	8	6	6	9
Garden Blue	10	10	6	8	9	9
Callaway	7	9	8	8	10	10
Tifblue	9	9	7	10	10	10
Homebell	10	8	9	7	7	9
Menditoo	9	7	9	6	8	10
Woodard	8	10	9	10	9	7

Source: Darrow and Moore, 1962.

[a] For bush size, 1 = smallest, 10 = largest; for season, 1 = latest, 10 = earliest; for other characters, 1 = poorest, 10 = best. Relative values in this table are not comparable with those in Table XV for the highbush varieties; e.g., 10 for bush size here does not indicate the same size as 10 in Table XV.

southern markets, home use and processing. Though seedlings mature through a long season, the named varieties mature mostly in early and midseason. Their bush size, season, and some fruit characteristics are rated in Table XVI (Darrow, 1947, and Brightwell and Darrow, 1950).

Tests have shown that the rabbiteye varieties are nearly self-sterile. Hence it is essential that two rows of one variety should alternate with two rows of another. All have a relatively long ripening season and some fruit can be picked for four to six weeks, making the rabbiteye well suited to home gardens. In Georgia, Tifblue and Woodard are now recommended.

Description of Varieties

Callaway. Myers × Black Giant. Introduced 1950. Bush vigorous with low chill requirement; fruit cluster loose; berry large, ovate, light blue, medium firm, high dessert quality, scar excellent; season early.

Coastal. Myers × Black Giant. Introduced 1950. Bush very vigorous; fruit large, ovate, dark blue, medium firm, good dessert quality, scar excellent; season early.

Garden Blue. Myers × Clara. Introduced 1958. Bush very vigorous largest of all and very productive; berry medium size, light blue, firm, good flavor, dry scar; season early.

Homebell. Myers × Black Giant. Introduced 1955. Bush vigorous, very productive; berry large, medium light blue, firm, good dessert quality, scar excellent; season late.

Menditoo. Myers × Black Giant. Introduced 1958. Bush vigorous, large, and productive; berry large, dark blue, free of grit cells, small, dry scar, good flavor; midseason to late, long season.

Tifblue. Ethel × Clara. Introduced 1955. Bush vigorous, erect; berry medium large, very light blue, firm, high dessert quality, scar excellent; season medium early.

Woodard. Ethel × Callaway. Introduced 1960. Plants have spreading habit, not so vigorous as Tifblue, productive; berries are large, round, light blue in color, medium size scar, excellent flavor, earliest ripening of all rabbiteye varieties.

Note: As this book goes to press, the New Jersey Blueberry Variety Council announced the naming of the highbush variety **Elizabeth**: (Katharine × Jersey) × Scammell. Berries are loose-clustered, medium blue, with excellent flavor. Season is mid to late, and production is high and dependable.

CHAPTER 6

Propagation and Planting

by Charles M. Mainland, Department of Horticulture and Forestry, Rutgers University, New Brunswick, New Jersey

Blueberry propagation methods vary with the type of blueberry being grown. The highbush blueberry is usually propagated by hardwood cuttings, but a few nurseries are using softwood cuttings. The lowbush blueberry can be propagated by hardwood or softwood cuttings, but usually is propagated by rootlike underground stems called rhizomes. The rabbiteye blueberry is propagated chiefly by softwood cuttings, since a poor rooting percentage is obtained from hardwood cuttings. Other methods such as budding, mound layering, and seed propagation are used for special purposes.

HIGHBUSH BLUEBERRY

Propagation by Hardwood Cuttings

Varietal Differences in Rooting. Some highbush varieties do not root as readily as other varieties with the methods commonly employed in propagation. Experiments in the state of Washington indicate that varieties such as Rancocas, Pemberton, and Rubel are easy to root from cuttings while June, Stanley, and Ivanhoe are difficult to root. Jersey, Scammell, Pioneer, and Concord are somewhat intermediate in this respect (Schwartze and Myhre, 1947). In Michigan and New Jersey, the Jersey variety is considered easily rooted. Newer varieties such as Blueray, Berkeley, and Coville are relatively easy to root; Earliblue, Collins, and Herbert are somewhat less so; and Bluecrop is more difficult.

Selection of Cutting Wood. Cutting wood is taken from well-matured, firm, and healthy shoots of the past season's growth. Slender shoots as well as poorly hardened, pithy growth of late summer should be avoided (Fig. 38). Browning of the pith is a sign of inferior wood and is most

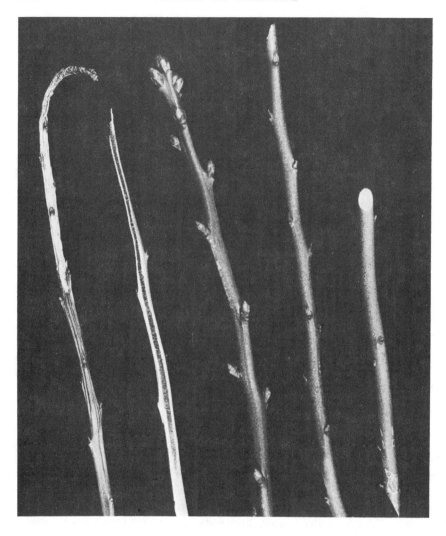

Figure 38. Selection of hardwood cuttings for propagation. Shoot at left shows winter injury; beside it is another shoot showing the extent of the injury. Shoot in center is undesirable for cuttings because of the abundance of fruit buds. Shoot at right illustrates best type of wood for hardwood cuttings. At the extreme right is a cutting ready for planting. (Stanley Johnston, Michigan Agricultural Experiment Substation, South Haven.)

likely to occur in poorly hardened growth. If virus diseases are present in a field and the location of every diseased plant is not exactly known, it is wise to look elsewhere for cutting wood. If a careful inspection has been made during the growing season to locate all diseased plants, it is probably safe to take cutting wood at a distance of 50 feet from the nearest diseased plant (Doehlert, 1953c).

Shoots from which cuttings are to be made are commonly 12 to 30 inches long and are referred to as "whips." These whips may or may not have fruit buds at the tip, but they must be well hardened. The portion bearing only leaf buds is selected. If cutting wood is scarce, sections with occasional fruit buds are used and the fruit buds are rubbed off.

The effect of flower buds on the rooting of hardwood cuttings of the highbush species *Vaccinium atrococcum* Heller was determined by O'Rourke (1942). Cuttings of three types: (1) with leaf buds only, (2) with one or two flower buds in addition to at least two good leaf buds, and (3) with one or more flower buds which were removed, were taken from the middle portion of the previous season's growth. A much higher percentage of rooting was obtained with leaf-bud cuttings than with flower-bud cuttings even if the flower buds were removed.

O'Rourke (1944) also studied the relationship between the position of the cutting on the original shoot and the success of rooting. Progressively more rooting occurred from tip to base of shoot, and vegetative wood rooted much better than flowering wood in all but the basal portion (Table XVII). It thus seems advisable to secure as many cuttings as

Table XVII. Per Cent Survival of Hardwood Cuttings of Three Varieties of Blueberries as Affected by Position of Cut and Wood Type

Variety	Basal		Hyper-basal		Sub-terminal		Terminal		Average	
	Veg.	Fl.	Veg.	Fl.	Veg.	Fl.	Veg.	Fl.	Veg.	Fl.
Pioneer	66	57	66	37	39	2	24	0	49	24
June	32	34	36	18	19	7	13	0	25	15
Cabot	63	58	44	27	11	7	6	1	31	23
Average	54	50	49	27	23	5	14	0	35	21

Source: O'Rourke, 1944.

possible from the basal portions of the shoot, where the factors which influence rooting apparently are most favorable. The groups of cuttings that had the highest survival also produced the most shoot growth.

Time to Take Whips. For small-scale operations, the cutting wood is collected in the spring just before bud growth begins, is cut into the desired lengths, and is set directly into the rooting medium. The time to take the whips is between March 15 and April 10 in New Jersey (Doehlert, 1953c), late March in Washington (Schwartze and Myhre, 1954a), early April in Massachusetts (Bailey, Franklin, and Kelley, 1950), mid-April in Michigan (Johnston, 1959), and March and April in North Carolina. In larger-scale operations the collection of cutting wood is started in February. The 800-to-1,000-hour chill requirement necessary for shoot growth and flowering is satisfied by February in most seasons. Some North Carolina growers claim a better and more uniform growth and rooting of the cuttings if the bundled wood is refrigerated for 30 days after cutting.

Storage of Whips. Whips collected before the proper time for setting into the rooting medium can be stored at 35 to 40 F, with adequate moisture and ventilation. Storage boxes for whips should have cracks in the bottoms and sides and should be elevated from the storage floor for ventilation. Sphagnum moss, kept moist but not wet, is used to line the boxes. The whips can also be packed in sphagnum moss and stored without boxes on a dry, clean floor. Whips that have been cut into cuttings may also be stored as described above, but the wood is likely to stay in better condition if cuttings are not made until planting time. Better results will usually be obtained if the storage period for whips or cuttings can be eliminated or reduced to a very short time.

Making Cuttings. Cuttings can be made with a sharp knife, a bench saw, or pruning shears adjusted for this work. The cut must be clean without compressing or crushing the tissue. The use of ordinary pruning shears in making blueberry cuttings is likely to result in injury to the cuttings. However, the blade of a pair of snap-cut shears can be removed and ground much thinner. The metal bumpers should be replaced with wooden ones. Cuttings can be made with these adjusted shears without injury and with greater comfort and speed (Johnston, 1959). Commercial propagators commonly use a bench saw to cut large quantities of whips into cutting size at one time.

Hardwood cuttings with the basal cut made just above a bud were compared with cuttings with the basal cut made just below a bud (O'Rourke, 1952b). In both groups the distal (or upper) cut was just above a bud. Economy of material and fewer snips were afforded where the basal cut was made just above a bud. Four-inch cuttings of both types were made from one-year shoots and were rooted for comparison of survival and subsequent shoot growth. Cutting survival was higher and mean shoot length was greater where basal cuts were made just below the buds, except in the Concord variety. The advantages of the above-the-bud method, however, probably out-weighed the disadvantages. This conclusion is supported by the successful use of the bench saw in commercial operations to make many cuttings at one time without considering the position of the cut with respect to the leaf buds.

The recommended length for cuttings is 3 to 4 inches in New Jersey (Doehlert, 1953c), 4 inches in Massachusetts (Bailey, Franklin, and Kelley, 1950), 4 inches in Michigan (Johnston, 1959) and 4½ to 5 inches in Washington (Schwartze and Myhre, 1954a). Cuttings less than ¼ inch in diameter but not of spindling nature root most readily. Cuttings with a diameter greater than ¼ inch root less readily but produce desirable plants (Bailey, Franklin, and Kelley, 1950). If wood of a variety is scarce and prospective prices for plants are favorable, cuttings can be reduced in length, thus obtaining a larger number of cuttings from a given shoot. Since such a cutting has less food reserves, additional care must be given in maintaining environmental conditions suitable for rooting.

Air drafts, which can cause dessication of the wood, should be eliminated while making cuttings. Cuttings are generally assembled into groups of 50, appropriately labeled, and replaced immediately in moss. A dip in Semesan (2 level tablespoonfuls per gallon) for two to five minutes may be used to prevent development of mold and blackening of the cuttings (Doehlert, 1953c).

Propagating Structures. Propagation of hardwood cuttings may be conducted in frames of various types. Johnston (1930) investigated four propagation structures at South Haven, Michigan. The rooting response of Rubel, Pioneer, Cabot, and Adams was tested in a German-peat rooting medium. Rooting percentages were above 75 per cent in the box frame and solar frame, and slightly less in the open frame. Less than 25 per cent rooting occurred in the cold frame.

Box frames (Fig. 39) may be constructed for sashes of varying sizes; suitable dimensions are 6 feet in length, 27 inches in width, and 16 inches in height. The trays for cuttings are made of 4-inch lumber, and hardware cloth of ⅛-inch to ¼-inch mesh is attached to the bottoms. The trays are set upon braces which are attached inside the frame 8 inches from the top. The trays are removable to permit easy handling. It is necessary to place the frames in full sunlight. A glass sash or a polyethylene cover and a burlap cover are placed over the cuttings until rooting occurs. The box frame has two important advantages over the solar frame: it does not require as much attention on hot days and it is about 50 per cent cheaper. The box frame with movable cutting trays (Johnston, 1959) is still used in Michigan.

The solar frame resembles a box frame with one side replaced by a glass sash sloping away from the frame. Infrared rays from the sun penetrate the glass and dissipate their energy by heating the air under the glass. This warm air circulates under the suspended trays and warms

Figure 39. Low propagating frame with glass sash and burlap shade in place. One of the trays in foreground is filled with rooted cuttings. These movable trays, which fit in the top of the propagating frame, provide a convenient way of handling cuttings prior to planting in the nursery. (Stanley Johnston, Michigan Agricultural Experiment Substation, South Haven.)

the rooting medium. On clear, hot days, it is necessary to open the lower sash to avoid excessively high temperatures in the bed.

The cold frame differs from the box frame only in respect to the position of the rooting medium. The rooting medium is placed directly on the ground in the cold frame rather than in suspended trays. With the medium on the ground, drainage becomes an important consideration.

The open frame differs only from the box frame in that no glass or plastic sash is placed over the frame.

Johnston (1958) compared clear polyethylene, clear polyethylene reinforced with wire mesh, and glass as covers for propagating frames. The average rate of rooting of Jersey blueberry cuttings for three years was 89 per cent under glass and 87 per cent under wired plastic, and the two-year rate under clear plastic covers was 84 per cent. Plastic covers have the advantages of lower cost, less weight, greater ease of handling, and less chance of breakage. But these advantages might easily be offset,

Propagation and Planting

in the case of clear plastic, by considerably higher temperatures within the propagating frame on hot days in June or early July, when the cuttings are in full leaf and are just starting to form roots. The use of colored polyethylene covers without burlap shade gave unsatisfactory results. Best rooting results were obtained when glass or clear polyethylene was used with burlap shade.

Ground beds of peat over sawdust are also used in Michigan. The boxes are 6 feet by 12 feet and approximately 15 inches deep. These are sloped to allow for drainage of water from the propagating sash. Five inches of sawdust goes into the bottom and then 5 inches of peat moss, which leaves 5 inches for growing space. The boxes should be placed in a well-drained location, or gravel should be placed under the sawdust. Hotbed sashes, 3 feet by 6 feet, and $7\frac{1}{2}$-ounce burlap cover the boxes.

In New Jersey, propagating beds are generally built 4 feet wide and 8 inches deep and vary in length (Doehlert, 1953c). For protection from grubs, moles, and mice, hardware cloth of $\frac{1}{8}$-inch or $\frac{1}{4}$-inch mesh or fly screen is placed in the bottom and secured to the sideboards. Drainage is provided by placing the bed on cedar poles, with the bottom several inches above the ground, in a well-drained location. Although Doehlert recommended placing the beds under a lath house, no sash or shade is placed over these beds. Commercial propagators have found that the rooting percentage is just as good and the cuttings make more growth in full sunlight.

In North Carolina, open beds under a lath house are used almost exclusively. Most beds have wooden framing down to the ground level. Hardware cloth or gravel is usually placed under the bed for drainage. Beds are occasionally raised above the ground surface.

In Rhode Island, rooting was tested in suspended trays in the Michigan type of box frame, in common cold frames, and in cold frames in which the propagating medium was laid directly upon a base of gravel in layers ranging from fine on top to coarse, with stones, at the bottom (Stene and Christopher, 1941). Rooting in the common cold frames was found to equal that in the more expensive trays.

Ordinary cold frames are used in Massachusetts (Bailey, Franklin, and Kelley, 1950). In order to prevent root injury by grubs, wire screen is attached under the bed. The shade is not applied until the buds break. The shading may be accomplished by use of burlap or slats supported about four inches above the sash.

Use of glass-covered frames with electric bottom heat gives the best results in Washington (Schwartze and Myhre, 1954a). A 3-inch to 4-inch layer of coarse gravel or cinders is placed on the bottom for drainage and is covered by a 1-inch layer of sand. Heating cables of 60 linear feet, such as those used in a standard frame of 3 by 6 feet, are then placed on the layer of sand. Care must be taken to fill in any air pockets around the cable to insure uniform heating. An inch of sand is placed over the cables and is covered by a sheet of ⅝-inch hardware cloth the size of the inside of the frame to hold the sand and wire in position. If rubber-covered cable is used it should be fastened to the hardware cloth with the cable side on the first layer of sand. The second layer of sand is then added over the wire. A layer of 3 parts peat to 1 part medium sand by volume is added to a depth of 5 inches for the rooting medium. Bottom heat is maintained at 70 F. A 7-foot lath house giving two-thirds shade on the south and west and one-half shade on the other sides is built over the frames.

In Washington, cuttings taken February 22 or March 5 and either planted immediately or stored until March 22 rooted almost equally well (Schwartze and Myhre, 1947). Better rooting was obtained by providing bottom heat for two or three months rather than for only one month. Cuttings supplied with bottom heat for two to three months made nearly twice as much average shoot growth as those without bottom heat. More plants of a salable size were produced with bottom heat than without.

Another type of propagating structure used in Washington depends on the availability of raw peat (Schwartze and Myhre, 1954a). It is possible to mound this material into raised beds. Such beds are bounded by 1-by-4-inch or 1-by-6-inch boards. Rooting is satisfactory but growth and root development are slower than under frames.

Propagating Media. In New Jersey, North Carolina, and Massachusetts, a mixture of half sand and half horticultural peat furnishes a good rooting medium. The peat is shredded by rubbing through a coarse mesh screen prior to mixing by some suitable means (Fig. 40). Hot or cold water may be used to wet the bed, the former used where speed is important.

Comparisons of various rooting media in Michigan have shown that German peat is superior to American peat, German peat and sand, or sand (Johnston, 1930).

The use of auxin and non-auxin type root-inducing substances on hardwood blueberry cuttings has not proved to be of sufficient benefit to warrant recommendation (Chandler and Mason, 1940; Doran, 1941; Schwartze and Myhre, 1947; and Johnston, 1928, 1930, 1939).

Methods of Striking. In New Jersey the cuttings are pushed into the bed in a vertical position (Fig. 40) in rows 2 inches apart, with the cuttings in the row also 2 inches apart (Doehlert, 1953c). A spacing of 1 inch by 2 inches results in a lower percentage of rooting and smaller plants but a greater number of plants for a given area. Closer spacing is often used, but this encourages fungus disease. Only the top bud of the cuttings is left above the surface of the soil or even with it. After the setting, the bed is thoroughly watered.

In Washington, 4½-inch to 5-inch cuttings are stuck vertically into the

Figure 40. Preparation of a commercial propagating bed in New Jersey. a. Boxes of rooting medium consisting of peat and sand are moved on a conveyor to the beds. b. The rooting medium is spread and air pockets are filled. c. The bed is leveled. d. and e. Cuttings are struck into the propagating beds. f. Skinner irrigation system for watering the beds. (Paul Eck, Rutgers University, New Brunswick, N.J.)

medium to four-fifths their length when bottom heat is not applied, but inserting them at a third to a half of their length is most satisfactory with bottom heat. The spacing may be 1½ inches by 4 inches, but if fertilizers are used the spacing should be 3 inches by 4 inches (Schwartze and Myhre, 1954a). The medium is pressed firmly around the base of the cuttings.

In Michigan, cuttings are inserted into the suspended trays on a slant to about two-thirds of their lengths (Johnston, 1959). The cuttings are placed at an angle to keep the butt of the cutting from going through the wire screen into the air, since the trays are only 4 inches deep. The spacing between rows is 2 inches and within the row about 1 inch. The cuttings are set vertically in the peat over sawdust beds.

Culture. Maintenance of moist conditions in the propagating bed is essential, thus requiring frequent waterings during periods of high evaporation (Doehlert, 1953c). Precautions must be taken against using too much or too little water. Water reaching a high temperature within a hose exposed to the sun frequently kills cuttings. May 1 to late June is the most critical time, since cuttings are in leaf and are without roots. The first greening of the tip bud is an indication that rooting has occurred.

Fertilizer is never mixed with the sand-and-peat rooting medium before the cuttings are placed in the bed, and no form of fertilizer is applied prior to the rooting of the cuttings. After rooting, characterized by a new burst of foliar growth, a liquid fertilizer is generally applied. Doehlert (1953c) recommended a weekly application of a 15–30–4 or 13–26–13 soluble fertilizer at the rate of 1 ounce of concentrate in 2 gallons of water until mid-August in New Jersey. The solution should be rinsed from the foliage after application. Johnston (1951a) recommended that a stock solution be prepared by adding 1 pound of 8–8–8 blueberry fertilizer to 4 gallons of water. Since the fertilizer dissolves slowly, the stock solution should be made up well in advance and should be screened to remove all undissolved matter that would otherwise clog the sprinkler head. A quart of this stock solution mixed with a gallon of water can be used to sprinkle about 25 square feet of propagating bed. In Michigan the first application is normally made around the middle of July, with subsequent applications at 10-day intervals until the third week in August. Applications are discontinued after August to allow adequate hardening of tissues.

The plants are usually left in the propagating beds over winter, although they may be transplanted to the nursery in September. Later planting does not allow for adequate root development to prevent winter heaving. If the plants are retained in beds through the winter, the soil should be mounded around the sides of the bed to prevent bottom ventilation and excessive freezing.

In Michigan, morning watering is recommended with water from which the chill has been removed (Johnston, 1943a). For disease control, roguing (the removal of infected plants) instead of fungicide application is practiced. Ventilation may be increased to check disease development. After rooting, the sash opening is gradually increased each day until there is full ventilation. At this time the glass sash is removed, but the burlap covering is retained until mid-September. The trays are removed in October or early November and placed in trenches 4 inches deep. Soil is filled around them and wire fencing is used to keep out animal pests.

Good ventilation was shown to be an effective prophylactic against fungus diseases in Rhode Island (Stene, 1938). The sash was kept tightly over one box frame until rooting occurred and was raised 1 inch on hot days in another box frame. A higher average of rooting and a lower incidence of disease were obtained in the ventilated frame. Ventilated-frame temperatures were found to be higher above than below the cutting trays, a condition contrary to the generally accepted optimum for rooting.

In Washington, the frames are kept closed until the first leaves appear (Schwartze and Myhre, 1954a). Continuous ventilation is then started, using a $3/16$-inch space between the frame and the sash. About July 1, or when second top growth is under way, ventilation can be increased gradually. Water is applied so as to keep the medium moist but not soaked. Ammonium sulfate applied at the rate of 2 grams per square foot or approximately $3/4$ ounce per 3-foot by 6-foot frame provides sufficient nitrogen. The cuttings should be kept in the frames until next spring, when they are sold or transplanted to nursery rows.

Growers in the Puget Sound area transfer rooted cuttings to open peat-filled frames, where they remain for a year before selling or transferring to the nursery.

The importance of fertilizer applications to hardwood cuttings was demonstrated by work at Puyallup, Washington (Schwartze and Myhre, 1948). Fertilizers consisted of ammonium phosphate (11–48–0), tankage (6–10–0), and VHPF, a commercial fertilizer equivalent to 5–25–15 and provided with vitamins, hormones, and minor elements. They were applied at rates equivalent to 300 and 600 pounds per acre both at the time of setting and at the time of rooting. The cheaper ammonium phosphate was found to be as beneficial as VHPF. The time of application was not critical, but the early application avoided the necessity of washing off the fertilizer caught on the leaves in the later application. On the basis of these tests, a 30-pound application of actual nitrogen per acre was tentatively recommended.

A later experiment tested the value of various forms of nitrogen used alone and with phosphorus and potassium (Schwartze and Myhre, 1949). The form of nitrogen or the presence or absence of phosphorus or potassium apparently was not important. However, nitrogen at 40 pounds per acre induced additional top growth and encouraged sufficient maturity to permit safe late-fall digging. It was indicated that the nitrogen should be in a soluble form, as in ammonium phosphate, ammonium sulfate, ammonium nitrate, or sodium nitrate. The use of bottom heat induced earlier rooting and thus a longer response to fertilizer.

Propagation by Softwood Cuttings

Selection and Handling. Softwood cuttings, in contrast to hardwood cuttings, are taken during the active growth period of the plant. The cuttings should be made at the proper vegetative stage of the plant and not at a certain date. Since plant development does not proceed at the same rate in different locations, cuttings taken at a certain date would be at different stages of maturity. The appropriate stage for taking cuttings is when secondary growth from lateral buds first appears in the new shoots (Webber, 1954; Johnston, 1930). This period occurs as the first fruit are ripening in Michigan. Although cuttings taken as early as June 10 have been shown to give a slightly higher proportion of rooting than cuttings taken later, it takes longer for these plants to reach a size suitable for field planting, and they are more susceptible to winter injury.

Types of Cuttings. A suitable cutting is about 4 inches long and has all the leaves but the upper two removed to reduce transpiration. When mist is used, only enough leaves are removed to insert the cutting to half its length.

In Washington, all varieties can be easily propagated with the heel type of softwood cutting (Schwartze and Myhre, 1954a). The best cutting material is 6-inch to 8-inch lateral shoots that are still growing but have matured to the extent that the base is firm but not brittle. The cutting is stripped from the plant so that a heel of bark and wood is attached. A 5-inch cutting is made by removing the soft upper portion. Just enough leaves are removed so as to insert the cutting vertically to half its length in the propagating medium. The peat and sand mixture should be moist when the cuttings are set. Ventilation is critical; excessive ventilation may cause wilting, but enough ventilation should be provided to prevent condensation. Soft rot may be controlled by weekly applications of Ferbam at the rate of 1 ounce in 4 gallons of water for three to six weeks.

The use of side-shoot cuttings was found to increase the number of cuttings obtainable from a single plant (Johnston, 1935). The shoots which ordinarily would have been used earlier for making hardwood cuttings were retained on the plant and permitted to form side shoots. These side shoots can be made into three types of cuttings: straight, heel, and mallet. The straight cuttings consisted of the side shoot only, the heel cuttings had a patch of year-old bark and wood attached, and the mallet cuttings had a complete segment of year-old stem at the base. The straight cuttings were found to root almost as well as the standard softwood cuttings.

A more rapid method of increasing plants of a new variety was afforded by the use of softwood cuttings given a forcing treatment (Meader, 1952). After the cuttings were rooted, they were forced during the winter in a greenhouse at a night temperature of 60 F. Additional lighting was used to provide 15.5 hours of light per day, and an ammonium sulfate fertilizer was applied. The average length of growth of forced plants during the winter, including the few inches of growth made prior to forcing, ranged from 67.4 inches for Stanley to 117.6 inches for Jersey.

Frames and Media. In Washington, the same structure used for hardwood cuttings may be used for softwood cuttings (Schwartze and Myhre, 1954a). No bottom heat is needed because of the warm air temperatures. A mixture of sphagnum peat and sand furnishes a suitable medium for rooting.

Mist has been used successfully for the propagation of softwood cuttings in Michigan (Fig. 41). The beds used are 8 inches deep and filled with 5 inches of a mixture of equal volumes of German peat and Perlite. The beds are placed in a well-drained location and underlaid with gravel. Concrete reinforcing wire is arched over the beds and polyethylene and Saran shade (45 per cent) is placed over the wire. An automatic mist control keeps the cuttings covered with a film of moisture. A fan on one end of the bed draws in fresh air and accelerates the evaporation of moisture, which keeps the cuttings cool on hot days. The polyethylene is taken off as soon as most of the cuttings are well rooted, but the shade is left until mid-September.

Root-Promoting Substances. Softwood cuttings treated with various root-inducing substances have been found to root more rapidly and yielded higher percentages of rooted cuttings than untreated cuttings. Successful root-inducing substances have included indolebutyric acid and indolepropionic acid in dusts and in solutions.

Softwood cuttings taken in Massachusetts in July and treated with indolebutyric acid (50 mg per liter for 20 hours) rooted as follows: Jersey, 100 per cent; Cabot, 80 per cent; and Wareham, 77 per cent. Untreated cuttings rooted 50, 33, and 42 per cent respectively (Doran, 1941).

Cuttings of Rubel taken in mid-July were treated with indolepropionic acid and potassium indolebutyrate (Doran and Bailey, 1943). After five weeks, when untreated cuttings had not even begun to root, there was 40 to 72 per cent rooting of cuttings treated with indolepropionic acid, and excellent root systems had already developed.

Figure 41. Softwood propagation with mist. a. and b. Polyethylene shade folded back for inspection. c. A mist nozzle over the cuttings. Note the nozzle is above the pipe to reduce dripping. d. A softwood cutting just beginning to root. (Charles M. Mainland, Rutgers University, New Brunswick, N.J.)

When the cuttings were taken July 1, two to three weeks before the first berries ripened, untreated cuttings rooted well. The rooting of cuttings taken after July 1 was sufficiently hastened by root-inducing substances to warrant their use.

Rooting has been increased markedly in *Vaccinium australe* with the use of a dust of indolebutyric acid in talc, but for *V. angustifolium, V. ashei, V. atrococcum* and *V. pallidum* rooting was only slightly increased (O'Rourke, 1943). The Concord, Jersey, and Rubel varieties did not respond to indolebutyric acid (Schwartze and Myhre, 1947).

Culture. Watering is critical before softwood cuttings are rooted because of the high summer temperatures and the resulting rapid water loss through the leaves of the cutting. Excessively moist conditions at high temperatures, however, increases the incidence of disease. A suitable balance must therefore be maintained between ventilation and watering. The high percentage of succulent tissue in softwood cuttings will require more exacting conditions for rooting.

Other Methods of Propagation

Budding. Although cuttings are most generally used for highbush propagation, budding (the insertion of a single detached bud into a plant) can be used to increase plants of a new selection rapidly (Johnston, 1944). A much higher percentage of bud takes were obtained when the wood was removed from the bud shield than when it was left. Budding, therefore, is best performed at a time when the bark slips, permitting easy separation of the wood from the bark. Rubber budding strips have been found to be the best wrapping material to bring about the bud-stock union.

Mounding. Mounding can be used to propagate varieties that are difficult to root by hardwood cuttings (Eaton, 1951). Varieties rooted by mounding have included Pioneer and Kengrape. After an initial severe pruning and a liberal spring application of fertilizer, a frame was constructed around the plants and filled with sawdust or peat. The frames were removed at the end of the third year and vigorous but one-sided root systems were produced near the base of the stems. These stems were severed below the new roots, and the new plants were then ready for transplanting. The uneven root systems were temporary, however, and well-balanced plants were obtained when the cuttings were set in the field. Plants propagated by this method required much less care than plants propagated from cuttings and were ultimately equal in size. One particular advantage of mounding is that plants too small for use as cutting material could be satisfactorily propagated by this method.

Seed. Highbush propagation by seed is necessary in breeding programs where plants with more desirable horticultural characters are sought. Such a method could not be used in the propagation of a variety, since seedlings would not reproduce identically the characters of their seed-parent. (See Chapter III for methods of seed removal and information on seed life and dormancy.) A medium of dried, shredded sphaghum was found superior to a peat-sand medium for growing seedlings (Childs, 1946). Seedlings grown in the sphagnum had 67.3 per cent more linear growth and shoots of greater diameter than seedlings grown in the peat-sand mixture. Some difficulty was experienced, however, in separating seedlings from the sphagnum material. No significant differences in amounts of linear growth were found in seedlings grown in sphagnum shredded by hand or shredded through screens of various mesh, but screen-shredded sphagnum was found to be superior in its handling qualities (Childs, 1947). The greater ease of shredding through a coarse screen as compared to a fine screen warranted the use of a ½-inch-mesh screen.

Applications of liquid fertilizer to seedlings at three-week intervals were found to be superior to applications at shorter intervals (Childs, 1950). One pound of a soluble 4–12–4 fertilizer was dissolved in 8 gallons of water and applied at the rate of 1 gallon to 133 plants.

Planting

Nursery. In the spring the rooted cuttings are removed from the propagating beds and planted in the best available piece of suitable soil (Fig. 42). The rooted cuttings are planted in rows 18 inches or more apart, depending on the method of cultivation, and about 6 to 10 inches apart in the row (Johnston, 1959).

In Washington, a well-drained site with facilities for irrigation is recommended for the nursery (Schwartze and Myhre, 1954b). If the plot is not adequately supplied with organic matter, it is suggested that one quart of loose, damp, horticultural-grade peat for each plant be placed in a 3-inch-deep furrow. A part of the peat is first tamped down in the furrow, and then the roots of the plant are spread over the peat. The remaining peat is placed over the roots and tamped again, and finally soil is added.

Wooden frames 8 to 12 inches in depth to which a 6-inch to 8-inch layer of peat has been added and firmly pressed may be used for growing rooted cuttings (Schwartze and Myhre, 1954b). The cuttings are set in the peat at spacings of 6 by 6 inches or 6 by 8 inches; the latter is recommended for plants that are not to be moved for two years. An additional inch of peat is then placed over the roots of the cuttings and tamped again. The best location for the frames is on

Figure 42. Handling rooted cuttings. a. One-year-old rooted cuttings in the propagating beds. b. and c. Separation and sorting of the cuttings. d. Cuttings packed for transport to the nursery. (Paul Eck, Rutgers University, New Brunswick, N.J.)

well-drained soil with facilities for irrigation and fully exposed to the sun. Blossoms should be removed from nursery plants to conserve plant vigor. After one year in the nursery, plants are known as two-year-olds. Well-grown plants of this age are very suitable for field planting.

Field. The time of planting varies with the climatic conditions of the region. Some early fall planting is done in New Jersey, whereas in North Carolina, planting may extend into the late fall and winter. Northern regions with early freezes are more safely planted in early spring. Planting as early as plants are obtainable in the fall is advisable in western Oregon, but the planting may be conducted in winter or early spring (Boller, 1951). Fall planting permits the control of weeds before the plants are set and gives the plant more time to establish a good root system. A well-developed root system would be better able to fulfill the requirements of opening buds in the spring.

The plants are usually set 4 or 5 feet by 8 feet, 4 or 5 feet by 10 feet, or 6 by 8 feet in the field. The close spacing will produce a higher tonnage the first few crop years but will require a higher initial investment and may eventually result in a crowded condition that is likely to increase production costs. Planting 8 by 8 feet permits cross cultivation. Under good growing conditions this spacing will give maximum production per acre after about eight years (Boller, 1956). The fertility of the soil may be used as a gauge to the planting distance since more fertile soil will generally grow a larger plant. In Washington (Schwartze and Myhre, 1954b) the recommended spacing is 8 by 8 feet or 7 by 9 feet on peat and black muck and 6 by 8 feet on the less fertile mineral soils.

LOWBUSH BLUEBERRY

Methods of Propagation

Since the lowbush blueberry industry is dependent upon wild plantings, few commercial attempts have been made to propagate the lowbush vegetatively. Plants in the wild may arise from seed. However, when once established, the lowbush plant produces underground stems or rhizomes, from which new shoots arise at intervals (Fig. 43). These shoots develop into plants that are like the original plant. Such a family of plants is known as a clone. One clone may cover a smaller or greater portion of a field, depending on its ability to compete with other clones.

Seed. The lowbush blueberry can be propagated by seed, but this method will not produce plants identical with the parent. When the berries are ripe, they are chilled at about 50 F for several days in order to break the rest period. The seeds must not be allowed to go dry as they may enter a prolonged secondary rest period that will take two years to break (Latimer and Smith, 1938). After the seeds have been chilled, the skins, are removed from the fruit and the seed-containing pulp is planted ¼-inch deep in flats containing a mixture of sand and fine peat moss. When about six to seven weeks old the seedlings are separated with a needle and transplanted to other flats. Fine peat is sprinkled over the seedlings when they have developed several leaves. They may be grown indoors over the winter or placed in a frost-protected pit. Shredded sphagnum fertilized with 1 pound of 4-12-4 fertilizer in 8 gallons of water is a satisfactory growing medium.

Rhizome. Underground rhizomes of the lowbush blueberry exhibit polarity. That is, the end of a cutting farthest from the mother plant tends to produce shoot growth while the other end tends to produce root growth (Hitz, 1949; Eggert, 1955). Cuttings treated on the basal end with root-inducing substances have been shown to produce more root growth than untreated controls, but treatment of the tip ends severely

Figure 43. A two-year-old blueberry plant grown from a rhizome cutting. (G. F. Dow, Maine Agricultural Experiment Station, Orono.)

inhibits root growth. Therefore it is most important to keep like ends of cuttings together.

Cuttings taken closest to the terminal end of the rhizome tended to have a higher proportion of shoots, whereas, cuttings from the basal end exhibited better rooting. Cuttings 6 inches in length produced twice as many shoots and survived twice as long as 3-inch cuttings. Although rooting did not vary greatly between cuttings of large and small diameter, the numbers of shoots per cutting and numbers of leaves per cutting was significantly greater for the larger diameter cuttings. Cuttings stored under refrigeration for five to nine weeks rooted better and produced a higher number of shoots than did cuttings that were not refrigerated.

Propagation of the low dryland blueberry, *Vaccinium vacillans* Torrey, has been accomplished by the use of either rhizome or shoot cuttings (O'Rourke, 1952a). Rhizome cuttings 3 inches in length survived better when set vertically with the distal end up than when placed horizontally. Hardwood cuttings made from the basal 4-inch sections of one-year shoots had a higher survival rate than cuttings from the same sections of older shoots and from the median or terminal sections of one-year shoots. The hardwood cuttings were considered to be superior to the rhizome cuttings because of the savings in time, labor, mother plant, and space.

Softwood Cuttings. Another method of propagating the lowbush blueberry involves the use of small side shoots as softwood cutting material (Johnston, 1935). Instead of using the year-old shoots for hardwood cuttings, side shoots were permitted to develop until the leaves had reached almost normal green color (in early June in Michigan). Shoots allowed to develop further are of little value as propagating wood because of the development of fruit buds. The shoots were removed and made into heel, mallet, and straight cuttings of 2 to 4 inches in length. The heel type of cutting rooted best, 81.7 per cent, while only 61.7 per cent of the mallet cuttings rooted. Root formation started earlier in the side-shoot cuttings than in hardwood cuttings.

Transplanting

Barren areas can be replanted with a golf-hole cutter, which removes a volume of blueberry sod 4 inches in diameter and about 6 inches deep (Hitz, 1949; Eggert, 1955). The same cutter can be used to make holes to receive the clump of rhizome sod. Since new rooting and shoot emergence depend on the cutting of rhizomes, adequate numbers of rhizomes must be present in the transplant sod. The plants remaining in the area from which the sod is removed will not be injured if the distance between holes is at least 8 inches and if the holes are refilled.

RABBITEYE BLUEBERRY

Methods of Propagation

Selection of Cutting Wood. Rabbiteye blueberry plants may be produced from either softwood or hardwood cuttings, but poor results have caused the hardwood cutting method generally to be discarded. Softwood cuttings may be taken throughout the season, but higher rooting percentages are obtained when cuttings are taken just after the first flush of growth in the spring (Brightwell, 1962). Large numbers of cuttings may be taken at this time, since most of the shoots are in the same stage of growth. Very tender cuttings should be avoided in order to reduce wilting to a minimum. A longer period is required for rooting and the percentage of rooting is lower when cuttings are made from mature wood.

Frames and Media. In Georgia, a higher percentage of cuttings of the Hagood variety were found to root in open beds than in beds under glass (Brightwell, 1948). The greatest average number of cuttings were rooted in a mixture consisting of equal parts of peat and sawdust. Mixtures consisting mainly of peat supported the greatest root growth and the smallest amount of top and root growth was found in the sawdust medium.

Good results were obtained in Georgia with cold frames (Brightwell, 1962). Drainage is improved by placing 2 inches of cinders in the bottom of the frame. Five inches of a sifted mixture consisting of 2 parts native peat, 1 part coarse sand, and 1 part old sawdust is placed over the cinders. The frames are covered with hotbed sashes to maintain a high humidity and heavily shaded to reduce temperatures.

Intermittent mist has been successful under partial shade. Imported peat and mixtures of peat and Perlite have also deen used successfully as rooting media. Attempts to root cuttings in full sun have not been successful.

Planting

Nursery. The rooted plants should be grown one summer under partial shade such as is provided by Saran cloth or a similar material. There is usually a high mortality of cuttings if planted in the field directly from the rooting bed. Plants may be spaced 12 by 18 or 18 by 24 inches in a shadehouse. Liberal amounts of peat moss worked into the soil prior to setting plants into the shadehouse increases the growth (Brightwell, 1962).

Field. Plants are generally planted 6 feet apart in rows 12 feet apart. Rows should not be spaced closer than 10 feet and plants should not be spaced closer than 5 feet (Brightwell, 1962). As much as 1 bushel of peat moss should be mixed with the soil before planting each plant. Plants should be planted about 1 inch deeper than they grew in the nursery and should be planted promptly to prevent excessive drying of the fine fibrous root system.

CHAPTER 7

Soil Management, Nutrition, and Fertilizer Practices

by Walter E. Ballinger, Department of Horticultural Science, North Carolina State University at Raleigh

HIGHBUSH BLUEBERRY

Soil Management Systems

The blueberry is a shallow-rooted plant with a root system characterized by a lack of root hairs (Fig. 44). The fine, fibrous roots of the blueberry require an open, porous soil for ease of growth (Cain and Slate, 1953).

The spread of roots of the cultivated highbush blueberry during the first few years of growth is usually no greater than the spread of the branches. As the plants become older the roots may extend into the row middles and interlace with the roots of plants in adjacent rows. This condition is found only in mature plantings on deep, moist, and well-aerated peaty soils. The downward growth of roots is influenced by drainage, soil structure, aeration, moisture, and organic matter content. The lateral spread of roots may be checked by the deep cultivation of row middles (Doehlert, 1937).

The nature of the blueberry plant's root system and the specific climatic and edaphic requirements of this crop make exacting soil management practices necessary.

Clean Cultivation. Clean cultivation is generally practiced with the highbush blueberry, that is, it is protected from competing plant growth. As early as 1921, Coville (1921) suggested that the soil be tilled to prevent the development of all competing vegetation, particularly in young plantings. Results obtained with wild highbush blueberries in Massachusetts (Bailey and Franklin, 1935) indicated that plant growth increases when competing vegetation is removed.

Figure 44. The exposed root system of a mature blueberry bush. Note the fibrous nature and the shallow penetration of the roots. (C. A. Boller, Oregon State College, Corvallis.)

Tillage of the soil apparently stimulates blueberry plant growth. Doehlert (1937) observed that leaf and shoot growth was checked if cultivation was discontinued in early or mid-August. When cultivation was continued the plants grew through October.

Bailey, Franklin, and Kelley (1939) recommended that cultivation in Massachusetts blueberry plantings be practiced from early spring to mid-August. These workers considered later cultivations to be undesirable because late growth would be susceptible to winter injury.

Cultivation in early spring is advisable for mixing fertilizers into the soil. It also assists in controlling mummy berry by creating conditions unfavorable for the development of fruiting bodies by the fungus.

Since blueberry roots are shallow and generally do not grow into the subsoil (Beckwith and Doehlert, 1933), it is essential that any cultivation be shallow. A once-popular misbelief was that cutting off the roots near the surface forced the plant to root deeper. The depth of root development, however, is fixed by the depth of the topsoil layer, and so this practice merely reduced the total root volume and thereby the plant's capacity to absorb nutrients and moisture. On soils low in organic matter, shallow cultivation avoids excessive aeration and decomposition of soil organic matter (Doehlert, 1937).

Beckwith and Doehlert (1933) found yields by plants in a two-year-old planting of Rancocas and Cabot varieties to be from 6 to 60 per cent higher when tilled to a depth of 2 to 4 inches than when tilled to a depth of 1 to 2 inches. These workers theorized that shallow cultivation left the soil beneath the tilled area hard and packed over the outer region of the root zone. Cultivation to a greater depth seemed to loosen the soil and induce a greater lateral spread of the roots. On the basis of these trials, Beckwith and Doehlert recommended a tillage depth of 3 inches in the region extending from beneath the fringe of the outer branches toward the center of the row.

Doehlert (1937) suggested that a blueberry tillage implement should be "less than 12 inches high where it must go under the branches. The tall parts of the equipment should be more than two feet from the outer edges of teeth or discs. It must pulverize the soil enough to kill weeds and yet not deep enough to injure roots or check plant growth. It should be adjustable to the slopes of mounds."

Since the roots require adequate soil aeration, many low, poorly drained soils must be mounded to provide raised rows in which to plant the blueberries. Many growers have observed that plants in poorly drained fields grow and produce better when planted this way (Doehlert, 1937). The mounds provide an area in which roots have adequate aeration for growth above the water-filled soil of the row centers. The row centers in such plantings are generally cultivated deeply to control weeds. Deeply rooted weeds may be eliminated by a small-diameter disc harrow early in the spring after the soil has become friable or by spring-tooth harrows with closely spaced teeth (Fig. 45). The mounds, however, are more difficult to till because of the mat of roots formed near the surface, and it is generally more expensive to control weeds in this type of planting. When mounds are used, a tillage tool that provides shallow cultivation and yet conforms and adjusts itself to the varying slopes must be used (Doehlert, 1937). One such implement used today in blueberry plantings attaches quickly to the rear of many tractors. It consists of an arrangement of discs decreasing in size from the middle toward the sides (Fig. 45).

Hoeing around the plants and pulling weeds by hand are expensive and time-consuming practices, but they are necessary to obtain the best growth, particularly with young plants. The costs of weed control are reduced in some states by using a tractor with a grape-hoe attachment (Fig. 46). Great care must be used to set the tripping bars to eliminate injury to the crowns of the bushes as far as possible. This implement must be carefully regulated to permit only very shallow penetration into the

Figure 45. Above. A tractor-drawn spring-tooth harrow clean-cultivating a new Wolcott blueberry planting in eastern North Carolina. *Below.* Specially designed cultivating discs for blueberries. Note the way the discs are tapered. The gangs may be adjusted to fit the contour of the row. (Walter E. Ballinger, North Carolina State University at Raleigh; Department of Horticulture and Forestry, Rutgers University, New Brunswick, N.J.)

soil beneath the bushes, where the feeding roots of the plant are located. These automatic weeders are used to a greater extent in Michigan and New Jersey than in North Carolina, where farm labor is more readily available (Carlton and Kampe, 1954; Doehlert, 1953a).

Cover Crops. The sowing of cover crops soon after harvest is recommended in many areas to allow for the timely hardening of the blueberry plants in the fall to decrease susceptibility to winter injury. Johnston (1951a) recommended shallow cultivation throughout the harvest season, followed by the sowing of an annual cover crop such as oats or Sudan grass or a mixture of these. A weed cover is also desirable if a

Figure 46. An automatic grapehoe attachment designed for blueberries. (Walter Fort, Pemberton, N.J.)

sufficient one can be produced. This practice has also been suggested for blueberry plantings in Ottawa, Canada (Eaton, 1949).

Use of a cover crop after harvest is also suggested for New Jersey blueberry fields. It increases organic matter in the soil, decreases soil erosion, and encourages the hardening of blueberry tissues before winter through the competition for nutrients and soil moisture. Cover crops have not been generally employed in the blueberry plantings of North Carolina.

Sod Culture. Relatively few commercial highbush blueberry plantings have been grown under sod culture. A 15-year-old planting under sod in southeastern Michigan was observed to be comparable in size and growth to that of an adjacent planting grown under clean cultivation. The Kentucky bluegrass sod used in this planting provided a thick mat of partially decomposed grassy materials which appeared in many respects to simulate a mulch. Yield records from 10-bush plots in each of the two plantings indicated that plants in sod plots produced at least as much fruit as those in adjacent clean-cultivated plots (Ballinger, unpublished data).

Johnston (1937) grew plants of the Rubel variety in Michigan under sod culture in which the planting was allowed to grow up in grass and weeds. The grass cover was cut twice and allowed to remain where it fell. In 1936, a season with a long drought period, 68 plants in sod yielded only 53 quarts whereas a comparable number of clean-tilled plants gave 156 quarts. Johnston suggested that on sandy soils, such as the soils used in his study, clean cultivation should be employed through the harvest season and cover crops provided in the fall to maintain soil organic matter.

A distinct advantage of sod culture in blueberry plantings is its convenience for workers. A commercial "pick-your-own" planting near Flint, Michigan, for example, has been particularly profitable to the grower. Customers are protected from the inconvenience of walking about in a highly organic soil. Disadvantages include the increased difficulty in controlling the mummy berry disease.

Mulches. The highbush blueberry is naturally adapted to a lowland, acid soil. Attempts to adapt it to drier soils of a higher pH and lower organic matter content have necessitated the employment of practices to maintain constant moisture content near the surface of the soil and optimum soil acidity. The use of mulches has been one of the more successful practices. Annual mulching has been found to reduce weed growth, maintain lower soil temperatures in summer, help maintain uniform soil moisture, maintain better soil structure, prevent heaving of the soil and subsequent root injury, control soil erosion, and reduce the costs of cultivation (Darrow, 1957; Shoemaker, 1955).

Highbush blueberry plants grown on an upland soil under clean cultivation were found to exhibit improved growth following the application of salt hay, leaves, straw, or Sudan grass as mulches (Clark, 1936). Clark found less weed growth with the mulch treatments and no need for cultivation, which had previously caused severe root injury. The elimination of this root damage may have been one of the factors responsible for the improvement in performance. Hay mulch was found to be particularly beneficial on this soil, which was naturally unsuited for blueberry production.

The effects of several mulches, cover crops and fertilizer treatments on the survival and yield of young highbush and dryland (*Vaccinium vacillans* Torrey) blueberry transplants grown in an acid loam soil were studied at Beltsville, Maryland (Kramer, Evinger, and Schrader, 1941). All mulching materials employed, with the exception of oak leaves, increased plant survival. Soil acidity was not affected greatly by mulch or fertilizer treatments. Soil moisture under all mulches was approximately twice as great as that under clean cultivation. Of particular note was the fact that lateral root growth in peat mulch was much greater than under clean cultivation. This greater root spread was associated with a doubling of yields and a reduction of soil erosion. Placing peat in trenches alongside the roots improved plant survival and yield, whereas the use of a lespedeza cover crop greatly reduced yields. As a result of this finding, the experimenters recommended that organic matter be placed in the hole at planting time and that a peat mulch and complete fertilizer be applied after the plants are established. Plants grown on a dry, sandy loam soil with hardwood sawdust mulch have been found to be more thrifty than those maintained under clean cultivation (Slate and Collison, 1942).

Chandler and Mason (1942) studied the effects of mulching and clean cultivation on blueberries grown in a sandy loam, a clay loam, and a very sandy soil. They reported that the soil in mulched plots contained a greater amount of soil moisture at both the 6-inch and 12-inch depths than did the soil in the clean-cultivated plots. The authors concluded that mulching maintained a lower soil temperature and increased the growth in clay loam soils but reduced the growth in sandy soils.

A mulch is reported to be necessary for the successful growth of highbush blueberries under the conditions of high temperature and soil-moisture deficiency that often exist in northern Georgia (Savage and Darrow, 1942). On a Clarksville gravelly loam, sawdust was by far the best mulch tested. Sawdust exhibited a greater capacity for reducing surface runoff and retaining soil moisture than any of the other materials tested. Mulches consisting of loose materials such as rye straw and oak leaves were found to be better than clean cultivation, but not as effective as sawdust. On a combined plant growth and survival basis, the mulching materials tested were rated as follows: sawdust 100, oak leaves 54, rye straw 41, and clean cultivation 0, with 100 representing best growth.

Griggs and Rollins (1947) planted three blueberry varieties in a mixture of soil with peat moss and also in a moderately drained Gloucester sandy loam. In the fall of the first season three cultural systems were initiated: clean cultivation, sawdust mulch, and hay mulch. After five years, analysis indicated that the soil phosphorus content had increased under all treatments. Potassium and magnesium were little affected, but nitrates increased under clean cultivation and hay mulch. Conversely, nitrates decreased while ammonia increased in the soil under the sawdust mulch. Little difference in moisture equivalent or organic-matter content of the variously treated soils was found. It was particularly noteworthy that plants grown with a sawdust mulch yielded 6 pints more than those with hay mulch and 8 pints more than those under clean cultivation. In these plots, a sawdust mulch was found to be more desirable than clean cultivation or hay mulch in regard to the convenience of pruning, harvesting, and weed control. Sawdust was also associated with a greater linear shoot growth. No differences in plant survival were associated with methods of soil management. However, greater yields were obtained from bushes planted in the peat-soil mixture, regardless of soil surface treatment, as compared with those planted in the natural soil.

The ascorbic acid and moisture content of fruit harvested from the bushes used in these soil management experiments were found to be unrelated to either the soil management systems or varieties (Griggs and Rollins, 1948). Some later pickings suggested that variations in ascorbic acid content may have been associated with temperature variations prior to harvest.

Four soil management programs were tested with Pioneer highbush blueberries grown on a Narragansett loam in Rhode Island (Christopher and Shutak, 1947). The programs consisted of (1) clean cultivation, (2) mulch of straw and hay, (3) mulch of sawdust, and (4) clean cultivation with a cover crop of buckwheat sown about August 1. Yields from the sawdust plots were twice as great as those from plots receiving clean cultivation or clean cultivation plus a cover crop. A severe infestation of quack grass in the clean-cultivated plots may have contributed to this yield reduction.

A higher soil-moisture content was found under straw mulch than under sawdust (Shutak, Christopher, and McElroy, 1949). The least amount of moisture occurred in soils that were clean-cultivated. Mulches reduced fluctuations in soil temperature compared with clean cultivation, and soil acidity was slightly lower under sawdust than with clean cultivation.

Yields were consistently higher from plants mulched with sawdust than from plants subjected to clean cultivation (Shutak and Christopher, 1952). Berry size was generally found to be larger on the sawdust-mulched plots, but ripening was slightly delayed. However, because of a greater total yield from the mulched plots, more berries were available at earlier harvests. Bush size and yield were found to be related. Softwood sawdust produced greater amounts of new growth than did hardwood sawdust. Hardwood sawdust was finer and apparently packed too firmly, whereas the softwood sawdust remained looser. The hardwood sawdust was also found to decompose more rapidly than softwood sawdust. Soil organic matter and total nitrogen content were not affected by the soil management practices.

Early-season soil temperatures were lower under the mulches but the difference decreased as the season progressed. In the fall the reverse was true; temperatures decreased more rapidly in the clean-cultivated soil. Fluctuations in soil temperature were only 2 per cent under mulch as compared to 12 per cent in cultivated soil and 40 per cent in the air. Soil moisture was greatest under sawdust mulch and lowest with clean cultivation. Root growth under the mulches was heavily fibrous. Roots grew within the sawdust mulch, but only on the surface of the soil beneath the straw; clean cultivation was associated with very poor root development (Fig. 47).

One study found that sawdust mulches settled ¾ inch per year and required annual maintenance, depending upon the specific rate of decomposition (Boller, 1956). Mulches of fir sawdust eliminated the cost of cultivation and improved the growth of blueberries on many soils not ideally suited for them, with the poorest soils receiving the greatest benefit. Most weeds, besides morning glory and Canadian thistle, were eliminated by mulching, and exceptions could be easily pulled, hoed, or sprayed.

The disadvantages of mulching include the potential fire hazard and the increased incidence of injury from mice (Bailey, Franklin, and Kelley, 1950). Mulching materials are generally difficult to obtain, and the hauling and spreading of the tremendous quantities needed per acre may be quite expensive unless efficient methods are employed (Shoemaker, 1955).

Greater amounts of nitrogenous fertilizer must be used to obtain good plant growth when mulches of leaves, sawdust, hay, or straw are used. As an example, with clean cultivation only 110 pounds of ammonium sulfate per acre are necessary, compared with two applications, six weeks apart, of 300 pounds each when mulches are employed.

Recommendations for thickness of application range from 2 to 4 inches or up to 6 to 9 inches in a solid mat. A sawdust mulch of 6 to 8 inches will

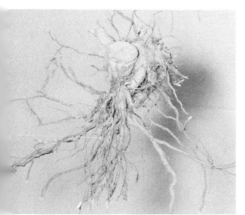
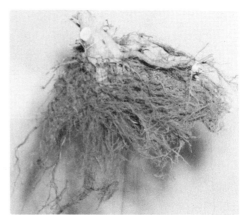

Figure 47. Above. Mulching versus cultivating is demonstrated by these two plants grown for three years on a silty clay loam. The plant on the left was cultivated and the one on the right had a 6-inch sawdust mulch. Fertilizer and water applications were the same for both plants. (C. A. Boller, Oregon State College, Corvallis.) *Below.* Blueberry roots from clean-cultivated plots (*left*) and sawdust-mulched plots (*right*). (Vladimir G. Shutak and E. P. Christopher, University of Rhode Island, Kingston.)

effectively control most weeds. Some investigators, however, believe that the mulch need be only 3 to 6 inches thick. The amount of mulch to be added yearly depends upon the rate of decomposition. One inch has been found sufficient under most conditions.

Strip mulching has been advocated in order to reduce costs. A strip $1\frac{1}{2}$ to 2 feet wide on each side of the plant is suggested in Washington state. Herbicides or cultivation may then be used to control weeds in the unmulched row centers.

Only relatively small areas have been mulched on a commercial basis in New England, New Jersey, Ohio, North Carolina, and northern Georgia. However, as the national demand for blueberry fruit increases, lands ideally suited for blueberry production will become increasingly more difficult to obtain. Consequently, the trend must be toward the use of less ideally suited lands which may require the use of special soil management practices such as mulching.

Chemical Weed Control

The possibility of chemical weed control in highbush plantings has been demonstrated (Hill, 1958). For Washington, a mixture of $1\frac{1}{2}$ quarts of Dinitro Weed Killer, 30 gallons of diesel oil, and 70 gallons of water per acre has been recommended (Schwartze and Myhre, 1954b). In Delaware, two sprays of CMU (3-chlorophenyl-1,1-dimethylurea) at 1 and 2 pounds per acre during the growing season give adequate weed control (Hitz and Amling, 1952).

Monuron (CMU) was not found to be safe with all blueberry varieties in Massachusetts. Diuron, which is less soluble but almost as effective as CMU, has been found to give good control of annual weeds when applied as a preemergence spray (Bailey, 1960). The Food and Drug Administration has granted a label for use of diuron in New Jersey and Massachusetts blueberry plantings.

Diuron has been recommended in New Jersey as a preemergent application (Schallock, 1962) at the rate of 2 pounds active diuron per acre (based on full coverage of the ground) in at least 40 gallons of water. In addition to diuron, simazine has been cleared for preemergent weed control at the rate of 5 pounds of the 80 per cent wettable powder (4 pounds active ingredient) in 40 or more gallons of water (Schallock, 1965). For specific weed problems such as winter annual weeds and dodder, chloro IPC is recommended as a commercial liquid containing 4 pounds active ingredient per gallon at the rate of $1\frac{1}{2}$ gallons of the commercial product in 20 to 40 gallons of water (Schallock, 1965).

Excellent results have been reported with diuron in Ohio (Hill, 1958). According to Hill, chemical weed control should be used as a means of preventing rather than overcoming a weed problem in small fruits.

Soil Management, Nutrition, and Fertilizer Practices 143

Trevett and Murphy (1958) suggested the use of Premerge and Sinox PE, applied at rates of 1½ to 2 gallons per acre in early summer or as annual weeds appeared. Care must be taken to apply these materials well beneath the foliage of the blueberry plant to prevent damage to new blueberry shoots.

As yet, no chemical herbicide has been found that is completely effective in controlling weeds but will not damage the blueberry plant or its fruit. It is important that the recommendations of the county agent and the manufacturer's label be followed exactly. Materials recommended for use in one state are not necessarily adapted for use in other areas.

Irrigation Practices

A uniform and adequate supply of soil moisture is essential for optimum growth of the blueberry. If this is not provided throughout the growing season by natural soil water, irrigation is required.

Blueberry yields are greatly affected by the amount of soil moisture available during the period of fruit development. In Michigan, fruit size and growth rate were retarded and some shriveling occurred during years of extreme drought (Young, 1952). Since the blueberry fruit matures over an extended period of time, the need for maintaining optimum soil moisture will extend from the time the first berries are expanding rapidly until harvest. This period may be as long as six weeks.

When to Irrigate. Fields should be irrigated when the soil becomes relatively dry and before wilting is evident. If several days are required to irrigate a large acreage, it is advisable to start irrigating the first field while 30 to 50 per cent of the available water still remains in the root zone. In this way the entire area can be irrigated before the last field becomes too dry.

In some fields it is difficult to judge soil moisture content. Black mucks are misleading because of their dark color; they may appear moist when actually they are quite dry. Sandy soils retain less water than heavier soils or soils containing high amounts of organic matter and require frequent checking. Field observations and soil tests are just as important following light showers as they are during extended periods of dry weather.

A number of methods are used to determine the moisture content of soil. Probably the one best understood and used most often by growers is the "soil feel" method. To obtain satisfactory results with this method, experience and skill are necessary in relating the feel of the soil to its available water content. A soil auger or spade may be used to obtain soil samples at various depths in the rooting zone. These samples are squeezed in the palm of the hand and the need for water determined by the feel or appearance of the soil. Generally, coarse-textured soils contain sufficient water when a ball of soil is formed after applying pressure to the sample. This ball of soil, however, is weak and easily broken. Under similar conditions a medium-textured soil often forms a ball that is very pliable. If the clay content is relatively high the soil with sufficient moisture will have a slick feeling when pressed between the fingers. Organic soils and mineral soils having a high organic matter content contain too much moisture for blueberries when water can be squeezed from them with the hand.

A comparison of monthly rainfall with the normal rainfall during the growing season may be of value in determining when to irrigate. In most blueberry areas in the United States there are Weather Bureau stations that can supply figures for the normal monthly rainfall. The larger blueberry growers may wish to purchase equipment to keep an account of rainfall in their specific areas. When a month's rainfall is 1 to 2 inches below normal, it may be advisable to irrigate, taking into consideration soil type, ground-water level, size of bushes, size and value of current crop, and the "soil feel."

A more accurate method that is widely used for measuring the soil moisture content is the electrical resistance or conductivity method. It consists of measuring the resistance of a block of gypsum permanently imbedded in the soil from 6 to 18 inches deep. Rubber-insulated wires lead from the block to the surface for connecting to the terminals of a test instrument. The electrical resistance of the block varies with its moisture content, which in turn depends on the moisture content of the soil around it. As the soil dries the block loses moisture and its electrical resistance rises.

In recent years, tensiometers have been widely used to measure the tension of the soil moisture. They consist of porous porcelain cups, which are filled with water and buried in the soil and are connected by tubing to vacuum gauges that indicate the tension on the system when the water in the cup is in equilibrium with the water in the soil.

In the final analysis, however, the grower's decision on when to irrigate is based on his experience with a particular blueberry soil and the time required to cover all of the area to be irrigated.

Method of Irrigation. In some blueberry areas, particularly in the East, the summer water table can be regulated by a system of ditches and control gates to provide subirrigation (Fig. 48). In extremely dry years, however, the use of water from reservoir ditches may be so great that the water table drops low enough to cause damage to the plants. During such times, water must be pumped from wells or nearby ponds.

The level of water in drainage ditches is not necessarily an indication of the water level in a field and should not be used as the only guide in determining when irrigation is necessary. The width of these ditches, the

Figure 48. Above. Pumping water from main drainage ditch to portable sprinkler-irrigation pipe lines. Note wire screen box around pipe to filter out large trash. *Below.* Control gate and irrigation ditch used in sub-irrigation of blueberries. (Department of Horticulture and Forestry, Rutgers University, New Brunswick, N.J.)

rate of lateral subterranean flow and the recentness of rainfall are factors which cause differences between field moisture conditions and the water level in drainage ditches.

On muck soils where the water table is controlled, a suitable set of tiled measuring wells may be employed. These will allow for a more accurate determination of the water level in a field.

Portable-sprinkler irrigation is the most widely used method (Fig. 48). This system distributes water evenly by spraying. The water is piped under pressure to the fields, and the pressure forces the water through sprinkler nozzles in pipelines and thus forms a spray. Sprinkler irrigation is the best method to use on soils that have high water-intake rates (more than 0.15 inch per hour), on fields that have steep slopes or irregular topography, and on soils that are too shallow to grade level.

Surface irrigation may be used where topography and soil are favorable. For best results, a uniformly smooth land is needed with a gentle slope of between 4 and 8 inches per 100 feet (Rubey, 1954). In North Carolina and New Jersey, furrow irrigation is being used successfully. Some growers are using portable aluminum pipe or sunken iron pipes across the ends of the rows with appropriate outlets to distribute the water down the row middles. One grower in New Jersey has several acres equipped with the usual 6-inch and 4-inch aluminum mains mounted permanently on 6-inch posts throughout the field. Experience and skill are necessary to insure an adequate lateral spread of water and a uniform moistening throughout the field. Poorly planned and executed surface systems may cause "plant skips" and soil erosion.

Source of Water. Streams, ponds, lakes, wells, and drainage ditches have been used as sources for supplemental irrigation (Fig. 49). It is desirable to select the most dependable source if more than one is available. It may be necessary to check the intended water supply for its salt concentration. Sources of water with a relatively high salt content (such as brine wells) should be avoided. Underground water sources in areas along the sea coast may become saline from the back-up of inlet waters during extended droughts when pumping of ground water is heavy.

Surface waters that contain an appreciable amount of silt and organic matter may cause clogging of sprinkler systems. These sources of water may be used, however, if the suspended material is screened out in the pumping operation.

Frequency and Amount of Sprinkler Irrigation. Irrigation should be frequent enough to prevent drying out of the soil. The amount of water to be applied during each irrigation depends upon the moisture-holding capacity of the soil. Enough water should be applied so that it penetrates as deep as most of the roots. The depth of penetration may be deter-

Soil Management, Nutrition, and Fertilizer Practices 147

mined by digging into the soil. In some soils the downward movement of water continues for a day or two after irrigation.

An excessive application of water, particularly on sandy soils, may leach fertilizer materials beyond the root depth and hence decrease the amounts of plant nutrients available to the roots. Light applications of water that do not reach root depth are just as undesirable.

Water should not be applied faster than the soil can absorb it. Puddling and erosion occur when water accumulates on the soil surface. Often water can be applied to a given field at a greater rate without puddling by using more sprinkler heads that deliver a smaller volume of water per hour.

Water may be applied to sandy soils at a rate of up to 2 inches an hour with little or no difficulty (Judkins, 1951). On sandy loam and silt loam the rate should be reduced to $\frac{1}{4}$ inch to $\frac{1}{2}$ inch per hour. An acre-inch of water is equivalent to 27,154 gallons of water, so a pump delivering 150 gallons per minute will cover an acre with about an inch of water in three hours.

Figure 49. Typical irrigation pond on a blueberry farm. (Norman F. Childers, Rutgers University, New Brunswick, N.J.)

Nutrition

Nutrient Deficiency Symptoms. Solutions high in nitrogen and low in phosphorus and potassium have been shown to be most favorable for the growth and yield of the blueberry in sand culture (Doehlert and Shive, 1936). Nitrate nitrogen appeared to be superior to ammonium nitrogen, but the nitrate solutions became more acid with time, indicating greater cation than anion uptake. The solutions that produced the best growth and yield contained 40 per cent of the total nitrogen as ammonium sulfate. The sensitivity of the blueberry to lack of manganese and boron was indicated by the relatively short time in which deficiency symptoms developed when plants were deprived of them.

Kramer and Schrader (1942) grew plants of the Cabot variety in sand culture under greenhouse conditions to study the effects of nutrient solutions deficient in a given element and the effect of rooting media on growth. Plant growth was excellent in sand but even better in sand covered with a layer of peat.

Nutrient deficiency symptoms appeared on plants grown in sand alone but often did not appear as quickly when the peat was used this way. The only treatments in which this was not the case were those in which potassium was deficient.

This experiment suggests that the order in which elements will become deficient in poor, sandy soils may be: nitrogen, phosphorus, sulfur, boron, calcium, potassium, iron, magnesium, and manganese. In a field soil of sand covered with peat the order of deficiency would be: nitrogen, potassium, phosphorus, magnesium, boron, calcium, sulfur, iron, and manganese. Peat apparently contains available sulfur, calcium, iron, and boron, since plants not supplied with these elements grew normally in sand covered with peat.

Nitrogen deficiency in these treatments was associated with the abortion of the terminal growing point. The first symptom on the leaves appeared as a uniform yellowing of the entire leaf, followed by reddening and necrosis (death). Older leaves were affected first, but all leaves were eventually affected and the plant was severely stunted.

Potassium deficiency was accompanied by marginal scorching and the appearance of necrotic spots on older leaves. Interveinal chlorosis, or mottling, appeared later on new growth from axillary buds.

Phosphorus deficiency symptoms were characterized by a slight purpling of the leaves and stems. The color of the leaves was duller than that of normal leaves.

Magnesium deficiency was noted as a uniform chlorosis of leaf margins while the area near the midrib remained green. In later stages the chlorotic areas became red and necrotic. Older leaves were affected first.

Boron deficiency symptoms appeared abruptly, first as a bluish coloring of growing points and later as a chlorotic spotting of leaves situated under the shoot terminals. Severe cases displayed blotched and misshapen leaves.

Calcium deficiency symptoms appeared first on younger growth as an interveinal chlorosis, with the green regions adjacent to the veins remaining narrower than in the case of potassium deficiency.

Sulfur deficiency symptoms were similar to those observed in the early stages of nitrogen deficiency. The younger leaves exhibited a bleached yellowing and later turned pink. The older leaves retained their green color.

Manganese deficiency symptoms were not apparent during the period of experimentation. The manganese-deficient cultures were continued until it appeared that a lack of manganese resulted in a breakdown of axillary buds.

Iron deficiency symptoms appeared first on the younger leaves as an interveinal chlorosis, quite similar to but less severe than potassium deficiency symptoms on younger leaves. In general, the interveinal chlorosis of leaves deficient in iron, potassium, and calcium was somewhat similar, suggesting a common maladjustment of the leaf process by any or all of these three elements.

In another experiment, a marked yellowing of young foliage on blueberry plants grown in sand culture and supplied with a high-calcium nutrient solution was temporarily corrected by the addition of iron tartrate (Cain, 1952a). However, the addition of ammonium nitrogen to this previously ammonium-free culture solution caused the chlorosis to disappear for the remainder of the season. Analysis of the foliage revealed that the foliar nitrogen and iron content varied with the ammonium nitrogen supply in the culture solution. The author concluded: "Iron deficiency symptoms (chlorosis) are not necessarily related to soil pH, calcium content, or the iron content of the foliage, since the healthiest plants had more calcium and sometimes less iron than those showing acute chlorosis and making very poor growth." Ammonium nitrogen was thought to be superior to nitrates for the growth of blueberries and may be involved in the internal iron nutrition of the plant. Therefore one of the factors for poor blueberry growth and foliar yellowing on marginal soils may be a lack of ammonium nitrogen. Ammonium nitrogen may be essential for normal blueberry growth.

In further studies of the relationship between nitrogen metabolism and yellowing of blueberry leaves, Cain and Holley (1955) compared the free amino acid and basic cation contents of chlorotic and green blueberry leaf tissue. Amino acids, especially arginine, were found to increase tremendously with the appearance of chlorosis. Green leaves showed greater amounts of dry weight and basic cations than chlorotic leaves. The authors concluded that a detailed interpretation of these relationships awaits further research.

Nutrient Disorders under Field Conditions. Iron chlorosis characterized by a yellowing or interveinal chlorosis of the leaves has been often observed in blueberry plantings on marginal soils (Fig. 50). A reddish-brown interveinal chlorosis developed on leaves of several varieties in a Massachusetts planting (Bailey, 1936). The veins of these leaves remained green, which is also characteristic of iron deficiency. As the symptoms became more severe, an interveinal yellowing developed which eventually covered the entire leaf. In advanced cases basal leaves on new shoots were stunted and both leaves and shoots were yellow, but generally the terminal leaves of the shoot showed the symptoms first. The symptoms occurred mostly on plants growing in soil that had a low organic matter content and was quite dry, but no relationship was found between soil pH and chlorosis.

Figure 50. Iron chlorosis on highbush blueberries. (Walter E. Ballinger, North Carolina State University at Raleigh.)

Soil Management, Nutrition, and Fertilizer Practices 151

Treatments of manganese sulfate, sodium nitrate, ammonium sulfate, ferrous sulfate, German peat placed in trenches, a complete fertilizer, magnesium sulfate, and zinc sulfate were made in attempts to correct the observed conditions. Of all the materials, ammonium sulfate produced the greatest recovery (Bailey and Everson, 1937). Approximately one month was required for leaves sprayed with ferrous sulfate solution to turn green again, but this recovery was only temporary. Manganese sulfate treatments proved toxic to the leaves. The authors attributed the symptoms to a deficiency of iron because when the ferrous sulfate spray was used the green color reappeared only in the spots where the spray contacted the leaves. A soil analysis indicated that the iron content of the topsoil and the subsoil beneath chlorotic plants was approximately one-third that of the soil beneath healthy plants.

To further test the diagnosis, blueberry plants were grown in crocks to which from 5 to 40 grams of lime had been added. Chlorosis appeared on all lime-treated plants, increasing in severity as the quantities of applied lime increased. An increase in pH of the limed soils from pH 4.2 to 6.4 may have decreased iron solubility. Ferric citrate crystals placed in a slit made in the stem of one plant caused the chlorotic leaves above and below this point to turn green. The conclusion was that the chlorosis had resulted from an iron deficiency (Bailey and Everson, 1937).

Kramer and Schrader (1945) theorized that frequent reports of iron deficiency in plants may be explained if it is assumed that the amphoteric proteins in the blueberry act as cations. The authors suggested that this may explain the differential absorption of anions in blueberries. If anion radicals are absorbed in excess, it may be difficult to adequately maintain iron in the blueberry in the reduced, available form, since anions would serve as oxidizing agents.

Iron chelates have been used to correct iron chlorosis in blueberries (Hill, 1956). One hundred grams of iron chelate per bush, evenly distributed and worked into the soil about the base of a four-year-old plant, eliminated the chlorosis within 30 days. During the next season the treated bushes displayed vigorous shoot growth with dark green foliage accompanied by heavy bud development. Hill's recommendations for fertilizing blueberries included the use of 1.5 ounces of iron chelate for small, newly established plants and up to 4 ounces for larger, well-established ones.

Soil calcium in excess of 10 per cent of the soil cation exchange capacity (CEC) was associated with poor growth of blueberry plantings growing on a marginal upland soil of pH 5.2 in Michigan (Ballinger,

1957). The CEC of a soil represents its storage capacity for plant nutrient cations and is measured in terms of me/100 g of soil. Soil in which chlorotic bushes were growing was found to have a calcium content as high as 48 per cent of the soil CEC. Analyses of foliar tissue from bushes, displaying a complete yellowing as well as less advanced stages of interveinal chlorosis, revealed a leaf iron content approximately half as great as the standard established for normal plantings in Michigan. Concurrently, the potassium content of these leaves was 123 per cent of the standard value. This may lend support to Cain's theory that excessive bases in the blueberry leaf may interfere with the utilization of iron within the leaf.

Magnesium deficiency symptoms were reported by Mikkelsen and Doehlert (1950) in high bush blueberry plantings in New Jersey on a Leon sand with a pH of 4.0 to 4.5 to which 600 pounds of a 7–7–7 fertilizer had been applied in the regular fertility program. The symptoms began at berry-ripening time as a marginal and interveinal pale green coloring on lower leaves of rapidly growing shoots. Later the affected areas turned a yellowish, olive green and grew a vivid orange and red in advanced stages. The application of 70 pounds per acre of magnesium oxide in the form of magnesium sulfate (Epsom salts) and 300 pounds per acre in the form of dolomitic limestone in September corrected the deficiency the following season. The magnesium content of leaves from plants receiving 300 pounds of magnesium oxide per acre was 0.24 per cent as compared to 0.14 per cent for untreated chlorotic bushes.

Popenoe (1952) studied an abnormal foliar condition of the Rancocas variety of highbush blueberry grown on a Sassafras gravelly loam at Beltsville, Maryland. Analyses of leaves from the affected plants indicated a low level of magnesium (0.13 per cent) and suggested that these symptoms, similar to those described by Mikkelsen and Doehlert, were related to a magnesium deficiency. The symptoms appeared as a marginal reddening of the basal leaves and covered nearly half of some of the more severely affected leaves. The symptoms were not present in the second flush of growth following fruit harvest but reappeared during the period of fruit maturation the following year. Unlike the symptoms described by Mikkelsen and Doehlert they did not increase in severity as the season progressed. Applications of Epsom salts at the rate of 200 pounds per acre did not lead to recovery. Popenoe suggested that the magnesium-potassium ratio might play an important role in the expression of these symptoms.

Magnesium deficiency symptoms were found on leaves of highbush blueberry plants grown on a Gloucester sandy loam of pH 3.8 to 4.2 (Bailey and Drake, 1954). The older leaves of the plants displayed

Soil Management, Nutrition, and Fertilizer Practices 153

typical yellow and red coloration between the veins and were found to contain as low as 0.04 per cent magnesium. Magnesium oxide at a rate of 25 to 150 pounds in the form of Epsom salts and 100 to 600 pounds as dolomitic limestone increased magnesium in leaves but had no effect on their potassium, calcium, or nitrogen content. Particularly noteworthy was the fact that 1.5 tons per acre of limestone did not cause leaf chlorosis and increased the soil pH from 4.0 to only 5.2, which is slightly higher than that of an ideal blueberry soil. As little as 25 pounds of magnesium oxide as Epsom salts and 200 pounds as dolomitic limestone virtually eliminated the magnesium deficiency symptoms.

A dieback on highbush blueberries was observed in western Washington by Woodbridge and Drew (1960). Pioneer and Pemberton were the varieties most seriously affected. Pioneer's symptoms included a failure of both leaf and fruit to develop. On some bushes only a few branches were affected and on others only a few buds failed to develop. The buds of the Pemberton variety broke normally, but after a few weeks of satisfactory growth the terminal bud shriveled, turned brown, and died. Stanley showed slight dwarfing and Jersey showed no abnormalities even when growing where the other varieties showed severe symptoms.

The symptoms were correlated with a low boron content of tissues from affected bushes. Since there was adequate boron in the soil, the condition apparently was induced by low soil moisture. The spring and summer of 1957 were unusually dry, and even with irrigation many of the bushes had wilted from time to time. Boron sprays were applied the fall of 1958 and there was no dieback in 1959 and 1960.

Boller (1956) has suggested that several conditions or combinations of conditions might lead to typical mineral deficiency symptoms on blueberries in the field, including: (1) insufficient or poor distribution of soil moisture, (2) a small or weak root system caused by poor drainage, insect injury, fertilizer burn, disease, or excessive packing of the soil, (3) insufficient quantities of available ammonium nitrogen, and (4) a deficiency of some mineral element other than nitrogen.

Plant and Fruit Composition. From an analysis of highbush blueberry leaf tissue (Table XVIII), one can interpret that the required amounts of cations: potassium, magnesium, and calcium, are lower in the blueberry than in many other pomological fruit crops (Childers, 1966). Many workers have emphasized the importance of nitrogen to blueberry growth and production (Bailey, 1958a; Ballinger, 1957; Boller, 1956; and Doehlert, 1953b). In one study of highbush blueberries in Michigan, yields were found to be directly proportional to the nitrogen content of the leaves up to 2.1 per cent (Ballinger, 1957). Higher content was associated with decreased yields. It is interesting to note that in the reported surveys presented in Table XVIII, the per cent found in blueberry leaf tissue

Table XVIII. Mineral Composition of Blueberry Leaves

Per Cent Dry Weight					ppm					Sampling Remarks	Source
N	P	K	Mg	Ca	Fe	Mn	B	Cu	Zn		
Highbush Blueberry											
1.98	0.16	0.53	0.28	0.74	150	168	49	15	20	mid-July; mid-position of fruiting shoots; survey of Michigan plantings	Ballinger, Kenworthy, Bell, et al., 1958a
2.02	0.18	0.53	0.22	0.32	—	—	—	—	—	July 1; mid-position of fruiting shoots; Rubel	Bailey, Smith, and Weatherby, 1949
—	0.20	0.50	0.26	0.51	60	120	—	—	—	August 2; 3rd to 7th apical leaves of actively growing shoots; Rubel	Mikkelsen and Doehlert, 1950
2.02	0.17	0.37	0.12	0.37	—	—	—	—	—	July 7; plants under a sawdust mulch receiving 800 lb ammonium sulfate; Rancocas	Popenoe, 1952
Lowbush Blueberry											
2.18	0.17	0.64	0.12	0.39	45	1993	15	9	32	mid-July; sprout fields (burn year) in Nova Scotia; *Vaccinium augustifolium* var. Laevifolium	Lockhart and Langille, 1962
1.53	0.10	0.52	0.51	0.40	118	—	33	22	—	mid-July from sprout fields in Maine. *V. angustifolium* var. Laevifolium	Trevett, 1962
Rabbiteye Blueberry											
1.50	0.05	0.13	0.12	0.04	—	—	—	—	—	plants receiving full nutrients in sand culture; Walker variety	Minton, Hagler, and Brightwell, 1951

from plants grown in three different geographical areas came close to the optimum value suggested by Ballinger.

Little difference in phosphorus content can be observed from the analysis of leaf tissue from different areas. In the absence of reported deficiencies or response to phosphorus, one might assume that the optimum phosphorus level in the highbush blueberry leaf ranges between 0.16 to 0.20 per cent.

Potassium deficiency has not been reported in the field although Merrill (1944) has demonstrated potassium response by the blueberry on muck soils. In the surveys reported in Table XVIII, leaf potassium values in field grown plants without mulch were very similar (0.50 − 0.53 per cent), but in plants grown under mulch which received only ammonium sulfate, potassium content in the leaf was only 0.37 per cent.

Magnesium deficiency (less than 0.12 per cent magnesium) has been reported in field grown plants in most commercial highbush blueberry growing areas (Bailey and Drake, 1954; Ballinger, Kenworthy, Bell, et al., 1958b; Mikkelsen and Doehlert, 1950). A survey of 58 New Jersey plantings (Eck, 1964) showed that 27 per cent of the plantings were deficient in magnesium; 58 per cent were in a critical range (0.12 to 0.20 per cent); and only 15 per cent of the plantings analyzed had magnesium contents greater than 0.20 per cent. Results by Amling (1958) would suggest possible varietal differences in the critical magnesium level at which deficiency symptoms would appear. Foliar symptoms of magnesium deficiency were associated with a much higher medial leaf concentration of magnesium in Rubel than in Jersey.

Calcium level in blueberry leaves appears to vary the greatest. Values reported from Michigan were much higher than values reported by other workers. The relatively high plant calcium values reported for Michigan may be related to Ballinger's discovery that poor blueberry production and vigor were associated with calcium saturations greater than 10 per cent of the exchange complex. The wide range of calcium values found in blueberry tissue suggests the need for additional information on calcium needs of the blueberry.

Biweekly leaf sampling indicated that leaf concentrations of all nutrient elements varied less during the three weeks prior to, and including, the first week in which 35 per cent of the crop is harvested than at any other period of the growing season (Amling, 1958). This would suggest that leaf samplings for diagnostic purposes might best be taken during this period. Nitrogen, potassium, phosphorus, and copper decreased, while magnesium, calcium, iron, manganese, and zinc increased as the season progressed. Late in the season potassium increased and zinc decreased in the leaf. The concentration of all nutrient elements in the fruit declined with increased maturity with the greatest decline occurring in the manganese concentration and the least decline in the potassium concentration.

In a recent publication Watt and Merrill (1964) have compiled information on the composition of the blueberry fruit (Table XIX). Their analyses closely compare with the analysis of blueberry fruit made by Chandler in 1944, although vitamin A contents differed considerably. Chandler had observed, however, that the vitamin content in the blueberry varied with location and season. The vitamin C content of fresh blueberries ranges from 14 to 27 mg per 100 g of fresh berries (Chandler, 1944; Watt and Merrill, 1964). Bessey (1938) observed that the vitamin C content is about two and one-half times greater in highbush than in lowbush blueberries.

Blueberries contain relatively high concentrations of manganese and iron. Hodges and Peterson (1931) determined that a serving of blueberries contained more manganese than did a serving of any other one of 120 foods tested except wheat bran. Blueberries are low in calcium, magnesium, and phosphorus but have a fair amount of potassium.

Most of the acid in blueberry fruit is citric and malic (Chandler and Highlands, 1950; Nelson, 1927). The benzoic acid content in blueberries has been measured by Merriam and Fellers (1937) and found to be low (0.002 per cent).

Ballinger, Bell, and Kenworthy (1958) reported a range of soluble solids in blueberry fruit from approximately 10 to 18 per cent. High leaf nitrogen and high yield appeared to be associated with low soluble solids percentages, and conversely.

Fertilizer Practices and Recommendations

The blueberry plant must produce vigorous growth if fruitfulness is to be maintained. This requires the optimum employment of cultural practices, including fertilization. Berry size and other responses have been related to the vigor of shoot growth (Shutak, Hindle, and Christopher, 1957). Bailey, Franklin and Kelley (1939) have stated that "since success with blueberries depends on growing large berries, the plants must be kept highly vigorous. The need for strong growth is all the greater because of the severe pruning required." Some blueberry soils are naturally fertile and contain adequate proportions of organic matter and nitrogen to maintain vigorous plant growth. Many other soils that are not as well endowed with organic matter and have been continuously cropped for years require the replacement of depleted nitrogen. Thus blueberries have been found to respond more to nitrogen than to any other material (Ballinger, 1960).

Soil Reaction. Blueberry soils should be checked occasionally, and measures should be taken if necessary to adjust the soil reaction. If the

Table XIX. Composition of 100 Grams of Raw, Canned, and Frozen Blueberries

Type of Product	Water (%)	Food Energy (cal)	Protein (g)	Fat (g)	Carbohydrate Total (g)	Carbohydrate Fiber (g)	Ash (g)	Minerals Ca (mg)	Minerals P (mg)	Minerals Fe (mg)	Minerals Na (mg)	Minerals K (mg)	Vitamin A (IU)	Thiamine (mg)	Riboflavin (mg)	Niacin (mg)	Ascorbic Acid (mg)
Fresh	83.2	62	0.7	0.5	15.3	1.5	0.3	15	13	1.0	1	81	100	0.03	0.06	0.5	14
Canned																	
in water	89.3	39	0.5	0.2	9.8	1.0	0.2	10	9	0.7	1	60	40	0.01	0.01	0.2	7
in syrup	73.2	101	0.4	0.2	26.0	0.9	0.2	9	8	0.6	1	55	40	0.01	0.01	0.2	6
Frozen																	
unsweetened	85.0	55	0.7	0.5	13.6	1.5	0.2	10	13	0.8	1	81	70	0.03	0.06	0.5	7
sweetened	72.3	105	0.6	0.3	26.5	0.9	0.3	6	11	0.4	1	66	30	0.04	0.05	0.4	8

Source: Watt and Merrill, 1964.

soil is light in texture and slightly acid and if the subsoil is low in lime reserve, further acidification may be easily effected. There are a number of ways of doing this. Peat worked into the soil will reduce and maintain the desired pH and is valuable in other ways. If the pH is too high or if the soil contains some reserve of lime, sulfur or aluminum sulfate may be used. Crude tannic acid is also effective. If sulfur is used to increase soil acidity, it should be applied the year before the planting. It should be used sparingly, for too much acidifying material may be quite harmful to blueberry plants or may even kill them outright.

Table XX will serve for calculating roughly the amount of sulfur or aluminum sulfate required to attain certain pH levels in sandy or loam soils. Generally, it is recommended that to acidify sandy soils ¾ pound of sulfur per 100 square feet be applied for each full point that the soil registers above pH 4.5, and that to acidify medium loams 1½ to 2¼ pounds be applied for each full point above pH 4.5. If a soil has a very appreciable lime reserve, it is impractical to try to acidify it. Fertilizers which leave acid residues such as ammonium sulfate are generally used in order to maintain the desirable acid soil reaction.

Plants growing on soils below pH 4.0 respond to the addition of lime (Merrill, 1939; Harmer, 1944; Johnston, 1951b). Harmer found that 8 tons of ground limestone per acre increased growth by 20 per cent in a muck soil at pH 3.3 but did not cause an increase with muck at pH 4.8. Merrill showed that soils below pH 4.0 can be made suitable for blueberry production by applying limestone at the rate of 1 to 4 tons per acre, depending upon the degree of soil acidity to be corrected. If possible, the limestone should be worked into the soil before planting.

Manures. As early as 1910 (Coville, 1910), manures were thought to be detrimental to the blueberry. Coville (1921) reported that stable manure stimulates vegetative growth but may cause later injury.

In Michigan, spring applications of manure for three consecutive years did not produce a significant increase in yield from 18 Rubel bushes (Johnston, 1943b), although more berries from the manured bushes ripened for the first picking. Johnston attributed this effect to the extra nitrogen contained in the manure. Foliage of the manured bushes was a darker green color than that of the plants not receiving manure. Contrary to earlier reports, however, no injury occurred from the use of the manure. Johnston postulated that manure would be more beneficial on poor, sandy soils low in organic matter.

Bailey (1944) applied up to 10 tons of manure per acre to blueberry soil with no apparent detrimental effects to the plants. Horse manure was applied in 1941 at the rate of 10 tons per acre and cow and poultry manures were used at rates to provide equivalent quantities of nitrogen. In 1942 and 1943 the amounts of manure applied were doubled. Yield

Table XX. Changing Soil pH with Sulfur or Aluminum Sulfate [a]

Present pH of Soil	4.0		4.5 (Blueberry Range)		5.0		5.5		6.0		6.5		7.0		7.5	
	Sand	Loam	Sand	Loam	Sand	Loam	Sand	Loam	Sand	Loam	Sand	Loam	Sand	Loam	Sand	Loam
4.0	0.0	0.0														
4.5	0.4	1.2	0.0	0.0												
5.0	0.8	2.4	0.4	1.2	0.0	0.0										
5.5	1.2	3.5	0.8	2.4	0.4	1.2	0.0	0.0								
6.0	1.5	4.6	1.2	3.5	0.8	2.4	0.4	1.2	0.0	0.0						
6.5	1.9	5.8	1.5	4.6	1.2	3.5	0.8	2.4	0.4	1.2	0.0	0.0				
7.0	2.3	6.9	1.9	5.8	1.5	4.6	1.2	3.5	0.8	2.4	0.4	1.2	0.0	0.0		
7.5	2.7	8.0	2.3	6.9	1.9	5.8	1.5	4.6	1.2	3.5	0.8	2.4	0.4	1.2	0.0	0.0

Pounds of Sulfur per 100 Sq Ft for pH of

Source: Slate and Collison, 1942.

[a] To substitute aluminum sulfate for sulfur, multiply the number of pounds of sulfur required by 6. Example: If the present pH of the soil is 6.5 and blueberry culture at pH 4.5 is contemplated, then from the table, 1.5 lb of sulfur per 100 sq ft would be required for a light soil and 4.6 lb for a medium loam. If aluminum sulfate were used instead of sulfur, 6 times these amounts would be necessary, or 9 and 27.5 lb, respectively.

The amount required can also be calculated easily for any pH. The present pH of the soil must first be known fairly accurately, as well as whether the soil contains appreciable lime reserve, which may be true if the pH is much in excess of pH 7.0. Sandy soil: for every 0.1 pH over 4.5, apply 0.075 lb sulfur per 100 square feet. Example: A sandy soil with a pH of 5.8.

5.8 − 4.5 = 1.3 or 13 tenths.

13 × 0.075 = 0.975, or nearly 1 lb of sulfur per 100 sq ft.

Example: A loam soil with a pH of 5.2.

5.2 − 4.5 = 0.7 or 7 tenths.

7 × 0.25 = 1.75, or 1¾ lb of sulfur per 100 sq ft.

Wherever no figures are given, lime applications would be required.

0.25 lb sulfur per 100 square feet. Example: A sandy soil with a pH of 5.8. loam soil: for every 0.1 pH over 4.5 apply and fruit size from manured plants were similar to plants receiving chemical fertilizer.

Types of Fertilizers. Early research by Beckwith (1920) on the use of mineral fertilizers for the highbush blueberry indicated that applications of sodium nitrate produced very little increase in yield over controls but that a complete fertilizer increased yields about 40 per cent. Organic sources of nitrogen produced greater yields than did sodium nitrate.

On a sandy soil in New Jersey, yields were tripled, compared with unfertilized plants, by spring additions of 600 pounds per acre of a mixture of sodium nitrate, dried blood, steamed bone, phosphate rock, and potash (Coville, 1921). Coville concluded, however, that if a blueberry soil contained enough peat, fertilizer was not needed.

In western Washington, Crowley (1933) was able to double yields in the first year by using a mixture of 100 pounds of sodium nitrate, 200 pounds of rock phosphate, and 50 pounds of potassium sulfate. During the second season the yields of fertilized plots were increased $2\frac{1}{2}$ times over those of control plots. Increased berry size resulting from a greater amount of new wood growth was responsible for the greater yields.

The following complete fertilizer mixture was recommended for blueberries on the basis of long-term fertilizer studies: 450 pounds of sodium nitrate, 450 pounds of dried blood (an organic nitrogen source), 800 pounds of rock phosphate, and 300 pounds of potassium sulfate, applied at the rate of 600 pounds per acre per year in two applications three weeks apart (Beckwith, 1933; Beckwith and Doehlert, 1933). The response to nitrogen at the rate of 33.6 pounds of nitrogen per acre in the form of sodium nitrate or sodium nitrate plus dried blood was found to be superior to dried blood alone. Split applications of sodium nitrate were associated with increased yields, but split applications of dried blood, which acts more slowly, were not superior over a single application. No additional benefit was derived from the application of 480 pounds of rock phosphate per acre in addition to the annual application of complete fertilizer.

In later studies, Beckwith, Coville, and Doehlert (1937) showed that yields were doubled over a nine-year period by the use of a complete fertilizer but that ammonium salts, dried blood, and acid phosphate alone did not give satisfactory results. The fertilizer mixture utilized by these workers was 450 pounds of sodium nitrate, 450 pounds of calcium nitrate, 800 pounds of rock phosphate, and 300 pounds of potassium sulfate. This was applied at the rate of 300 pounds per acre in early May and another 300 pounds three to four weeks later to bushes producing an average of two quarts of fruit. Smaller bushes received reduced amounts.

On Michigan soils of higher organic matter content, Johnston (1934) found that 335 pounds of superphosphate per acre gave very good results

Soil Management, Nutrition, and Fertilizer Practices   161

with plants of the Rubel variety. The same quantity of a 5-10-12 fertilizer also gave good results, indicating that applications of potash were beneficial. The lack of response to nitrogen applications might have meant that the soil was already rich in nitrogen.

Slate and Collison (1942) grew highbush blueberry plants under both clean cultivation and sawdust mulch. Potassium chloride killed many of the clean-cultivated plants but did not adversely affect plants under mulch. The injury was attributed to the chloride, since sulfate applications were not injurious. They recommended that chlorides not be incorporated in blueberry fertilizers.

Instances of harmful effects from the use of muriate of potash (potassium chloride) have been reported from Michigan and New York (Merrill, 1944; Slate and Collison, 1942), but no preference concerning the potassium fertilizer form is made in New Jersey and North Carolina. From the results of sand culture tests in the greenhouse on the Wolcott blueberry, Ballinger (1962) concluded that chloride content in excess of 0.5 per cent of the total foliar dry matter resulted in leaf injury. A survey of commercial plantings in eastern North Carolina revealed that the only instance of excessive foliar chloride occurred on bushes that had been overfertilized when the fertilizer was applied unevenly in lumps about the base of the plants.

Applications of nitrogen and phosphorus were found to be most effective in increasing growth of plants in sand under Michigan conditions, and nitrogen and potassium treatments were found more beneficial with a muck soil (Merrill, 1944). A direct relationship was found between the accumulation of nitrogen and phosphorus in the wood of highbush blueberry plants grown in sand. Concurrently, a direct relationship between the nitrogen and potassium content of plants was found in the muck. Merrill recommended that a complete fertilizer should be used until evidence exists that a given nutrient element is limiting plant growth.

Massachusetts blueberry plantings were found to respond to applications of nitrogen (Bailey, 1958). Responses to other fertilizer elements, however, were not as evident. Bailey applied nitrogen as ammonium sulfate at rates of ½, 1, and 2 pounds per bush. Potassium and magnesium were supplied in the form of Sul-Po-Mag (0, ½, and 1 pound per bush). All possible combinations of these were employed and all treatments were spread in one application just prior to bloom. The 2-pound application of ammonium sulfate stimulated considerable late fall growth which was susceptible to winter killing. Bailey speculated that, over a period of years, even the 1-pound rate might prove too great. No response from the Sul-Po-Mag applications was obtained.

In a North Carolina test, Ballinger, Kushman, and Brooks (1963) found that increasing the rates of nitrogen application increased bush vigor and fresh weight, dry weight and nitrogen content (per cent of dry weight) of the leaves of the Wolcott blueberry. Fruit yield in both pounds and numbers per bush and pH of the fruit were increased. There was no effect upon date of peak harvest or the berry's size or sugar content, although the total acidity and the percentage of dry weight and soluble solids in the juice were decreased. Nitrogen, therefore, has a profound effect on the blueberry plant and fruit quality.

Fertilizer Ratios. Complete fertilizers appear to be necessary for maximum production of highbush blueberries, particularly on soils low in organic matter. Although recommendations vary widely, the trend seems to be toward a 1–1–1 ratio. Mixtures of 7–7–7, 8–8–8, 10–10–10, or 11–11–11 are available. It is often more economical, because of handling, to use the higher analysis mixtures.

Blueberry recommendations have included those of Blasberg (1948), who suggested the use of a 5–10–10 mixture in Vermont; Shutak and Christopher (1952), who advised applications of a 5–10–10 in Rhode Island; and Johnston (1951) and Kenworthy, Larsen, and Bell (1956) in Michigan, who urged the use of a 1–1–1 mixture on mineral soils and a 1–2–3 or 1–3–4 mixture on organic soils. Liebster (1960) recommended the 1–1–1 and 1–2–3 mixtures for German blueberry plantings on mineral and organic soils respectively. In North Carolina an 8–8–8 is recommended (North Carolina State University Agricultural Extension Service, 1964). Blueberry fertilizer recommendations in New Jersey call for the use of an 8–8–8 mixture. Doehlert (1953b) advised the addition of 2 per cent magnesium oxide to this mixture. Darrow (1957) suggested the use of an 8–8–8 mixture (not neutralized) for locations where satisfactory practices are still unknown. Eaton (1950) at Kentville, Ottawa, Canada, reported that a 5–10–5 mixture was used prior to World War II but that a standard 9–5–7 previously suggested for apples had been used during the war years and has given favorable results.

Recently a trend away from the 1–1–1 ratio recommendations has been indicated (Bell and Johnston, 1962; Kenworthy, Larsen, and Bell, 1962). As a result of field trials in which there was little response to phosphorus and potassium applications, fertilizer ratios of 2–1–1 for mineral soils and 1–2–2 for organic soils are presently being recommended for Michigan plantings where leaf analysis indicates that phosphorus and potassium are low. Where phosphorus and potassium levels are sufficient only nitrogen applications are recommended.

Supplemental applications of nitrogen fertilizers are often recommended for use in conjunction with complete fertilizer applications, particularly on mineral soils low in organic matter (Kenworthy, Larsen, and Bell, 1956; North Carolina State College Agricultural Experiment Station, 1956). The ammonium form of nitrogen is generally recommended for use on soils in which the pH is higher than 5.0 or 5.5. The

Soil Management, Nutrition, and Fertilizer Practices

relative merits of the ammonium and the nitrate forms of nitrogen remain a matter of dispute, however.

Time of Application. To stimulate vigorous growth, particularly from the time of leaf and blossom appearance until fruit ripening, New Jersey recommendations for bearing fields call for the application of one-half of the fertilizer during the last week in April and the remainder during the first week in June (Doehlert, 1940, 1941a, 1944). A third application may be employed in October if signs of nitrogen or other types of deficiency are apparent.

For North Carolina (North Carolina State University Agricultural Extension Service, 1964), a complete fertilizer is recommended for application when the first plants begin to bloom. Four to six weeks later the plants should be top-dressed with an ammonium form of nitrogen fertilizer. For Michigan, two applications are suggested, the first in early spring and the second in early June (Kenworthy, Larsen, and Bell, 1956). A supplemental application of ammonium sulfate in late June was also suggested for use on mineral soils low in organic matter. Bailey, Franklin, and Kelley (1950) suggested the use of split applications of fertilizer for soils susceptible to nitrogen loss by leaching, weeds or mulches. In general, fertilization of highbush blueberries should correspond to periods of greatest growth.

Amounts of Fertilizers. Blueberries are extremely sensitive to excessive quantities of fertilizer. It has been suggested that, since highbush blueberries are grown on acid soils naturally low in exchangeable bases, a low cation requirement is necessary for growth under these conditions (Kramer and Schrader, 1945). Ballinger, Kenworthy, Bell, et al. (1958b) compared the cation content of soils beneath poor and vigorous bushes and observed that the total of the three major cations (potassium, calcium, and magnesium) on the soil exchange complex was much higher beneath the less vigorous plants. They concluded that such conditions in Michigan may cause reduced plant growth in relatively dry soils with low organic matter contents. This would lend support to an earlier suggestion that several integral applications of fertilizer be made rather than a single application (Doehlert, 1940). This reduces the possibility of excessive concentrations of cations occurring on the soil exchange complex at any one time.

Increases in the concentration of the nutrient solutions applied to blueberry plants in sand culture resulted in a reduced growth rate (Ballinger, 1962). With another ericaceous plant (azalea) Colgrove and Roberts (1956) found growth and intensity of foliage color to increase as the total base content of the rooting medium was reduced. Similarly, blueberries have been found to grow best in acid soils low in nutrients (Doehlert, 1957).

General recommendations for amounts of fertilizers to apply to blueberries vary with the type of fertilizer used, age and location of plants, type and fertility of the soil, and the general vigor of the plants (Table XXI). Trials using various levels of the locally recommended materials should be made to determine the optimum amount of a given fertilizer, particularly those containing nitrogen.

Table XXI. Fertilizer Recommendations for Commercial Blueberry Production

	Recommendation		Source of Recommendation
lbs/A.	Fertilizer	Application	
	HIGHBUSH BLUEBERRY [a]		
400–600	8–8–8	early spring	United States Department of Agriculture (Darrow, 1957)
100	ammonium sulfate	6 weeks after complete fertilizer; 6-week intervals thereafter if needed	
500	8–8–8	last week in April	New Jersey (New Jersey Agricultural Experiment Station, 1960)
500	8–8–8	first week in June	
500	8–8–8	October, where there are signs of nitrogen hunger	
408 (or)	16–8–8–4	in April just before bud break; 16–8–8–4 for mineral soils and 8–16–16–4 for organic soils; 4% MgO is standard	Michigan (Bell and Johnston, 1962)
408	8–16–16–4		
300–400	8–8–8	at bloom;	North Carolina (North Carolina State University Agricultural Extension Service, 1964)
30	nitrogen	4–6 weeks after bloom;	
30	nitrogen	4–6 weeks after harvest if plants have not made good growth	
700	10–10–10	split application in early May and a month later; nitrogen application must be increased 50 to 100% if fresh mulch has been added	Massachusetts (Bailey and Kelley, 1959)
6–16 oz/ bush	7–7–7; 10–10–10, with at least ½ nitrogen in organic form	lowest rate for newly set plants and highest rate for mature plants, ⅔ application rate applied late April, the remainder in June	Rhode Island (Christopher and Caroselli, 1958)

Table XXI. Fertilizer Recommendations for Commercial Blueberry Production (Continued)

	Recommendation		Source of Recommendation
lbs/A.	Fertilizer	Application	
6–12 oz/ bush	0–20–20	equivalent to 1–1½ lb/100 sq ft applied at planting	Vermont (Blasberg, 1961)
2 oz/ bush	ammonium sulfate	equivalent to ¼ lb/100 sq ft; double rate of nitrogen if sawdust mulch is used	
340–425	ammonium sulfate	applied to the surface of the sawdust in a broad ring	Oregon (Boller, 1956)
		LOWBUSH BLUEBERRY	
300 (or) 600 (or) 160	10–10–10 5–10–10 ammonium sulfate	for stems less than 4 inches tall, applied in the spring of the burn year; reduce rates to approximately ½ if stems are greater than 6 inches	Maine (Trevett, 1955)
500 (or) 1,000 (or) 250	10–10–10 5–10–10 ammonium sulfate	spring of crop year if poor second crop is anticipated and if poor growth was made during the previous year	
		RABBITEYE BLUEBERRY	
500–800	4–8–6, 8–6–4	rate depends on age and spacing of bearing bushes	Florida (Florida State Department of Agriculture, 1941)
600–1,200	8–8–8	rate depends on age and spacing of bearing bushes	Georgia (University System of Georgia Agriculture Extension Service, 1957)

[a] Recommendations based on mature plants producing at least 6 pints per bush and at least 1,000 plants per acre.

Methods of Application. For effective utilization, fertilizer must be applied in the region of the plant's absorbing roots. This area generally extends as far out as the branches. The spread of roots may often be increased by spreading the fertilizer farther out into the row each year, but few blueberry plants exist in soils or receive cultural practices that permit root growth across the row middles.

A broad strip application (Fig. 51) of fertilizer has proven best (Doehlert, 1953b). The placement of fertilizer in narrow bands results in poor distribution and increases the plant's susceptibility to drought. It is recommended that the fertilizer be scattered well and not dropped in lumps.

Nursery Fertilization. Blueberry cuttings placed in the nursery are not fertilized at planting time (Doehlert, 1953b). After the second flush of growth starts, the established plants may be side-dressed at the edge of the root balls at the rate of 5 pounds of an 8–8–8 fertilizer on each side of 1,000 feet of row, and the application may be repeated six weeks later. During periods of drought, fertilizer should not be applied to young plants.

Figure 51. A tractor-drawn mechanical fertilizer distributor in eastern North Carolina. (E. B. Morrow, North Carolina State University at Raleigh.)

Soil Management, Nutrition, and Fertilizer Practices

Fertilizing Newly Planted Fields. Newly planted bushes should never be fertilized until the roots have become reestablished and second growth has been initiated. The application of 100 pounds per acre of an 8-8-8 fertilizer (1.25 ounces per bush) in a band at least 4 inches away from the crown of young plants, followed by a similar amount one month later if needed, has been recommended (Doehlert, 1944). As the plants grow, the rate may be increased by 200 pounds per acre per year until the fourth year in the field. It is stressed that fertilizer is not a cure-all, and during the early years of a blueberry planting other cultural practices such as cultivation and weed control must be accomplished faithfully to insure the efficient utilization of the fertilizer.

LOWBUSH BLUEBERRY

Soil and Plant Management

The lowbush blueberry is produced in areas that were formerly forests but that have been cut and burned. Clean cultivation is not commonly employed. It has been reported, however, that the lowbush blueberry may be cultivated by providing cover crops and by mixing peat into the soil (Chandler, 1947). Plants selected for cultivation should spread well, should be affected as little as possible by three-year burnings, and should produce fruit of a good character on a large number of stems.

Lowbush blueberry plantings that are declining in vigor and production may possibly benefit from the restoration of soil fertility which results from the application of mulches of peat, sawdust, or hay. However, the mulches may be hard to stabilize on the surface and difficult to obtain (Trevett, 1956). Commercial fertilizers are needed to force new blueberry stems up through the thick mulch layer, when it is applied.

Irrigation has resulted in increased blueberry yields when moisture was limiting in lowbush fields (Struchtemeyer, 1956). The results of seven years of trials on lowbush blueberry fields indicated that irrigation was highly beneficial, particularly on first-crop land with a high yield potential.

Lowbush blueberry plants may persist, but will seldom fruit, on forest floors where the light is dim. Chandler and Mason (1946) found that more than 80 per cent of full sunlight is required every year for large yields. The clearing of the trees and brush from a forest generally causes the multiplication of scattered "ghost" blueberry plants because of the more favorable conditions of light and other factors. However, this land must be burned occasionally to prevent growth that would again crowd and shade out the blueberries (Smith, 1946). The land used for lowbush blueberry production is generally burned every two or three years to stimulate growth of new shoots and to eliminate many weed plants. However, many other weed plants survive the burnings and must be controlled by some other means. This is one of the most difficult problems of the lowbush blueberry grower.

Weed Control

A blueberry weed may be classified as any plant other than the blueberry plant. Weed species harbor disease and insects and interfere with the proper application of fertilizers. Chandler and Mason (1946) classified blueberry weeds according to their effects on the lowbush blueberry industry. The first category consisted of "plants which have fleshy fruits that may be harvested with the blueberries and are an adulteration in the pack, such as bunchberry, sugar-pear, huckleberry, wintergreens, bearberry, mountain cranberry, rose and chokeberry." The second was "weeds which have windborne seeds such as spreading dogbone, goldenrod, fireweed, milkweed, orange and yellow hawkweeds, kind devilweed, wild fall astor and willows." The third category consisted of "weeds which form dense masses and crowd out the blueberry plants, such as bush honeysuckle, sheep-laurel, bunchberry and wintergreen." A fourth category included woody weeds normally occupying newly cleared land, such as alder, birch, sweet fern, willow, hazel, and sprouting oak.

Soil type and fertility greatly influence the kinds of weeds found in a blueberry field (Chandler and Mason, 1946). In the blueberry barrens, alder, birch, sheep laurel and sweet fern are prevalent, but not bunch grass and poverty grass. In old fields, however, grass is more abundant and is generally accompanied by cinquefoil. Alder, poplar, birch, bayberry, and scrub oak shade blueberries in cut-over woodlands, and old pastures contain weeds found in both fields and cut-over woodland.

The following weeds are common in blueberry plantings (Eaton, 1950): common brake or bracken (*Pteris* sp. Linnaeus), sheep laurel or lambkill (*Kalmia angustifolia* Linnaeus), bayberry (*Murica colinensis* Miller), sweet gale (*Myrica gale* Linnaeus), wild spirea, hardback or meadow sweet (*Spires latifolia* Borlch), and wild roe or brier (*Rosa blanda* Aiton).

Weed control practices fall into three general classes: mechanical, such as hand-pulling and mowing; burning; and chemical (Trevett, 1952). Burning has been recommended to eliminate many weed plants and cutting to control those that survive. Cutting leafy weeds in July hinders their development and may eliminate them in two or three years (Smith, 1946).

Annual mowing in July, August or September has been recommended for the control of sweet fern in Yarmouth, Nova Scotia (Eaton, 1950). For many other plants with growth requirements similar to those of blue-

Soil Management, Nutrition, and Fertilizer Practices 169

berries, midsummer mowing to slow their spread was suggested until proven control measures could be offered.

Trevett (1952) recommended mowing for control of brake fern in lowbush blueberry fields of Maine. Mowing has also been used successfully in July to remove the weeds that survived chemical treatment. Hand pulling is economically justifiable only for removing what less costly weed-control practices have missed. Fall or spring burning does not control many weeds. For evergreen types, however, fall burning has some possibilities.

Chemicals appear to be the most efficient (Trevett, 1952) and acceptable means for control of woody weeds in Maine. Herbicides should be applied with extreme caution since they usually will affect blueberry plants as well as weeds. However, if used with care they may cause only minor damage.

Herbicides recommended by Trevett (1952) were 2,4-D, 2,4,5-T and Ammate. These materials are plant hormones which, upon entering the plant, become systemic and kill the roots as well as the tops. The 2,4-D may be obtained as a powder (sodium salt) or as a liquid (amines and esters). The 2,4,5-T may be acquired only in the liquid form (amines and esters). A mixture of 2,4-D and 2,4,5-T is sold under the trade name Brush Killers. Trevett (1953) grouped the more common weeds occurring in lowbush blueberry plantings into three classes on the basis of susceptibility to 2,4-D.

Ammate contains 80 per cent ammonium sulfamate and is obtainable as a powder. It is effective against more kinds of weeds than either 2,4-D or 2,4,5-T. Blueberry plants are more easily damaged by it, however, and it corrodes spray equipment. For red maple, one pound of Ammate per gallon of water, with a sticking agent, may be used (Trevett, 1952).

Lambkill can be controlled (Trevett, 1961) by sprays of 3 pounds per acre of the low-volatile ester forms of 2,4-D (*not* the amine) in midautumn, after blueberry leaves have fallen. The leaves of the lambkill absorb the 2,4-D, which is then translocated to the roots to kill the plant. Little 2,4-D is taken up by the bare blueberry stems. Sprays should be applied only prior to autumn burning of the field. If the fields are not burned, or burning is delayed until late spring, the blueberry stand will probably be thinned by half.

Chemical weed killers may be applied either as a foliar treatment or as a stub treatment, that is, after mowing. The stub method can be used throughout the year and gives better control of some weeds, while the foliar method requires the presence of leaves.

Foliar applications of chemicals may be made in several ways. First, the entire field or area may be covered rapidly and economically with a power sprayer, using a 2,4-D type of weed killer. This method is used primarily on thick, extensive growths of sweet fern and bayberry. Under these conditions the spray wets only the protective umbrella of bushy-topped weeds, not the lower layer of blueberry plants. Area spraying is generally limited to relatively new fields.

Spot spraying is more common than area spraying. This method consists of applying the herbicide to individual clumps of weeds with a hand sprayer in the summer (Trevett, 1952). A hand boom with a single nozzle from a garden type of knapsack sprayer, or a power sprayer, is used to limit the spray to the weed clumps and to minimize contact with the blueberry bushes. This method is useful in controlling alder, willow, birch, maple, and other clump weeds.

The brush method of herbicide application eliminates the disadvantage of wind drift of aerial applications. A film of a herbicide solution is deposited on the leaves of susceptible weeds using a brush 8 to 12 inches wide wrapped in an absorbent cloth. The brush is dipped into the herbicide solution and applied with jabbing or sawing strokes from the base of the weed upward, thoroughly covering both sides of the leaves and stems.

The glove method is a good substitute for hand-pulling of sweet fern. A cotton glove, worn over a rubber glove, is kept moist by dipping into a pail of herbicide solution. Weeds are grasped as closely to the blueberry bush as possible. As the glove is pulled lightly upward the weed is thoroughly wetted. Great caution must be taken to prevent the herbicide from coming in contact with the blueberry plants.

Spraying stubs of woody weeds after mowing is an effective means of control at any time of the year, although less resprouting of treated clumps occurs if the ground is not frozen. It reduces the resprouting of weeds such as birch, alder, and red maple, which send shoots only from the clump. Stub treatments are less effective on plants such as poplar, which send up shoots from roots at a distance from the clump.

When a concentrated spray solution is used, caution must be exercised during placement. The use of a protective "shoe" on the end of the spray wand and the use of low spray pressures hold down spray splattering, which might contact and injure blueberry plants.

Toxic vapors are given off by the esters of 2,4-D and 2,4,5-T used for stub treatments. Injury to blueberry stems 3 to 4 inches away from treated stubs has been observed, and it is therefore best to use these materials when the blueberry plants are not in leaf. Ammate is nonvolatile.

The contact method of applying herbicides is mainly used for controlling weeds such as poplar. A blanketing material on a wooden frame, moistened with weed killer solution, is dragged as low as possible across the tops of weeds without touching the blueberry plants.

Soil Management, Nutrition, and Fertilizer Practices   171

The costs of weed control are difficult to calculate because of the variation in size, kind, and number of weeds in a given field (Chandler and Mason, 1946). Woody shrubs must be sawed or cut, whereas herbaceous weeds need only chemical sprays. In Maine, the costs during 1946 ranged from 50 cents to $50 per acre, depending on the method of control used and the numbers and kinds of weeds encountered (Chandler and Mason, 1946).

Chemical weed control has the potential of saving the grower a great deal of labor and time. However, precision in the selection of the chemical, measurement of the exact quantity for use, and correct timing and method of application are required. It takes but a short time to apply the chemicals; their adverse effects can last a long time.

Nutrition

Nutrient deficiency symptoms were described by Lockhart (1959) and, as a whole, were similar to those reported for the highbush and rabbiteye species. Leaves of calcium-deficient plants, however, had red to dark flecks that later turned to dark brown blotches. These coalesced in more advanced stages as the leaves curled and died. Leaf composition of the essential mineral elements (Table XVIII) has been reported by Lockhart and Longille (1962).

Fertilizer Practices

For many years the lowbush blueberry industry consisted primarily of wild blueberry stands on cut-over woodland. Periodic burning every second or third year was the principal means of pruning and maintaining vigor. As the industry grew, however, a decline in soil fertility and productivity in older fields was recognized. Consequently, the means of maintaining and increasing production, including fertilization, have assumed added importance.

The objectives of lowbush blueberry fertilizer practices for the lowbush blueberry have been summarized as follows: "Tall stems ordinarily produce more fruit buds than short stems. Consequently the objective of the grower during the year of the burn is to produce a tall stem early so as to insure abundant fruit bud formation for the first fruit crop. The objective during the year after burning is to produce numerous side branches without decreasing the production of the fruit crop. These side branches are essential for obtaining a high-yielding second crop if a three-year cycle is followed [Mason, 1950]." Although fertilizers may improve yields, they should be used with caution. Excessive use may cause too much growth during the first year after the burn, resulting in tall, thin stems with few fruit buds. Fertilization must not upset the delicate balance between vegetative growth and fruitfulness. The most desirable result is the best blueberries with a minimum stimulation of wood growth.

Chandler and Mason (1933) found that a complete fertilizer increased yields of the native Maine lowbush blueberry by 128.6 per cent compared with untreated plots. Nitrogen increased growth, numbers of fruit buds per stem, and yields. Plots receiving phosphorus and potassium showed no appreciable increase over plots receiving no fertilizer. All treatments decreased the content of reducing sugars in the fruit. The acidity of the fruit varied from pH 3.63 to 4.11 and was found to be highest from bushes fertilized with ammonium sulfate and lowest from bushes which received applications of fish meal. A plot treated with a manganese fertilizer yielded fruit whose acidity was similar to that of the check, but the amino acid and total nitrogen contents were higher than those of fruit from the untreated plants. It was concluded that differential nitrogen levels affected maturity and indirectly the berry constituents.

Plantings that had been burned one, two, and three years before application of fertilizer failed to benefit from phosphoric acid, muriate of potash, hydrated lime, potassium sulfate, sulfur dust, lamp black, charcoal, or wood ashes (Smith, Eggert, and Yeager, 1946). Plants receiving applications of a 7-7-7 fertilizer at 200, 500 and 1,000 pounds per acre in May and June produced higher yields and better shoot growth than those that received sodium nitrate at the same rates. Sodium nitrate overstimulated stem growth of weeds, grass and blueberries. Ammonium sulfate increased shoot growth, the number of blossoms per shoot, and the fruiting area of the shoots, but produced less fruiting area than the 7-7-7 fertilizer. No differences in ripening or fruit size were found to be related to treatments. Increases in yield were due primarily to greater numbers of fruit.

Excessive weed growth resulting from fertilization of lowbush blueberry plantings is frequently responsible for reductions in yield due to shading (Chandler, 1943). Fertilizers may also cause low-growing weeds and grasses to crowd out the blueberry plants (Mason, 1950). The amounts applied, therefore, should be between what will produce the best blueberry bushes and what will cause excessive weed growth.

Applications of complete fertilizers are not always superior to those of nitrogen alone (Mason, 1950). Responses of the lowbush blueberry to nitrogen applications often have been most striking, whereas the use of phosphorus and potassium has not been shown to be generally beneficial.

Fertilizers are not generally included in lowbush blueberry management practices in Canada (Eaton, 1950). Nitrogen, phosphorus, and potassium fertilizers were tested in Yarmouth County, Ottawa, separately and in all possible combinations for three years. Increased growth of grass was the only consistent response observed in plots receiving nitro-

Soil Management, Nutrition, and Fertilizer Practices 173

gen. It was noted that the contact of fertilizer with blueberry foliage may result in damage, therefore, applications should be made before the buds open.

Extensive field fertilizer tests with the lowbush blueberry have been conducted in Maine (Trevett, 1962). Since the yield per stem was found to be closely correlated with the number of stems per square foot and the yield per square foot, attention was given to the effects of fertilizers on stem length, density per square foot of planting, branching, and the number of fruit buds per inch of stem. Both the year of fertilizer application and date of application within a given year modified responses. The results are presented accordingly.

For preemergence fertilizing (after burning, but before stems appeared above ground), an increase in nitrogen application rate (17.5 to 35 to 70 pounds per acre) decreased the number of fruit buds per inch of stem but increased stem length, the total number of fruit buds, and the yield of berries. Additions of phosphorus and potassium to the fertilizer did not change this response appreciably. A standard rate of 35 pounds per acre of nitrogen was chosen. The use of more or less nitrogen depended on such factors as weed competition, genetic differences from clone to clone, and plant vigor prevailing in a given field. The use of preemergence nitrogen fertilizers also increased the number of blossoms per fruit bud, increased the stem length, and extended the growing period of the emerged unbranched stems during the burn year. The extended growing period did not increase winter injury.

The date of the burn did not affect plant response to nitrogen applied after emergence if the nitrogen was always applied to bushes of the same stage of development. Nitrate and ammonium sources of nitrogen produced similar plant responses.

The initiation and growth of rhizomes in plots mulched with sawdust and peat were increased by application of nitrogen. When the majority of the stems were ready to branch (when 50 to 100 per cent of the stems had ceased to elongate due to terminal meristem abortion, termed "dieback" by Trevett), applications of nitrogen fertilizer produced the greatest percentage of branched stems by fall. The report stressed, however, that rainfall patterns must be considered to insure that the nitrogen reaches the root zone. Complete fertilizers did not produce more branching than nitrogen alone. The rate of nitrogen application determined the amount of branching. For unbranched stems, nitrogen applied after 100 per cent of the stems had died back did not increase stem length but did increase the number of fruit buds per stem. However, whether or not the yield is increased depends on the effects of this late-applied nitrogen on disease, the number of blossoms per fruit bud, pollination, winter hardiness, and weed growth.

Of the various minor elements tested, only copper and zinc applications increased the number of buds per stem.

Trevett attempted to increase the second-crop yield by fertilizing with nitrogen at the rate of 35 pounds per acre in the spring of the first-crop year. The second-crop year yield increased by 41.3 per cent, but this increase was made at the expense of the first-crop yield. As a result the overall three-year totals were not materially increased. This first-crop reduction may have been due to detrimental effects of early nitrogen applications upon either pollination and fruit development or the increase in blossom and twig blight. Early applications of nitrogen also may have stimulated vegetative growth so that subsequent moisture stresses resulted in smaller berries. The date of nitrogen application appears to be important since early applications have, in some cases, reduced fruit set.

Fields were heavily fertilized at the precluster bud stage of the second-crop year in an attempt to increase the food reserves of the rhizomes. This, it was thought, might enhance growth in the year of the burn. However, results were disappointing and first-crop yields increased only 18.9 per cent. This insignificant carry-over of nitrogen's effects from year to year indicated that more frequent applications should be made at opportune times when specific effects, previously mentioned, could be influenced. Primarily, the nitrogen content of the leaves should be maintained at a minimum level to prevent leaf drop. The importance of leaf-to-fruit ratios in the production of maximum yields of many fruit crops is commonly known. Trevett theorized that a failure of the leaves to provide photosynthates for fruit development results in a drain of reserves from the rhizomes. In Trevett's words, blueberries should be fertilized "not only to insure stems of adequate length for maximum fruitfulness, but also to supply nutrients in effective amounts at critical stages of plant growth and to reinforce the stimulus from fertilizer applied in preceding years."

Fertilizers may cause injury to blueberry foliage or newly opened buds. Therefore it is desirable to apply them in the spring, seven to 10 days before the initiation of growth.

No two blueberry fields are alike and the grower must determine the optimum amount of fertilizer to be applied. Small sections of each field should be tested to determine the effects of fertilizers on weed growth as well as the desirability of fertilizing the blueberries.

The fertilizers should be spread uniformly over the entire field by hand or by machine. Hand broadcasting is made more accurate by dividing the field into strips 10 feet wide and 100 feet long, and then determining the number of handfuls of fertilizer that must be spread over this area to provide the desired rate of application.

RABBITEYE BLUEBERRY

Soil Management Systems

The rabbiteye blueberry responds to clean cultivation (Darrow, 1957) and benefits from the elimination of competing weeds. These plants possess a fibrous root system that penetrates more deeply into a well-drained soil than the root system of the highbush species, but it is still relatively shallow. Accordingly a cultivation system similar to highbush practices may be followed.

The rabbiteye blueberry responds well to mulching (Darrow, 1957); however, very few data concerning the mulching of this species are available.

Nutrition

Nutrient deficiency symptoms were produced on the rabbiteye blueberry grown in sand culture by Minton, Hagler, and Brightwell (1951). Deficicncy symptoms were very similar to those found for the highbush blueberry (Kramer and Schrader, 1942) but were sufficiently different to present here (Fig. 52).

Nitrogen deficiency symptoms were noticeable 40 days after the initiation of differential treatment. Leaves were smaller, turned yellow and reddish, and exhibited small necrotic pinhead spots in later stages. The plants were stunted.

Potassium deficiency symptoms appeared after 70 days, first as an interveinal chlorosis of young leaves and later as a complete blanketing of leaf surfaces with pinhead spots which developed into a severe necrosis. Marginal scorching and rolling appeared during more advanced stages of the deficiency.

Phosphorus deficiency symptoms appeared after 90 days as a darker green leaf color than normal and smaller leaves than normal.

Magnesium deficiency was characterized, after 75 days, by a distinct interveinal reddening followed by an upward cupping of older leaves. Affected leaves were smaller than normal and later dropped, leaving the basal areas of the shoots bare. Poor growth resulted.

Calcium deficiency, after 90 days, was observed as a scorching and upward cupping of both old and new leaves. The leaves and plants were of moderate size.

Sulfur deficiency was noted after 65 days. It was characterized as a chlorosis of the leaves which progressed to a mottling and completely bleached appearance. Affected leaves were only of medium size and plant growth was reduced.

Figure 52. Leaf deficiency symptoms of the rabbiteye blueberry when grown in sand culture. (N. A. Minton, ARS, USDA, Auburn University, Auburn, Ala.)

a. Complete mineral solution
b. Minus sulfur
c. Minus magnesium
d. Minus calcium
e. Minus nitrogen
f. Minus potassium
g. Minus phosphorus
h. No nutrients

Fertilizer Practices

Although the available information on fertilizer effects on the rabbiteye blueberry is limited, Darrow (1957) reported that rabbiteye plantings have responded to applications of 4–8–4, 4–8–6, and 4–8–8 fertilizers in Florida. Formulations containing 6 to 8 per cent potash have been found to benefit older plantings. An 8–6–4 formulation applied at the rate of 100 pounds per acre every two or three years also has been found to be satisfactory. Half a pound to a pound per plant has been suggested in first-year plantings, 1½ to 2 pounds per plant in the second year, 2½ to 3 pounds per plant in the third year, and 500 to 800 pounds per acre in mature plantings.

Fertilizer recommendations for the rabbiteye blueberry in Georgia (University System of Georgia Agricultural Extension Service, 1957) include the use of an 8–8–8 fertilizer at 400 to 600 pounds per acre together with an annual supplemental side dressing of 200 to 300 pounds per acre of ammonium sulfate.

Wynd and Bowden (1951) found that chlorotic rabbiteye blueberry bushes grown on a Cecil clay loam with a topsoil pH of 5.2 near Athens, Georgia, responded to a very insoluble iron-containing glass frit. Ferrous sulfate sprays at the rate of 1 pound per 25 gallons of water had produced only temporary improvement of leaf color on five-year-old plants. Periodic application of the sprays was necessary in order to provide a continual supply of iron. A finely powdered glass frit containing 5 per cent ferric oxide was applied at the rate of 5 pounds per bush and thoroughly mixed into the upper 12 inches of the soil. After 194 to 445 days, complete recovery of leaf color was obtained.

CHAPTER 8

Plant and Fruit Development

by **Vladimir G. Shutak,** *Department of Horticulture, University of Rhode Island, Kingston, and* **Philip E. Marucci,** *Cranberry and Blueberry Research Laboratory, Rutgers University, Pemberton, New Jersey* *

HIGHBUSH BLUEBERRY

Fruiting Habit

Highbush blueberries produce fruit from buds on one-year-old wood. Flower buds are initiated during the summer, and bud development continues throughout the fall and winter. The flower buds are found near the tip of the new growth. Under good growing conditions every new shoot, including tiny laterals, usually set one or more flower buds (Fig. 53).

Shutak, Hindle, and Christopher (1957) found that the thickness of new growth affected the size of berries. They divided new growth into three groups—thin (less than 0.1 inch in diameter), medium (0.1 to 0.2 inch in diameter) and thick (over 0.2 inch in diameter)—and found that the largest berries were produced on the thickest wood and the smallest on the thinnest wood. Measurements were made on Pioneer, Pemberton, and Dixi varieties, but general observations indicate that the same relationship between wood thickness and size of berries also exists in other varieties.

Pollination

A properly pruned blueberry bush which receives good cultural care is capable of setting almost 100 per cent of its blossoms. To yield a good commercial crop a set of about 80 per cent is required. In this respect

---

* Parts of this chapter were originally written by Charles A. Doehlert, formerly in the Department of Horticulture and Forestry, Rutgers University (retired, 1959). The pollination section was written by Philip E. Marucci.

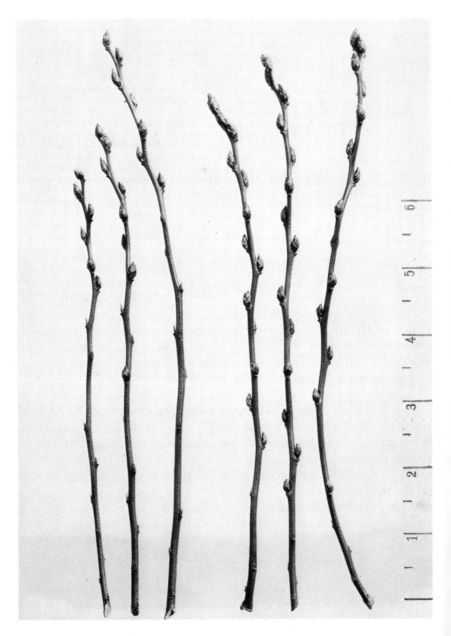

Figure 53. Terminal shoots showing arrangement of flower and leaf buds. Note the differences in size between the two types of buds. (John S. Bailey, University of Massachusetts, Amherst.)

blueberries are quite dissimilar to apples, peaches, and other fruits that may produce large yields with only 20 per cent blossom set. The failure of Earliblue and Coville to produce adequate commercial crops consistently in New Jersey has been attributed to poor pollination by Filmer and Marucci (1963). Thus, pollination is a very critical phase of blueberry growing.

Floral Characteristics. The blueberry blossom has all of the characteristics of an entomophilous flower, one adapted for insect pollination. The corolla is well developed and fragrant, and at its base are located glands, or nectaries, which secrete the nectar bees seek. The pollen grains are comparatively heavy, tend to adhere in masses, and are not easily propelled by wind. Bees and other insects are guided to the flower by its conspicuous color and odor.

Coville (1910) described the peculiar structure of the floral parts, which tend to facilitate cross-pollination by bees while rendering self-pollination difficult or improbable. The blossom is in a pendant position with the opening of the corolla lowermost (Fig. 54). Ten stamens hang downward in a close, tight circle around the pistil, which extends well beyond the stamens almost to the opening of the blossom. The nectar is welled up at the base of the pistil and may be procured by the insects only by pushing through the clustered stamens. A mere touch of any one of these structures, which are under a slight springlike tension, communicates movement to all of them and dislodges pollen from all of the sacs, showering the bee's body. Pollen released from the stamens and not intercepted by the bee has little chance of landing on the stigma. The stigma's angled sides deflect the falling pollen like an inverted funnel. The prominent central position of the stigma at the opening of the corolla and the sticky liquid on its surface enhance the probability of pollination with each bee's visit. In its search for nectar it is not possible for a pollen-laden bee to avoid brushing against the stigma and transferring pollen to it. The pistil of an unpollinated blossom continues to grow outward until it may extend beyond the orifice of the corolla, enhancing further the probability of its contact with insects. Merrill (1936) found that 79 per cent of the pistils of highbush blueberry remained receptive to pollen four days after the opening of blossoms and that 63 per cent were still functional after five days.

The pollen of the blueberry, like that of its close relative the cranberry, is tetrad, divided externally into four cells. The cranberry pollen can produce four germ tubes, but Merrill (1936) obtained no evidence that more than one was produced by a single blueberry pollen grain. He found that fresh blueberry pollen from a wide selection of varieties germinated readily and in an agar culture media obtained better germination in a 9 per cent sugar solution than with 3 per cent or 6 per cent concentrations.

Figure 54. Cluster of flowers of *Vaccinium angustifolium*, Washington County, Maine. Note the pendant position of the individual flowers in the raceme and the prominent pistil. (Walter J. Kender, University of Maine, Orono.)

Histological observations of self-pollinated material showed no retardation of pollen tube growth, which is frequently associated with self-infertility. Pollen tubes were found to traverse the length of the style in about two days and enter the ovule and accomplish fertilization in about three days.

It has been observed that blueberry varieties may vary in the ability of the flower to attract insects (Filmer and Marucci, 1964). In New Jersey the failure of the Coville and Earliblue varieties to produce good crops was found to be related to an innate unattractiveness of their blossoms to honeybees (Filmer and Marucci, 1963). These varieties were considered to be producing good crops only in a few isolated fields that had a superabundance of bumblebees. The wild bees worked blossoms at random while the honeybees preferred other varieties to Earliblue and Coville. Filmer and Marucci (1964) showed that there is almost a direct relationship between the attractiveness of a variety to bees and its productiveness. Rubel, June, and Rancocas were the most attractive to bees and are among the most productive of the cultivated blueberry varieties. Earliblue, 1316-A, Coville, and Stanley had relatively poor attractiveness to bees, and these are the varieties that recently have been failing to set good crops in New Jersey. The basic reason for the difference in attractiveness to bees is not known. In limited observations it was noticed that the sugar content of the nectar of attractive and unattractive varieties differed very little but that the quantity of nectar produced by the attractive varieties was notably greater (Filmer and Marucci, 1963).

Cross-Pollination vs. Self-Pollination. In his *Directions for Blueberry Culture*, Coville (1921) stated: "When blueberry flowers are pollinated with pollen from their own bush, the berries are smaller and later in maturing than when pollen comes from another bush. Some bushes are almost completely sterile to their own pollen." He advocated alternating rows of different varieties. Beckwith (1930) also concluded that cross-pollination was a requirement for good blueberry production. Based on three years of pollination experiments, using 15 highbush varieties, Meader and Darrow (1947) showed that cross-pollination usually increased the crop sufficiently to warrant interplanting of two or more varieties. This emphasis on the value of cross-pollination resulted in the use of a system in the early cultivated blueberry fields which alternated varieties every two rows. Some New Jersey fields still have this planting pattern.

In Michigan, Merrill (1936) made a comprehensive and thorough study of highbush blueberry pollination. To determine whether blueberries were self-fruitful, Merrill employed three methods: bagging of blossoming twigs before opening of blooms throughout the period of bloom, hand pollination, and isolation. Data procured in each of these tests clearly showed that the blueberry is self-fertile. Perhaps the isolation test was the most impressive. Forty-eight Rubel plants in their fourth growing season, growing two miles from the nearest blueberry field, all produced very good crops. These findings had an important impact on blueberry culture, particularly in Michigan. Growers were encouraged to plant standard varieties in solid blocks to simplify cultural and harvesting operations (Johnston, 1947). Practically all fields in all blueberry areas are now planted in solid blocks. In North Carolina several fields of 100 acres or more contain only the Wolcott variety.

The question of the value of cross-pollination in blueberry production is still unsettled. Recent investigations by Filmer and Marucci (1963) confirm the findings of both Merrill and Coville. Their data on cross-pollination by hand, shown in Table XXII, reaffirm Merrill's contention regarding the self-fruitfulness of blueberries. However, the data also support Coville's opinion that cross-pollination results in larger, earlier-maturing berries.

Solid block plantings of all standard varieties are yielding satisfactory crops, but it is doubtful that they are producing at their full potential. No data are available on the comparative production of solid block planting and mixed block arrangements. Solid block plantings will probably

Table XXII. Cross-Pollination of Coville Variety [a]

Pollen Used for Cross	% Blooms Set	% Ripe by July 6	Wt. of Berries (g)	Cup Count
Earliblue	51	57	2.4	57
Rancocas	83	35	2.4	57
Weymouth	86	47	2.8	49
Burlington	83	60	3.1	41
Concord	83	57	—	—
Jersey	100	59	2.4	57
June	79	57	2.6	53
3850-A	100	33	2.5	55
Cabot	87	43	2.4	57
Rubel	91	70	2.8	49
wild	84	62	2.7	51
average cross-pollination	84	53	2.6	53
self (Coville)	95	5	1.5	91
open pollination	46	6	1.8	76

Source: Filmer and Marucci, 1963.
[a] Crosses made May 8, 1962; counts made July 6, 1962.

Plant and Fruit Development

continue to be standard practice unless it can be demonstrated that cross-pollination offers overriding advantages.

Bees as Pollinating Agents. The fact that bees are required in blueberry pollination was discovered very early in the development of the cultivated blueberry industry (Coville, 1910; Beckwith, 1930).

Merrill (1936) considered honeybees to be incapable of pollinating highbush varieties with deep narrow blossoms such as Cabot and Pioneer and believed that blueberries in Michigan were pollinated largely by bumblebees, which have longer tongues. These wild bees are of inestimable value in promoting blueberry set. The high esteem growers have for these busy creatures is quite justified; they work at much lower temperatures and apparently with greater vigor than honeybees. Unfortunately, the populations of these insects fluctuate greatly from year to year, and there can be no assurance that the populations of wild bees will be sufficient to give adequate fruit set. There are no data to back this, but almost all growers of long experience in New Jersey feel that there has been a general decline in bumblebee population in most of the blueberry areas. Frequent forest fires, the destruction of nesting sites by intensive cultivation, and the constant elimination of wild areas, made easy by improved mechanical equipment, have undoubtedly reduced bumblebee populations. The increased use of potent new insecticides may also be a factor. However, it is interesting to note that observers in Maine were recording the inadequacy of bumblebee populations long before the advent of the modern organic insecticides (Chandler and Mason, 1935).

In 1963, Filmer and Marucci (1964) compared bumblebee and honeybee populations in nine representative New Jersey cultivated blueberry fields, two of which were thought to be inhabited by exceptionally large numbers of bumblebees. They found that the honeybees greatly outnumbered the bumblebees in every field. The ratios of domestic to wild bees varied from 240 to 1 to 4 to 1, and the average was 33 to 1.

The conservation of wild bees and protection of honeybees are of utmost importance to fruit growers. An effort should be made to preserve such wild bee nesting sites as fallen logs, ditch banks, and small wild areas close to blueberry fields. The use of insecticides during the open blossom period amounts to wanton destruction of beneficial wild creatures and should be avoided. After the pollinating period, when spraying or dusting is necessary the beekeeper should be given sufficient notice to remove his hives, and they should not be placed in fields early in the spring until at least five days after an application of insecticide.

Despite the relatively long period of pistil receptivity, the concentration of bees during the blossoming period is an important consideration. Research workers in highbush blueberries have given bee concentration scant attention, and until recently most growers believed that bumblebees were more efficient than honeybees and needed little help. However, high concentrations of bees have been shown to induce high yields by the highbush blueberry. A hive confined in a cage constructed over two bushes in a field resulted in extremely high production (Filmer and Marucci, 1963). Very large crops were obtained whether the bushes were of the same or different varieties and whether or not bouquets of blossoms from several varieties were placed in the cage. Caged Earliblue plants yielded 14 pints of berries, compared with only 2 pints from uncaged bushes in the same field. Caged Coville bushes outproduced comparable bushes in the open by 18 pints to 3. Plants of the Stanley, Weymouth, Jersey, and 1316-A varieties also bore abnormally large crops when isolated in cages with a hive of bees. These observations suggest the use of more honeybees, particularly in areas where wild bees are scarce or where production appears to be less than optimum. To date, there are no data available which would give a reliable basis for estimating the number of hives per acre necessary to achieve optimum pollination.

Recently, growers in New Jersey have been using higher concentrations of honeybees with encouraging results. In one area it has been observed that with one hive per acre, much more than previously used, a considerable reduction in the "June drop" of Weymouth resulted as well as an increased set of Coville. In areas where there are large adjoining blueberry fields the success of bee concentration in promoting better set depends on a cooperative community effort. Unless each grower uses an adequate number of hives, the bees will travel great distances to work on the more attractive varieties in other fields. With many bees about, the blossoms of the attractive varieties become pollinated and fall more rapidly, forcing the bees to work sooner on the less attractive varieties.

Weather Influences. The period of blossoming is one of the most critical periods of the year for the blueberry grower, since adverse weather at this time may greatly affect pollination. The effect of unfavorable weather on bee activity can cause a very substantial reduction of the blueberry crop. Honeybees are not very active below temperatures of 55 F. A long spell of cool weather could easily occur during the blossoming period in any of the large blueberry growing areas. Winds of more than 15 miles an hour, which are common in the spring, affect pollination by restricting the flight of bees. Rain also suppresses bee activity. In New Jersey, in early May of 1958, a constant drizzling rain for six consecutive days during the peak of bloom of early varieties so severely limited bee activity that production was one of the poorest ever recorded for a non-frost year. Rain may also ruin the pollen already on the stigma of blossoms (Kinsman, 1957). The full impact of bad weather can be lessened

Plant and Fruit Development

by placing hives in sheltered areas so that entrances have a southern exposure.

In some cases the weather may not seriously restrict bee activity and yet may be unfavorable for a good fruit set. This can happen when the average temperature is so cool that the germination of pollen or the movement of pollen tubes may be inhibited enough to prevent optimum fertilization (Kinsman, 1957).

Extremely hot or dry weather also affects the fruit set of blueberries. It is not unusual in most blueberry areas to have extremely high temperatures during blossoming of midseason and late varieties. During such spells, bees are apt to be considerably less active than normal. Dry weather may reduce the flow of nectar, which could reduce the rate of bee visits to blueberry blossoms. Severe hot and dry weather may also limit the period of receptivity of the pistil and may bring about a premature drop of the corolla.

Effect on Berry Size. Merrill (1936) found no relationship between fruit size and seed count; however, Darrow (1958) reported that small Stanley berries averaged only 2.4 seeds as compared to a 28.7 average for large ones of the same variety. Filmer and Marucci (1963) graphically presented data to show a direct relationship of berry size and seed count. Although data were given for only Coville and Weymouth, it was observed to hold true for other "weak" varieties such as Stanley, 1316-A, and Jersey, as well as for all other varieties. This may have practical implications to the blueberry grower. Since the blueberry pollen grain produces only one pollen tube which can result in only one seed, many pollen grains must be transferred to produce large berries. The use of higher concentrations of honeybees to place the greatest possible number of pollen grains on the stigma in the shortest possible time may have the effect of producing larger berries as well as more of them.

Berry Growth

Berry growth and rate of ripening is influenced by many environmental and nutritional factors, but on the average 50 to 60 days are needed from the time the blossom drops until the berry ripens (Bailey, 1947). The growth of the berry occurs in three separate stages (Hindle, Shutak, and Christopher, 1957; Young, 1952). The first stage is characterized by a period of rapid growth lasting about a month. During this period the corolla and the stamens abscise. The stigma turns brown and within a day or two the style also falls. The calyx remains permanently attached to the ovary and berry. During the second growth stage the berry makes very little growth. This period may last from five to 56 days. The final growth stage is again characterized by a period of rapid growth continuing until maturity, lasting 16 to 26 days. During this period the calyx turns purplish and the green berry takes on a translucent appearance. The berry then turns to a light purple and finally to a dark purple in most species. The increase in size during this stage is greater than in any other period of berry maturation. Representative growth curves are shown in Fig. 55.

Pruning

Before pruning blueberries, it is essential that the fruiting habit be thoroughly understood. The main objective is to balance fruiting and the production of new fruit wood for the following year. Excessive pruning will increase the size of berries but will decrease total production. Very light pruning, or none, will result in more but smaller berries and a reduction of the new vegetative growth essential for future fruit production (Doehlert, 1941; Brightwell and Johnston, 1944).

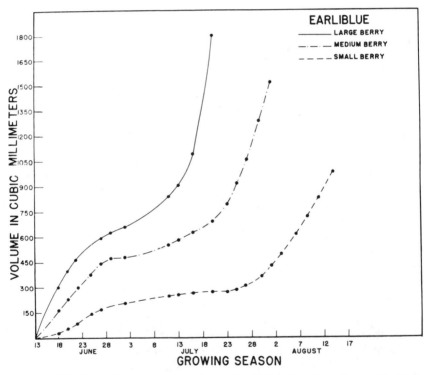

Figure 55. Growth curves for three Earliblue berries that ultimately attained a different size. Note that the large berry developed in considerably less time than the small berry. (From Shutak, Hindle, and Christopher, 1957.)

Time of Pruning. Highbush blueberries may be pruned at any time while they are dormant, but it is best to prune during the early spring just before buds begin to swell. There are two reasons for this delay in pruning: it decreases the danger of winter injury at cut surfaces and provides an opportunity to remove winter-injured parts of the plant. Pruning after growth starts is undesirable because many of the developing buds may be injured or rubbed off in the process. Late pruning may actually cause a delay in bud opening.

If considerable acreage is to be pruned and labor is short, the pruning period may have to be extended. Starting earlier than recommended in the spring is preferred to fall or early winter pruning.

Pruning Tools. For small plantations, the only pruning tools needed are loppers and pruning shears. It may be advisable to try out different sizes of loppers until one finds the size best suited. Power pruning equipment may be prudent for large plantations. Pneumatic pruners, which are really power loppers, come with blades of different sizes and handles of different lengths. Again, an individual must decide for himself what size and weight will suit him best. Power tools speed the pruning process, but one should be an experienced pruner before using this equipment or overpruning may result.

Pruning Cuts. It is generally accepted that for most plants, pruning cuts should be close and smooth. In pruning highbush blueberries this is less important, since most of the main canes do not remain a part of the bush for more than four to six years. The home grower may be justified in spending additional time in pruning to improve the appearance of the bush, but this is not profitable for a commercial grower.

In removing large canes it is best to cut them close to the ground, although some growers prefer to leave a short (2-inch to 3-inch) stub. New shoots usually develop from these stubs; however, new shoots developing from the roots are frequently stronger and more vigorous.

With heavy-bearing varieties, tipping back is frequently practiced. Tipping refers to clipping off the tips of some of the vigorous bearing terminals, thus reducing the number of berries per shoot. Most commercial growers ignore this practice. Some, however, believe that the number of fruit buds per shoot should be reduced to prevent overbearing. Tipping will do this without resorting to a heavy pruning. Tipping should be delayed as long as practical to allow winter injury to show but should be done before the buds swell in the spring. If done later than this, an undesirable checking of leaf growth and a slow, weak development of berries will result.

Soft autumn growth originating from the base should be removed entirely. This type of growth often winter-kills and is especially susceptible to fungus diseases. It is important to recognize wood of this type so that it is not saved in the belief that it is as valuable as strong wood. Soft wood is noticeably more limber, does not have the stiff, springy quality of well hardened wood, and is usually irregularly flat-sided instead of round and well-filled.

Twiggy or bushy growth occurs in a cluster toward the end of canes. This growth is weak and is not very productive. Removing this growth with a single cut saves time and shortens the cane. This procedure will involve an occasional loss of strong shoots, but if sufficiently strong fruiting twigs occur lower down on the cane this is not important. The habit of cutting out large clusters of small twigs with a single cut saves much time and labor in pruning blueberries.

Other small twigs can be removed more quickly by rubbing off than by clipping. Ordinary leather-palmed canvas gloves greatly facilitate this procedure. A skilled pruner uses the shears with one hand and rubs off small twigs with the other. A combination of the two procedures can cut the pruning time in half.

General Pruning Procedure. In pruning a full-grown bush a regular pruning procedure should be followed. Although it is impossible to give specific instructions applicable to every situation, there are some general guides that fit most of the normal situations.

1) Remove all diseased or injured canes.
2) Remove or cut back some of the oldest and least vigorous canes.
3) Remove low branches and soft wood.
4) Thin out and tip. This includes removing bushy or twiggy growth, rubbing off smaller shoots and tipping fruiting wood if needed.

If time is limited it is much better to give some pruning to all of the bushes than to spend all available time on thoroughly pruning half of the plants. An inexperienced commercial grower may benefit by seeking advice and help before attempting any large-scale pruning operation.

Pruning Young Bushes. It is recommended that two-year-old bushes be used to establish a new plantation. In exceptional cases large, rooted cuttings may be substituted. After planting and before spring growth starts, all the tips bearing flower buds should be removed. The removal of weak, twiggy growth is recommended so that a few strong shoots will be produced during the first summer instead of a great number of small, weaker laterals.

After one growing season in the field many of the bushes will be large enough to bear a small crop. On these it is only necessary to remove weak, thin wood. This reduces the number of fruiting tips to a favorable balance between fruit and foliage. On the smaller plants the weak, twiggy growth and all fruit buds should again be removed.

After two seasons in the field, all plants should be large enough to produce a small crop of 1 to 2 pints. Again, the pruning procedure is to remove all nonvigorous growth and concentrate potential crop on a small number of thick, fruiting shoots. By limiting the crop to only the strong shoots, continued rapid enlargement of the bush is assured. Allowing the bushes to produce a heavy crop at this time tends to dwarf the plants and delay the time of full production.

The same pruning procedure is followed for the next two or three seasons, depending on the vigor of the plant. After this period the bush is considered mature and should start producing a full crop.

Pruning Mature Bushes. In pruning mature bushes, the suggested general pruning procedures should be followed. After removing all defective wood, large canes are thinned out. The oldest and the least vigorous canes are cut out first. They may be cut to the ground or to a strong side shoot near the ground. Then additional large canes are removed until one in every five or six main canes is eliminated. In this manner a bush will never have any canes more than five or six years old. This is very desirable, since even on a well-managed, productive bush the profitable life of most canes is no more than six years. Older canes seldom produce vigorous shoots, and thus berry size becomes smaller each year. Cutting out large canes stimulates sprouting of vigorous shoots from the base and keeps the bearing wood young, vigorous, and productive. Lack of new shoots frequently indicates that the bush is too crowded with old canes.

The bush is then thinned by removing clusters of twiggy growth and rubbing off small, weak laterals. The erect-growing varieties such as Earliblue, Pemberton, and Collins definitely should be thinned out in the center. More spreading varieties require more pruning of the lower drooping branches. A properly pruned bush should admit sufficient sunlight to all parts of the plant to encourage new growth and flower-bud formation.

If there are too many buds on a bush, blossom buds will open first and the leaf buds will open late, resulting in insufficient early foliage. Such bushes are unable to support both the full development of the current fruit crop and the growth of enough vigorous shoots and fruit buds for next year's crop. Tipping may be done to bring about a better balance. Only the conspicuously heavy-budded shoots need tipping.

In commercial plantations pruning may be reduced to comparatively few cuts. This operation may actually consist of just two steps. First, some of the large canes are removed, and second, a number of smaller cuts are made, removing clusters of small twigs. In such a simplified procedure some of the long, well-budded twigs are removed and a number of small, poorly producing twigs are left; however, this produces a satisfactory commercial job and greatly reduces pruning time.

A typical example of blueberry pruning is shown in Fig. 56. Note that both loppers and shears were used. Commercially, it may be advisable to

Figure 56. Pruning recommendations. a. A 15-year-old June bush before pruning. Right top section of the bush shows poor current wood growth, resulting from insufficient pruning and heavy crop production the previous year. Center part of the bush shows new vigorous growth caused by severe pruning a year ago.

b. The same bush after the bulk of poor wood has been removed by 5 lopper cuts and 14 cuts with the small shears. Scattering of small fruiting twigs is not sufficient to require further work. In fact these smaller twigs will increase the total crop if they are not too numerous to compete for soil moisture during hot, dry weather.

c. The same bush after a detailed finishing job was done by removing weak twigs and making a few additional small cuts. (Norman F. Childers, Rutgers University, New Brunswick, N.J.)

have one man make all large cuts with loppers, followed by a second man making smaller cuts with shears. It is also recommended that all bushes of one variety be pruned before moving to a different type of bush. A pruner becomes accustomed to a variety's growth characteristics and can do a better and faster job this way.

LOWBUSH BLUEBERRY

Fruiting Habit

The fruiting habit of the lowbush blueberry is very similar to that of the highbush. Flower buds are initiated in the summer on the current year's growth, and fruit is produced the following year (Bell, 1950; Eggert, 1957; Trevett, 1955). Some progress in bud development can be detected throughout the winter, the so-called "dormant season" (Bell and Burchill, 1955). Vegetative growth originating from buds located on the underground stems (rhizomes) is more vigorous and productive than growth developing from buds on existing shoots above ground.

Pollination

The lowbush blueberry does not have the capability of setting as high a percentage of its blossoms as the highbush blueberry. Lowbush fields contain large numbers of clones of various inherent degrees of self-sterility. Aalders and Hall (1961) examined 21 clones of *V. angustifolium* in Nova Scotia and found that 12 were completely self-sterile while nine were partially self-fertile; the self-fertile clones set from zero to 52 per cent of the blossoms when they were selfed, while the range was from 81 to 90 per cent when they were cross-pollinated. Lee (1958), working in New Hampshire, caged three pairs of clones with honeybees and under these conditions of high bee densities and cross-pollination the fruit set ranged from 36 to 98 per cent, with an average of 68 per cent. Wood (1961) states that in New Brunswick lowbush blueberry set is rarely over 50 per cent and is generally considerably lower.

Pollination of the lowbush blueberry is complicated by the fact that the stand often shows a high percentage of the diploid species *V. myrtilloides*, which has pollen incompatible with the tetraploid species *V. angustifolium* (Aalders and Hall, 1961). These investigators found that cross-pollination of these two species resulted in reduced set of the important lowbush species *V. angustifolium*. Neither honeybees nor wild pollinators discriminated between species in their foraging habits. This resulted in dilution of the pollen of *V. angustifolium*, which adversely affected fruit set. Following pollination with *V. myrtilloides* pollen the *V. angustifolium* ovules aborted. When only *V. myrtilloides* pollen was involved, no berries were formed on the *V. angustifolium* plants, but when there was a mixture of pollen of both species, berries were produced that were reduced in size, had lower seed counts, and were delayed in ripening. Wood (1965) found that this interspecific pollen incompatibility was also a factor militating against fruit set in New Brunswick fields. He presented data to show that both species were common in fields, that their periods of blossoming overlapped, and that pollination by bees carrying mixed pollen loads was a common occurrence.

Considerable emphasis has been placed on the importance of bee concentration in lowbush blueberry fields in Maine and Nova Scotia. As early as 1930 Phipps used cages to demonstrate that bees were necessary to produce commercial crops from lowbush fields. Later, Phipps, Chandler, and Mason (1932) found that where several clones were represented in a single cage enclosed with bees, the percentage of fruit set was more than double that recorded in similar cages containing bees and plants of only one clone. In 1935 Chandler and Mason demonstrated that a single hive of bees promoted a better set of berries for a radius of only about 150 yards from the hive and that bees must visit a lowbush blueberry blossom within three days after it opens if a fruit is to be produced.

Kinsman (1957) considered inadequate pollination as probably the most common cause of poor yields of lowbush blueberry in Nova Soctia. He pointed out that the wild bee population fluctuates from year to year and that in order to insure good production of lowbush blueberries in Nova Scotia honeybees should be used to supplement the wild bee. As a rule of thumb, he suggested a minimum bee population of one per square yard under sunny, calm conditions with the temperature over 60 F. He suggested the use of at least one strong colony of bees (30,000 workers) for each two acres in Nova Scotia. To circumvent the diversion of honey bees from blueberries to other blossoms for pollen, he advised the rotation of hives and making pollen or pollen substitutes available to the bees. The rotation of hives entails moving colonies at least two miles to new sites, and allowing the bees to remain at one site as long as they work the desired area. The success of this procedure is based on the fact that the field bees of a newly moved hive will work only a short distance from the hive during the first few days.

Recently, Boulanger (1965) found that although 41 species of wild bees are associated with lowbush pollination in Maine, their total numbers are insufficient to give adequate pollination. Very good increases in fruit set were achieved by increasing the number of hives from the average of one or two to nine hives per acre. Rotating hives has proven to be an efficient management practice. One to two hives per acre are set out at 10 to 20 per cent bloom, followed by the addition of three to four colonies at peak bloom. The dilatory colonies are permitted work for two or three days, after which they are relocated. This can be done at least twice during a seven- to eight-day period and thus affords more effective use of available hives.

In New Brunswick the influence of honey bees on pollination has not been as marked as in the other lowbush areas. In a three-year study, from 1957 to 1959, honeybees induced a significant increase in set in only one of the years (Wood, 1961). The relatively high percentage of the diploid *V. myrtilloides* and the inherent fertility of the stand of *V. angustifolium* in this region imposes an upper limit on fruit set. This apparently can be accomplished by the work of wild bees in years of long periods of bloom. The service of honeybees is of more pronounced advantage in years when the blossoms stay open for a relatively short time.

Lowbush blueberry blossoms generally remain in good condition for seven to ten days and data obtained by Wood (1962a) make it reasonable to assume that pollination could continue until petal fall. The pistil receptivity declined with the age of the blossom but at the age of seven days from 15 to 58 per cent of the blossoms on the five clones observed set berries.

Pruning

In order to obtain a good crop, a large number of strong new shoots are needed. Severe pruning is necessary to produce them (Chandler and Mason, 1939, 1943).

A number of different pruning methods have been tried, but burning appears to be the best one. In Maine, burning is the only commercial method of pruning lowbush blueberries presently in use (Abdalla, 1964a). Burning not only prunes the plants by removing all the tops but also removes most of the competitive growth of weeds and small brush.

Experimental data of Kender, Eggert, and Whitton (1964) showed that it is important to remove all top growth, thus forcing new shoots to come from rhizomes. In other pruning methods, such as the use of herbicidal oil or mowing, parts of the old stems remain and new growth originates from buds on these stubs. This growth is much weaker and tends to branch. The yields are lower and harvesting (raking) is more difficult.

Blueberry fields should be burned every two or three years. If a two-year burn cycle is selected the field will produce only one crop between burnings. Plants grow their fruiting wood the first growing season after burning and the first crop is produced the following year.

In a three-year burn cycle, plants will produce two crops before the field is burned again. The first crop is always the best but the second may be profitable. If the field is left unpruned for a longer period, weed growth results in shading and competition that reduces new vegetative growth and flower-bud formation to the point where it is uneconomical to harvest the crop. The decision on which pruning cycle to use depends on the condition of the field, the vigor of the plants, and, of course, the owner, who must decide which is more profitable. Burning is expensive, and all factors should be considered. After a grower decides on the burning cycle he usually divides his field accordingly into two or three sections. By burning over one section a year he avoids fruitless years for the whole plantation.

The time of burning is quite important. Plant growth starts early in the spring, and burning must be done before any growth takes place. Either the late fall or the early spring is satisfactory, but fall burning is recommended unless the field is on a steep slope and removing vegetation by burning might result in more soil erosion during the winter.

Eaton and White (1960) conducted an experiment to study the effect of the time of burning on subsequent growth and production. They concluded that the number and length of sprouts and the total number of flower buds were greater when burning was done in the early spring or the late fall. Late spring burning decreased the flower bud formation.

For best burning results, the soil should be thoroughly wet or frozen. Either condition protects the topsoil, humus, and blueberry. There is always danger of fire spreading into adjacent fields or woods, and all necessary precautions should be taken.

To obtain a good burn, some kind of inflammable material is usually spread over the field. Straw is most frequently used, but hay or shavings are also satisfactory. One ton of hay per acre is considered adequate. The material is spread evenly just before burning. The objective is to produce a hot, fast-moving fire that destroys all above-ground vegetation without affecting organic matter in the soil or injuring blueberry rhizomes. Some growers mow the field before the straw is spread. This mowed material when dry will help produce a better burn.

Recently, because of the rising cost of hay and labor, most of the growers are changing from the hay-burning method to a large tractor-drawn oil burner. It is faster, safer, and generally quite satisfactory on large fields that are relatively smooth and not too stony. On rough fields burners may produce an uneven burn. At present most growers use burners similar to one shown in Fig. 57.

The most recent introduction is a gas burner. A representative type is shown in Fig. 57. These burners are constantly being redesigned and altered to suit specific field conditions. It is believed that in the near future there will be better burners and that nearly all of the fields will be pruned with this equipment.

Figure 57. Above. Burning equipment for lowbush blueberry fields. Woolery oil burner, presently the most used piece of burning equipment. *Below.* LP-gas burner. Nozzles are raised for a hotter flame and swing on individual joints for easy bypass of rocks, stumps, and other obstacles. (Dennis A. Abdalla, University of Maine, Orono.)

RABBITEYE BLUEBERRY

The rabbiteye blueberry has a fruiting habit similar to that of the highbush blueberry (Darrow, 1962). Because the bushes are sufficiently vigorous to produce a heavy crop and still produce enough new fruiting shoots at the same time, very little pruning is required (Brightwell, 1962). Many bushes are never pruned, but it is desirable to do some pruning every year to keep the bushes from becoming too dense and tall.

On young bushes, pruning consists mainly of removing low branches to prevent fruit from coming in contact with the soil and thinning the center of the bush.

Mature bushes require only light thinning to prevent crowding and increase light penetration and the occasional removal or heading back of older canes. Excessive pruning will produce undesirable and excessive water-sprout growth.

CHAPTER 9

Insects and Their Control

by **Philip E. Marucci,** *Cranberry and Blueberry Research Laboratory, Rutgers University, Pemberton, New Jersey*

HISTORY OF INSECT CONTROL

The control of blueberry insects is one of the most important phases of blueberry culture. Without attention to the insects that attack the roots, stems, leaves, and fruit of the plant, blueberry growing would soon become unprofitable. Some of these insects seriously reduce the productivity of the bush while others impair the quality of the berries, lowering their value or making them entirely unmarketable.

The evolution of the insect problem in the commercial production of the blueberry was quite rapid. The industry developed in regions where wild blueberries were indigenous. Thus, a simple adaptation from the wild species to the cultivated varieties of the same species was all that was necessary for a native insect to become an economic pest. C. R. Phipps (1930) catalogued 292 species of insects, representing 9 orders and 57 families, which were found in wild lowbush blueberry fields in Maine. Most of these are present throughout the blueberry areas of the country, and a few have become important pests requiring control measures (Chandler, Mason, and Phipps, 1932). Many of the insects that attack the lowbush blueberry are also important pests in highbush plantings.

The first reference to the importance of blueberry insects to the highbush blueberry was made by C. S. Beckwith in 1932, when he noted that four insects—cranberry rootworm, blueberry stem borer, blueberry stem gall, and blossom weevil—required attention in New Jersey's blueberry fields. At that time the only recommendation for checking these insects was to keep the wild bushes destroyed and the ground cleared around the field. This remained the only insect control practiced until the seizure and condemnation of several shipments of New Jersey cultivated blueberries infested with blueberry maggot in July 1935 (Beckwith, 1935).

Table XXIII Blueberry Pest Control

Jointly Prepared by Research and Extension Personnel, College of Agriculture, Rutgers – the State University, New Brunswick, New Jersey, and U. S. Department of Agriculture Cooperating.

This chart has been prepared for pest control **UNDER AVERAGE CONDITIONS**. Problems are markedly influenced by seasonal and local conditions. Exact timing of each treatment should be based on close observations by the grower, aided by information in Extension notices.

PEST	WHEN TO TREAT	WHAT TO DO*	REMARKS
SCALE INSECTS	March 1 to first bloom	**Spray:** 3 gal. of superior-type oil.** (300-400 gal. of spray per acre are necessary for control.) Pruning out old canes will prevent heavy scale infestations.	Best results are obtained with hand boom, using 300-400 lb. pressure. Dieldrin may be combined with oil for blossom weevil control if applied just prior to bloom.
MUMMY BERRY	When mummy cups first appear, about April 1	**Spread:** granular cyanamid at 200 lb. per acre along the rows; or destroy cups by cultivation. Cultivation helps control blossom weevil, cranberry fruitworms, and plum curculio. Disk between rows and rake, sweep, and hoe under plants.	Use cyanamid before leaves show or blossoms open for best disease control and to avoid plant injury. Do not wait until cups are fully formed to make application.
BLOSSOM WEEVIL	When LEAF BUDS SHOW GREEN and BEFORE BLOSSOMS OPEN	**Spray:** parathion, 1.5 lb. of 15% wettable powder; or dieldrin, 2 lb. of 25% wettable powder; or **Dust:** 1% parathion, 25-35 lb. per acre; or 2% dieldrin. Do not use dieldrin after fruit forms.	Treat edges of fields and adjacent brush carefully since weevils are most abundant there. Protect pollinating insects by not treating plants during bloom stage. Dieldrin is preferred where curculio is a problem.
PLUM CURCULIO	When egg punctures appear on Cabot or Weymouth berries that are about ¼ inch in diameter; this is when 75% of petals have fallen on Rubel.	**Dust:** 40 lb. of 5% DDT and 1% parathion per acre. Do not use DDT within 21 days of harvest.	Treatments at this time will give some control of fruitworms and leafhoppers. Failures in curculio and fruitworm control result when less than 40 pounds per acre are used. REMOVE BEES BEFORE TREATMENT.
CHERRY FRUITWORM SHARP-NOSED LEAFHOPPER	First application May 25 to June 1.	**Spray:** parathion, 2 lb. of 15% wettable powder; or malathion, 4 lb. of 25% wettable powder; or **Dust:** 1% parathion; or 4% malathion, 40 lb. per acre.	Fruitworms are active for about 5 weeks, and they cannot be controlled with only one application. These applications control sharp-nosed leafhopper, the only known carrier of stunt disease.

		harvest. Sevin may be used the same day of harvest and malathion 1 day before.
	Dust: 5% Sevin or 4% malathion at 40 lb. per acre.	
SHARP-NOSED LEAFHOPPER		
STUNT AND OTHER VIRUS DISEASES IN NONBEARING FIELDS	Mid-June and late September to mid-October	**Dust:** 40 lb. of 1% parathion or 4% malathion per acre. In late season, include 5% DDT in either dust. Remove all plants identified as having a virus disease. Symptoms of stunt are most apparent at these two periods. These applications are for non-bearing fields and are used only to control sharp-nosed leafhoppers and other leaf-feeding insects.
MAGGOT	Begin about June 28. Make applications at 10-day intervals as long as harvest continues.	**Dust:** 4% malathion at 25 lb. per acre. **Spray:** airplane or ground application of 10 oz. actual malathion and one of the following — 2 qt. Staley bait No. 2-FB, or 1 lb. of Hy-Case Powder, or 1 lb. yeast hydrolysate per acre. For aircraft sprays use at least 5 gallons of spray mix per acre. Airplane bait sprays have proven effective and can be used when conditions are unfavorable for dusting. *Malathion in a bait spray may not be used within 8 hours of harvest. Malathion dusts or sprays may not be used within 24 hours of harvest.*
CROWN GIRDLER	September 1	FOR CONTROL: Cut plants to ground level. Apply ½ oz. of PDB in trench 1½ inches deep and 1½ inches from base of plant. Mound and compact soil. FOR PREVENTION: Dust 1 oz. of 5% DDT in a band 1 foot wide about the base of plant and cultivate into soil. Control cranberry rootworms and other grubs by disking 25 lb. of actual DDT per acre into soil before planting. Use preventive treatment only about plants adjacent to the injured plants.
BLUEBERRY BUD MITE	September 15-30	**Spray:** 2 gal. superior-type oil** Add 1 pint of malathion emulsifiable concentrate. Usually this mite does not cause much trouble in New Jersey, but it increases in numbers after mild winters.

*Amounts of spray material are given for 100 gallons of spray mix. Apply sprays to give good coverage without excessive runoff. Good coverage may require up to *400 gallons per acre* in the scale spray and up to 200 gallons per acre in other sprays.

**The minimum specifications of a 100-second spray oil include: viscosity — 90 to 120 seconds; gravity — (A.P.I. degrees) 32 minimum; unsulfonatable residue — 92 percent.

The minimum specifications of a 70-second spray oil include: viscosity — 66-74 seconds; gravity — (A.P.I. degrees) 33-34; unsulfonatable residue — 92 percent or higher.

!!! WARNING !!!

BEFORE USING INSECTICIDES, STUDY THE PRECAUTIONS ON THE LABEL CAREFULLY.

PARATHION IS EXTREMELY DANGEROUS IF INHALED OR SWALLOWED OR ABSORBED THROUGH THE SKIN. REPEATED EXPOSURE MAY, WITHOUT SYMPTOMS, INCREASE SUSCEPTIBILITY TO PARATHION POISONING.

Source: Swift and Davis, 1965.

This experience pointed to the need for dusting and spraying blueberries, and since 1936 insecticides have been obligatory in highbush blueberry culture.

In North Carolina various fruitworms became troublesome soon after the first plantings. The problem had become so serious by 1939 that it elicited the following statement from the experienced blueberry grower H. G. Huntington (1939): "The conclusion of our experience on pests and diseases in North Carolina is that after a field has been out four years it will be necessary, among other things, to conduct a careful and expensive spray schedule of from 6 to 8 sprays per year."

For several years the blueberry fruitfly was the only insect requiring the use of insecticides in most New Jersey fields. In 1946 the first Blueberry Insect and Disease Control Chart was published by the New Jersey Agricultural Experiment Station. This chart included six insect pests, three of which were controlled by cultural methods. By 1965 the chart (Table XXIII) included nine important insects and a potential of eleven insecticide applications to control them if all are present in pest proportions (Swift and Davis, 1965). Various other experiment stations in states where blueberries are grown issue bulletins through their extension services advising growers how to recognize and control insect pests. The present status of the blueberry insects in the various growing areas, as rated by entomologists in the respective geographical regions, is shown in Table XXIV.

It is interesting to note that the cultivated blueberry is still relatively free of insect pests in the state of Washington, where only the black vine weevil has been a serious problem. The number of insect species attacking the rabbiteye blueberry is relatively small (Brightwell, 1962). Insects that are of economic importance in rabbiteye production include the fruitworms, stem borers, leaf rollers and mites. These insects will be described under highbush insects.

Growers must realize that the importance of the various blueberry insects will vary from year to year according to weather conditions, parasite and predator relationships, and other constantly varying environmental factors. Control measures discussed in this chapter will undoubtedly be modified as insects become resistant and better methods are discovered. Therefore, in order to reduce efficiently the losses from the depredations of insects, the grower must be in constant contact with the extension services of his experiment station.

Table XXIV. Importance of Blueberry Insects in the Various Growing Areas [a]

Insect	N.J.	Mich.	Me.	N.C.	Mass.	Canada	Wash.
Blueberry Maggot	1	1	1	3	2	1	
Cranberry Fruitworm	2	1	5	1	2	4	
Cherry Fruitworm	1	1	4	1	2	4	
Plum Curculio	1	2	4	1	4	4	
Putnam Scale	2	4	4		3	4	
Other Scales	3	3	4	3	3	4	
Blossom Weevil	2	4	4	4	2	4	
Bud Mite	3	4	4	2	3	4	
Sharp-Nosed Leafhopper	2	3	5		2	4	
Cranberry Rootworm	3	4	4	3	3	4	
Scarabeid Root Grubs	3	4	4		3	4	
Crown Girdler	3	4	4		4	4	
Black Army Cutworm	4	4	3		4	2	
Span Worms	4	4	3		4	2	
Blueberry Thrips	4	4	2		4	2	
Blueberry Flea Beetle	4	4	3	4	4	2	
Termites	4			3			
Prionus Larvae				3			
Gypsy Moth	4	4	3				
Leafroller	4	4	3				
Stem Borer	4	4		2	3		
Black Vine Borer							"serious"
Orange Tortrix							"harmful"
Tent Caterpillars							"harmful"

Source: Ratings made for Michigan, Maine, North Carolina, Massachusetts, and Canada through correspondence with the following: Ray Hutson, Michigan; F. H. Lathrop, Maine; B. B. Fulton, North Carolina; W. E. Tomlinson, Jr., Massachusetts; George W. Wood, Canada. Rating for Washington taken from Schwartze and Myhre, 1954b.

[a] Code for all areas except Maine:
    1—needs treatment virtually every year.
    2—almost always present but needs treatment only occasionally.
    3—rarely present in pest proportions.
    4—never present in pest proportions.

Code for Maine:
    1—present in most areas and needs treatment virtually every year.
    2—present in most areas and needs treatment in some areas virtually every year.
    3—often present in small numbers; needs treatment when outbreaks occur.
    4—has not been observed in pest proportions.
    5—has been observed in the state, but pest status uncertain.

HIGHBUSH BLUEBERRY

Factors Affecting Insect Infestations

Factors that have an important bearing on the abundance of insects in blueberry fields are weather, the environment surrounding the field, natural enemies, cultivation, pruning, resistant or susceptible varieties, and solid-block planting.

Weather. The most important factor affecting blueberry insects is weather. In any year the severity of the insect attack is directly dependent on favorable weather conditions. Since all of the important cultivated highbush blueberry insects, except the sharp-nosed leafhopper and blueberry bud mite, have only one generation a year, adverse weather conditions over a relatively short period can greatly reduce the insect attack. Cool, wet weather reduces the egg laying and increases the mortality of small larvae of plum curculio, cherry fruitworm, and cranberry fruitworm. Hot, dry summers shorten the life of the blueberry fruitfly and decrease its oviposition. Mild winters in New Jersey have tended to increase populations of plum curculio and cherry fruitworm while decreasing the emergence from the puparia of the blueberry fruitfly. Bud mite in North Carolina is especially severe after warm winters and may be troublesome in New Jersey under these conditions. Severe winters are injurious to Putnam scale.

In general, it may be stated that extremes of weather have deleterious effects on the blueberry insects. If these extremes occur during periods of oviposition and egg hatching the deterrent effect is most pronounced. Windy, wet weather just after or during spraying or dusting indirectly benefits the insect by reducing the efficiency of the application and diminishing the toxic residue.

Environment Surrounding the Field. Since almost all of the insect enemies of the blueberry originally lived on wild blueberries and closely related plants, the concentration of these plants around a field will directly influence the degree of infestation in the planting. Huckleberries (*Gaylussacia* spp.) as well as the true blueberries (*Vaccinium* spp.) harbor most of the insect pests. Abandoned cranberry bogs are important reservoirs of cranberry fruitworm, blossom weevil, and cranberry rootworm. Suppression of these wild plants around a field is quite helpful in reducing potential infestations in the cultivated fields.

Keeping the surroundings of the field clean of wild plants by regular burning has been found to be helpful in the control of blossom weevil, cranberry fruitworm, and plum curculio by destroying the hibernating cocoons and adults and by eliminating the host plants upon which they develop. The hibernating eggs of the sharp-nosed leafhopper in the fallen blueberry leaves tend to accumulate in windfalls along the edges of fields; therefore burning this area also should tend to reduce the leafhopper populations.

Natural Enemies. Predators and parasites of the insect pests of blueberries play an important role in keeping their populations within bounds. All of the blueberry pests have a series of natural enemies. Ants are probably the most important of the predators; their attacks upon the larvae and pupae of blossom weevil, plum curculio, cranberry fruitworms, and blueberry maggot undoubtedly have a considerable effect on their populations. Ladybird beetles are active predators of the scale insects in blueberry fields. This is shown by the fact that when excessive DDT is used, which decimates ladybirds, the scale populations build up very rapidly.

The very minute wasp parasite, *Trichogramma minutum* Riley, which spends its egg, larval and pupal stages within the eggs of the cherry fruitworm, the cranberry fruitworm, and the blueberry leafminer, is of inestimable value. As many as 50 per cent of the eggs of both the cherry fruitworm and the cranberry fruitworm may be destroyed by this parasitic wasp during some years in New Jersey. Unfortunately, in most years these natural enemies only reduce the insects to a point known as the natural balance. This population level is more than the grower can tolerate economically, so he must resort to other methods of control.

A parasitic fungus, *Beauvaria bassiana*, attacks the hibernating larvae of cherry fruitworm and cranberry fruitworm. During wet autumns and winters the destruction of these larvae by this fungus may be considerable. In New Jersey in the winter of 1951–52, approximately 48 per cent of the hibernating larvae of cherry fruitworm were attacked by this fungus.

Cultivation. Clean cultivation of blueberry fields is helpful in combating the crop's insect enemies. Cutworms cause damage only in very weedy or mulched fields. Blossom weevil, plum curculio, and cranberry fruitworm are nearly always serious problems in poorly cultivated fields. Their hibernating quarters are destroyed, and they are exposed to predatory birds and ants by thorough and repeated cultivation. They remain problems because of the difficulty of cultivating between the bushes and of cleaning out leaves and trash in the crowns of the plants.

In New Jersey the hibernating puparia of the blueberry maggot do not survive repeated discings between the rows to any great extent, but between the bushes, where the soil is only rarely hoed, the insects are able to overwinter successfully. The recent introduction of the automatic rotary hoe should militate strongly against all of these insects that spend a part of their life cycle on or in the soil.

Pruning. Pruning is an important aid in blueberry insect control. The blueberry stem gall is completely controlled by pruning. Putnam scales and other scales that attack the blueberry are principally stem feeders and do not thrive on strong, vigorous wood. Pruning out old, weak canes usually prevents their increasing to the point where spraying is necessary. Blueberry stem borer is also somewhat vulnerable to good pruning, as is blueberry bud mite. Cutting out old, dead pruning stubs at the crown and on the bush helps in eliminating the hibernating quarters of cherry fruitworm. Pruning also helps indirectly in controlling insects, since well-pruned bushes are easier to spray and dust thoroughly.

Resistant or Immune Varieties. The growing of resistant or immune varieties would be an ideal method of minimizing insect damage and eliminating costly and bothersome spraying and dusting. None of the cultivated blueberry varieties in use today exhibits complete immunity to any of the blueberry insects. However, some varieties show a few noteworthy differences in the degree of susceptibility to some insects. The cranberry fruitworm shows such an apparent preference for the Cabot that it was at one time called the Cabot worm. It also has a predilection for the Rancocas and Bluecrop varieties. All of the early-ripening varieties are more subject to attack by plum curculio and cranberry fruitworm than Jersey and later varieties. Conversely, the early varieties are less susceptible to blueberry maggot than the later ones. But this does not seem to be entirely a matter of the fruit's being available for attack at a certain period of development of the insect. Bud mite often seriously affects Weymouth, Cabot, Rancocas, Pioneer, June, Scammell, and Harding, while causing considerably less damage to other varieties and hardly any on Burlington. The Concord and Rancocas varieties seem especially susceptible to attack by various scales.

Solid Block Planting. The expense of spraying and dusting could possibly be reduced by planting blueberries in solid blocks of one variety. Under the system in use in some areas in which different varieties are planted in every other pair of rows to insure cross-pollination, it is often necessary to treat both early and late varieties to control an insect which may be a problem on only one of the two varieties. The efficiency of insecticide application could be enhanced in solid block plantings since the sprays or dusts could be timed better with respect to the variety's vulnerability to the insect. In solid block plantings consideration should still be given, however, to having an occasional row of a good pollinating variety to help achieve cross-pollination.

Scale Insects

Nomenclature and Distribution. The first insects to which the blueberry grower must give attention in the spring are the scales. The most

Insects and Their Control 207

important species that attack blueberries are the Putnam scale, *Aspidiotus ancylus* Putnam, terrapin scale, *Lecanium nigrofasciatum* Pergande, oyster shell scale, *Lepidosaphes ulmi* L., cottony maple scale, *Pulvinaria innumerabilis* Rathv., and the European lecanium, *Lecanium corni* Bouché. Of these only the Putnam and terrapin scales have occasioned appreciable losses to the grower. Oyster shell, cottony maple, and European fruit scales rarely become abundant in blueberry fields in the eastern United States on any variety other than Concord. They are potentially serious pests and may cause damage whenever their parasites and predators are unduly destroyed by the misuse of DDT, or when pruning has been neglected.

Life Cycle. The Putnam scale occurs in many fields that have bushes more than six years old. Bushes that are not pruned regularly to remove old wood are infested most heavily (Weiss and Beckwith, 1945). The scales are found most often and in highest concentration under loose bark and old canes, but they also migrate to leaves and berries. On the fruit, these scales appear as grey waxy dots about $1/16$ inch in diameter surrounded by small, circular, red discolorations, and thus harm the appearance of the product on the market. On the stems and canes they blend so well with the bark that they are often unnoticed until they form encrustations. Even a few scales will cause pitting of the wood, and a single one on a berry causes a slight dimple.

The scales pass the winter in New Jersey as fully developed adults. Crawlers begin to issue from beneath the adult scales after the middle of May in that latitude. Most of these immature scales settle on stems and old canes, but they also reach leaves and berries. They excrete a sticky fluid (honeydew) on the fruit and leaves, which later turns sooty black as the result of a mold that grows on it. Leaves so covered do not function as well as normal ones and the sooty, sticky berries are often undersized and unmarketable.

The terrapin scale is a dark brown, hemispherical-shaped scale, somewhat resembling a tiny terrapin, about $1/7$ inch in diameter. These scales feed almost entirely on twigs and are very prolific, one adult often producing more than 300 nymphs (Simanton, 1916). The feeding of these scales greatly reduces the twigs' vigor. Large excretions of honeydew cause blackening of leaves and twigs.

Control. Good pruning is the first step in control of scales on blueberries. All scales are well controlled by delayed dormant oil sprays applied before blossoming. The scales and blueberry bud mite are the only insects in blueberry fields that cannot be controlled adequately by dusting. Heavy spraying at high pressure is necessary; usually 300 gallons per acre and a pressure of at least 300 pounds is required to wet insects well under loose bark. Three per cent superior oil has been found to be significantly better than 2 per cent in New Jersey.

Blueberry Blossom Weevil

Nomenclature and Distribution. The blossom weevil, also called the cranberry weevil, *Anthonomus musculus* Say, may frequently become a problem in blueberry fields in New Jersey and Massachusetts. It is especially destructive in poorly cultivated fields and along wooded edges of fields, where it feeds on wild blueberries, huckleberries, cranberries, and several other wild ericaceous plants.

Life Cycle. Doehlert and Tomlinson (1947, 1951) have described the life cycle of the blossom weevil. This insect is a small, dark red snout beetle about $\frac{1}{16}$ inch long, including the snout. It emerges from hibernation early in the spring, at about the time the blossom buds are beginning to swell. The weevils hide between the clustered buds, and in small infestations they may be difficult to find. Their presence may be discovered by the small, brown holes in the unopened blossoms that result from their feeding and egg-laying. Infested blossoms turn purplish, fail to open, and eventually drop to the ground, where the grub consumes the floral parts and ovary. Adults of the second generation sometimes feed extensively on blueberry leaves. There is evidence that in New Jersey another generation is completed on cranberry and other ericaceous plants before the weevil hibernates in trash and weeds in and around the field.

Control. This insect is controlled easily by spraying or dusting immediately after punctures are observed. Parathion or dieldrin at rates of 0.3 pound of actual toxicant per acre gives excellent control. In fields where plum curculio is a problem, dieldrin is preferred. In the interest of protecting pollinating insects, care should be taken not to apply this insecticide when blossoms are open.

Plum Curculio

Nomenclature and Distribution. This general fruit pest, *Conotrachelus nenuphar* Herbst., likes blueberries and is common in blueberry growing areas. It is one of four larvae which may be found in marketed berries and cause condemnation of shipments (Table XXV). Plum curculio is particularly damaging on early varieties and in mulched or poorly cultivated fields. The Jersey and later varieties are hurt less frequently by this insect.

Life Cycle. A very good account of the life history of this insect is given by Snapp (1930). The plum curculio is a dark-brown snout beetle about $\frac{1}{4}$ inch long with light patches on its back and four knobs on its wing covers. The beetle hibernates in trash and in the soil under trash. It be-

Table XXV. Insect Larvae That May Be Found in Blueberry Fruit at Harvest Time

Insect	Adult	Eggs	Larvae or Worms
Plum Curculio	dark brown snout beetle, easily jarred off bushes	laid in crescent-shaped scars in green berries	white, legless grub with brown head; feeds on only one berry
Cranberry Fruitworm	inconspicuous dusk-flying moth	laid on inside rim of calyx cup of green berries	green caterpillar; webs together clusters of berries, may destroy as many as 4 berries
Cherry Fruitworm	inconspicuous dusk-flying moth	laid on leaves, usually under surfaces and on any place on green berries	red caterpillar; usually feeds on only one berry; does not web berries together
Blueberry Maggot	fly (about housefly size) with dark bands on wings; very active and hard to see in ordinary infestations	laid just underneath skin in green or ripe berries	white, legless, apparently headless maggot; causes berry to become soft and juicy; feeds on only one berry

Source: Marucci, unpublished data.

comes active on blueberries early in the spring at about the time that the early varieties are beginning to bloom. It immediately begins feeding on both leaves and blossoms. When the early berries are about $\frac{1}{4}$ inch in diameter, the female beetle begins to lay eggs in typically crescent-shaped depressions, which she gouges out with her snout. The egg hatches in about six days and the legless grubs feed on the pulp of only one berry. Almost all of the infested berries of midseason and late varieties drop to the ground, but a few of the late-hatching larvae may go to market in the ripe berries of early varieties. The grub spends about three weeks in the berry and then goes into the soil to pupate. After about a month a new generation of adult beetles appears and feeds on foliage. This type of damage is rarely very serious. The adult goes into hibernation in September or October.

Control. Thorough and frequent cultivation facilitates control of the plum curculio. A dust consisting of 5 per cent DDT and 1 per cent parathion or 2 per cent guthion at 40 pounds per acre gives good control. The treatment should be applied before too many oviposition scars are observed but after pollinating insects are gone. In New Jersey, waiting until 75 per cent of the Rubel blossoms have dropped before spraying or dusting gives adequate control of the curculio without causing appreciable loss of pollination. Bees should be removed from the field before this application. Dieldrin applied before blossoms open to control blossom weevil usually greatly reduces plum curculio populations in New Jersey.

Cranberry Fruitworm

Nomenclature and Distribution. The cranberry fruitworm, *Mineola vaccinii* Riley, the foremost pest of cranberries in Massachusetts, is also a formidable destroyer of blueberries. It is found throughout the eastern United States wherever wild blueberries grow (Beckwith, 1943a) and has been reported as an economic pest of blueberries in Michigan, North Carolina, New Jersey, Massachusetts, and Maine. Its effect on cultivated blueberries depends on the abundance of weeds, trash, or mulch in the field. It has been known to destroy more than 75 per cent of the berries on early varieties and up to 40 per cent of the berries of Jersey and Rubel in poorly cultivated New Jersey fields. The degradation of quality caused by its heavy webbing and frass and the appearance of crawling worms over packaged berries adds to the economic loss.

Life Cycle. The cranberry fruitworm hibernates as a fully grown green larva, about $\frac{3}{8}$ inch long, in a soil-encrusted cocoon. These cocoons are usually spun under weeds or trash near the surface of the soil, rarely more than $\frac{1}{2}$ inch below it. The night-flying moths, which have dark greyish-brown wings with two white splotches on each wing, emerge from the soil when the largest early berries are about quarter-grown and almost immediately begin laying eggs. The eggs are usually inserted along the inside rim of the calyx cup. Small green caterpillars hatch in about a week and immediately enter a berry, usually at the junction of the stem and berry. The first larval entries occur in New Jersey when the largest Rancocas berries are about 9 mm in diameter. The worm webs together as many as six berries and may destroy as many as four. Generally most of the core of the berry is eaten, and when the larva leaves it to enter another one, the consumed berry is filled with sawdust-like frass. The mature caterpillar, green and about $\frac{3}{8}$ inch long, crawls to the ground to form its cocoon about 24 days after hatching. There is only one generation a year.

Control. In the early days of blueberry culture, this insect was effectively controlled by picking off infested berries, which were easily detected because of the webbing and their early ripening. This method is still practical in small plantations with light infestations, but in larger fields insecticides must be employed for adequate control (Beckwith,

1941b; Fulton, 1946; Hutchinson, 1954). In New Jersey this pest has increased in importance with the increased tolerance of weeds in some areas. Repeated discings and thorough elimination of all weeds, mulch, and trash helps to reduce populations of this soil hibernator. The chemical control of this insect is the same as that for cherry fruitworm, given below.

Cherry Fruitworm

Nomenclature and Distribution. In New Jersey the cherry fruitworm, *Grapholitha packardi* Zell., advanced from the stage of being merely an entomological curio to a position of prime economic importance in the short span of about five years, from 1945 to 1950. It now poses a serious threat to the cultivated blueberry industry.

It is found in all the blueberry areas of the country but has been most abundant in New Jersey. In a 1951 survey of New Jersey blueberry fields, 18 of 23 fields sampled were infested with cherry fruitworm (Marucci and Fort, 1952). The presence of this large, conspicuous red worm in marketed berries reduces the high quality of cultivated blueberries and greatly depresses their market value. In several instances, shipments of berries were returned to growers or condemned when large numbers of red worms were observed crawling over the crates of packaged fruit. In addition, this insect sometimes leaves its highly repugnant frass in the berries and often causes them to shrivel. If the cultivated blueberry industry is to maintain its high prestige on the market, this insect must be closely watched and recommendations for its control carefully followed.

Life Cycle. The cherry fruitworm hibernates as a larva in a flimsy web which lines a burrow usually made in dead wood on the bush (Marucci, 1953a). Larvae have also been found in pithy weeds and in small prunings at the base of bushes, but not in the soil (Fig. 58). In New Jersey, pupation occurs toward the end of April and moth emergence starts about May 10, with a peak of emergence coming about the first week in June. The adults are small, dark grey moths; their wings are only about $\frac{3}{8}$ inch across and are marked by chocolate brown bands. The moths are rather inconspicuous dusk fliers and do not begin to lay appreciable numbers of eggs until the temperatures after sunset reach 70 F.

The first eggs, which are greenish-white, pancake-like and about $\frac{1}{50}$ inch in diameter, are not laid until after mid-May (Fig. 59), after which they may be readily found on the under surfaces of leaves as well as on fruit. After an incubation period of about a week, hatching begins toward the end of May, about the same time that entries of the cranberry fruitworm occur.

The newly hatched larvae are white with black heads and only about $\frac{1}{32}$ inch long. They usually enter the fruit through the calyx cup and after a few days of feeding turn pink, with brown heads. The larva only rarely feed on more than one berry, usually leaving only a slight amount of frass in it. After a few days of feeding the worm becomes red. About 24 days after hatching, the larvae, about ¼ inch long, leave the fruit and seek hibernating quarters. Unfortunately this migration often does not occur until the berry is ripe and on its way to market. However, the vast majority of the worms leave green or partly ripe berries before they are picked.

In Massachusetts there is only one generation a year, but in New Jersey a very small percentage of the larvae sometimes pupates and emerges in July for a partial second generation.

Control. Commercial cherry fruitworm control is not difficult when the treatments are well timed and when two thorough applications are made (Hoerner and List, 1952). Because of the long moth flight period and egg hatching period, at least two applications are ordinarily required. These should be made at 10-day intervals beginning at the time of the first larval entry. Control is obtained with dusts of 1 per cent parathion, 5 per cent Sevin, 4 per cent malathion, or 2 per cent guthion at the rate of 40 pounds per acre.

Figure 58. Larva, pupa, and moth stages of the cherry fruitworm, *Grapholitha packardi.* (Walter Fort, Pemberton, N.J.)

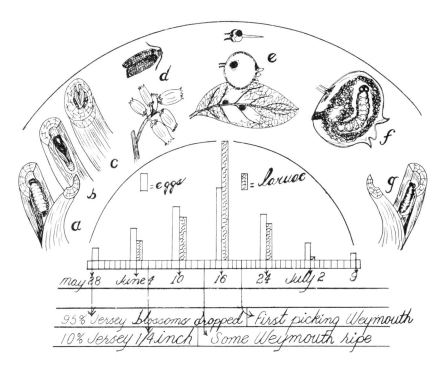

Figure 59. The life history of the cherry fruitworm on blueberries in New Jersey in relation to specific stages of fruit development in the Jersey and Weymouth varieties. At (*d*) the moth is a dark, inconspicuous night flier and rarely seen. Moth emergence begins about May 15, but egg laying (*e*) on green and ripe fruit and twigs may be delayed several days. In 1952 egg laying began May 29, reached a peak June 10-16 and continued until June 30. After an incubation period of about 7 days, eggs hatch and larvae enter fruit about June 3. The larva (*f*) is bright red and while it may feed on more than one berry, it does not web berries together as does the green cranberry fruitworm. Mature larvae begin to leave fruit about June 20 and seek a place to overwinter. Many bore into old pruning stubs (*g*) where about April 20 of the following spring they pupate (*a*) (*b*). At any time after May 15 pupal cases (*c*) may be found protruding from old pruning stubs indicating that moths have begun to emerge. (Ordway Starnes, Rutgers University, New Brunswick, N.J.)

Sharp-Nosed Leafhopper

Nomenclature and Distribution. The sharp-nosed leafhopper, *Scaphytopius magdalensis* Prov., is important in blueberry culture because it is the only known agent capable of transmitting blueberry stunt disease in the field (Tomlinson, Marucci, and Doehlert, 1950). It is common in all blueberry growing areas on the East Coast but is less abundant in Michigan and is not found on the West Coast. The geographic distribution of this insect and other leafhoppers which were trapped on yellow, sticky board traps (Marucci, 1947) in blueberry fields in the various blueberry growing areas is shown in Table XXVI. This leafhopper survey was an integral part of the Rutgers University blueberry stunt disease research program conducted from 1947 to 1952.

Table XXVI. Leafhopper Survey of Blueberry Fields, 1947 to 1948

Insect	N.J. 30 Traps	N.C. 10 Traps	Mass. 10 Traps	Mich. 10 Traps	Total
Graphocephala coccinea Forster	13,751	1,466	523	53	15,793
Scaphytopius magdalensis-verecundus	2,443	725	302	1	3,471
Macrosteles (divisus) fascifrons Staal	2,591	86	194	448	3,319
Empoasca fabae Harris	1,227	51	100	523	1,901
Paraphlepsius irroratus Say	873	36	134	454	1,497
Jassus olitorius Say	560	504	374	22	1,460
Graminella nigrifrons Forbes	1,167	130	13	0	1,310
Gyponana spp.	957	112	168	4	1,241
Graphocephala versuta Say	218	776	0	0	994
Agallia constricta Van Duzee	657	60	25	0	742
Scaphytopius angustatus Osborn	600	0	10	0	610
Scaphoideus spp.	251	3	5	5	264
Scaphytopius frontalis Van Duzee	124	4	14	120	262
Aulacizes irroratus Fabricius	9	241	0	0	250
Xestocephalus pulicarius Van Duzee	148	4	2	19	173
Scaphytopius acutus Say	59	10	10	42	121
Idiodonus kennicotti Uhler	97	2	16	0	115
Paraphlepsius collitus Ball	74	2	0	0	76
Deltocephalus flavicosta Staal	68	0	1	0	69
Oncometopia undata Fabricius	1	67	0	0	68
Erythroneura spp.	31	9	14	8	62
Osbornellus rotundus Beamer	38	0	18	0	56
Neokolla gothica Signoret	26	3	8	7	44
Penthimia americana Fitch	11	7	14	0	32
Draeculacephala spp.	19	4	3	5	31

Source: Marucci, 1947.

Life Cycle. The sharp-nosed leafhopper is dark brownish-black, with a distinctly sloped and pointed extension of the head in both the nymphal and adult stages. The adult is $\frac{1}{16}$ inch long and is drab in color. The nymph is brown or brownish-black with a notable white hourglass-shaped marking on the back. A closely related species, *Scaphytopius verecundus* Van D., which is not a vector of stunt disease, is indistinguishable macroscopically from the S. *magdalensis* as an adult, but has a red and yellow hourglass marking in the nymphal stages (Hutchinson, 1955).

S. *magdalensis* hibernates as an egg between the upper and lower epidermis of fallen leaves. Eggs begin hatching in mid-May in New Jersey, and after five sedentary nymphal periods adults are formed in late June. The adult leafhoppers are active flyers, and it is in this stage that blueberry stunt is transmitted. A second generation starts with eggs in July, producing a large flight of adults by early autumn.

Control. Since egg hatching is complete and very few migrating adults are on the wing by the time the second application for cherry fruitworm is made, the sharp-nosed leafhopper is controlled by the fruitworm treatments of parathion or malathion. In some fields where stunt disease was spreading rapidly, 25 pounds of 5 per cent DDT applied in early September greatly reduced populations of the adults of the second generation.

Blueberry Fruitfly or Blueberry Maggot

Nomenclature and Distribution. The blueberry maggot has been considered to be a biological form of apple maggot and both have been designated as *Rhagoletis pomonella* Walsh. Isolation in blueberry areas over many generations may have caused the blueberry maggot to become a distinct species, as demonstrated by the failure of the two forms to interbreed (McAllister and Anderson, 1935; Pickett and Neary, 1940). The blueberry maggot is sometimes given the status of a separate species and is referred to as *Rhagoletis mendax*, as suggested by Curran (1932). However, the approved list of common names of insects of the Entomological Society of America still records both blueberry maggot and apple maggot as *Rhagoletis pomonella* (Entomological Society of America, Committee on Common Names of Insects, 1960, 1961).

The blueberry maggot is the most serious pest of both lowbush and cultivated highbush blueberries in all of the important growing areas except the West Coast (Lathrop, 1945). In North Carolina the insect becomes a problem only after unusually cold winters when the cold requirement of the insect is fulfilled, and in very wet harvest seasons when frequent pickings are not possible.

The chief cause for concern about this insect is that the maggot stays in the berry for some time after it is ripe and picked and that it is not

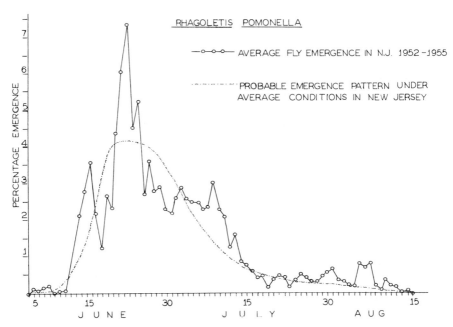

Figure 60. Average percentage emergence of *Rhagoletis pomonella* at various dates for the years 1952 to 1955 and probable emergence pattern under average conditions in New Jersey. (Philip E. Marucci, Rutgers University, New Brunswick, N.J.)

possible to wash out, winnow, or eliminate by hand all of the infested berries. Fruit containing maggots is liable to seizure and condemnation by the Pure Food and Drug Administration. Federal inspectors maintain a constant watch over interstate blueberry shipments, and processors and other buyers generally make laboratory tests to determine whether berries are contaminated. Maggoty berries are soft and mushy and leak over other berries, causing a sticky, wet, and unattractive fresh pack.

Blueberry maggot feeds on all kinds of huckleberries as well as all species of blueberries within its range. Cultivated plantings in areas containing dense stands of these wild plants are difficult to keep free of maggots because of continuous migration from the wild. If such areas are left untreated it is not unusual to find more than 100 maggots in a pint of berries.

Life Cycle. Lathrop and Nickels (1932) studied the life cycle of this insect on the lowbush blueberry in Maine in great detail, as did Porter (1928) on the apple in Connecticut. This fruitfly hibernates as a brown, seedlike, barrel-shaped puparium, about $\frac{1}{8}$ inch long and $\frac{1}{16}$ inch broad, buried in the soil from 1 to 6 inches deep. In New Jersey the flies begin to emerge from the soil about the middle of June and continue issuing through the month of July and usually at least half of August (Fig. 60).

The adult is about the size of a housefly, with black bands on the wings (Fig. 61). The female does not mature rapidly, and eggs are not laid until about seven to 10 days after emergence. In Maine it is reported that eggs are laid only in ripe berries of the wild lowbush blueberry, but the female flies have been observed ovipositing in green as well as ripe berries of the cultivated highbush blueberry in New Jersey and Massachusetts. The eggs are inserted just beneath the skin and hatch in from two to seven days. The tiny new larvae are colorless, and although they begin feeding immediately after hatching they generally cannot be seen

Figure 61. The blueberry fruit fly *Rhagoletis pomonella* is the most important enemy of cultivated and wild blueberries. (Walter Fort, Pemberton, N.J.)

until they are about a week old, after which the liquefying of infested berries makes their detection easier. The larva matures in about 20 days and then drops to the ground to pupate.

In Maine and Massachusetts about 80 to 90 per cent of the puparia that survive the winter and their natural enemies emerge the following summer, with a very small percentage emerging the second summer after pupation. In New Jersey and more southerly latitudes a larger proportion of the puparia emerge in the second summer. In all states a very few puparia sometimes do not emerge until the third summer after pupation.

Control. The control of this insect is complicated by its long emergence period, its migratory tendencies, and the fact that it does not start its attack on the fruit until harvest has already begun (Tomlinson, 1935). The use of residual insecticides is precluded, necessitating frequent applications of relatively nontoxic, nonresidual poisons.

In Maine, where almost the entire lowbush blueberry crop is processed and fruit may be thoroughly washed to remove residues, very good control is obtained with two applications of a dust consisting of 50 per cent calcium arsenate, 10 per cent monohydrated copper sulfate, and 40 per cent hydrated lime at 6 pounds per acre (Lathrop, 1952). In Canada the same formula is recommended in two applications, the first when berries first begin to turn blue and the second about two weeks later (Wood, 1962b).

Since much of the highbush blueberry crop is sent to the fresh market, washing is not feasible; and therefore residual toxicants may not be used. This makes it necessary for the grower of highbush blueberries to use as many as five applications of insecticide dusts to give protection over the long emergence period of the fly in the field and the extended migratory period of flies from the wild to the cultivated fields. Two per cent rotenone dust at 25 pounds per acre applied every 10 days after fly emergence until the end of harvest gives adequate control. Recently the use of 4 per cent malathion dust at 25 pounds per acre in New Jersey has given more efficient control than rotenone and has almost entirely replaced it. Bait sprays utilizing protein hydrolysates as the bait and malathion as the toxicant have proved very effective when applied by ground sprayers or by airplane. They have the advantage of reducing the amount of poison necessary; ½ pound of malathion with bait has given results equivalent to a full pound without bait (Marucci, 1958).

Experimentally, diazinon, guthion, cygon, ciodrin, korlan, and dylox have all given excellent control. These materials are not yet labeled for use against blueberry maggot, but at least one of them should be by the time the inevitable resistance of the fruitfly to malathion does occur.

In the eastern United States, two braconid parasites, *Opius melleus* Gahan and *Opius ferrugineus* Gahan, attack the larvae of the blueberry fruitfly. In New Jersey from 10 to 40 per cent of the maggots may be destroyed by these parasites. Biological control factors are apparently not important in New Jersey since wild blueberries are always very heavily infested even during and after years of high parasitism.

Blueberry Bud Mite

Nomenclature and Distribution. Frequently, blueberry bud mite, *Aceria vaccinii* Keifer, will destroy more than 25 per cent of the crop in North Carolina. It rarely does much damage in New Jersey or in other blueberry producing states. Its attack is very sporadic and the severity is related to the weather. Predacious mites and thrips may also influence its fluctuations. Recently eriophyid mites closely related to the blueberry bud mite have been incriminated as vectors of virus diseases, and it is possible that this species may be important in the transmission of the several new blueberry viruses.

Life Cycle. Blueberry bud mites are too small to be seen with the naked eye, being only $\frac{1}{128}$ inch in length. They do not have the typical oval shape of most mites but are elongated and arrow-shaped, pearly white in color, with four pairs of legs near the head end. All stages of the mites are present throughout the year, and they are confined to the buds and blossoms (Keifer, 1941). During the fall and winter, 50 or more mites may be present between the scales of a single bud. Feeding by the mites results in desiccated buds, flowers that may be distorted and fail to set, and berries that develop only partly or have roughened skins. Most commercial varieties are attacked, but serious damage occurs most often on Weymouth, Cabot, Rancocas, Pioneer, Scammell, and Harding. Usually no visible damage is found on Burlington.

Control. Two applications of spray consisting of 2 gallons of superior oil plus 1 pint of emulsion per 100 gallons, applied in late August and late September, control heavy infestations of this pest. In New Jersey, light infestations may be kept in check with one spray of 3 per cent oil in early October (Tomlinson, 1951). Recent tests in New Jersey show that thiodan gives excellent control of bud mite.

Blueberry Leafminer

Nomenclature and Distribution. Four species of leafminer occur on blueberries, but *Gracilaria vacciniella* Ely probably may be considered the most common. They are among the most numerous insects found in highbush blueberries, and have caused concern because of their great abundance in Michigan and New Jersey. They have also been found to occur on wild lowbush blueberries in Maine. *G. vacciniella* can be found in any field in New Jersey in any year; in some years it is easy to observe several hundred on almost any bush in most fields.

In 1955, populations of G. vacciniella were the heaviest ever observed in New Jersey. In one field it was estimated that 28 per cent of the leaves were infested. It was estimated that each miner damaged only 3 to 12 per cent of the leaf surface area, with 6 per cent the average. The total leaf surface area destroyed in this heavily infested field was therefore estimated to be less than 2 per cent. The bushes were not hurt ostensibly by this extremely heavy infestation and produced well the following year. While this insect can be considered innocuous to growers who do not sell plants, it can be quite important to nurserymen. On several occasions plant quarantine officers have condemned shipments of young blueberry plants because they contained the tents of G. vacciniella.

The principal blueberry leafminers can be determined by the mine type in the following simple key.*

1) Mine on lower surface of leaf — — — — — — — — — 2
   Mine on upper surface of leaf — — — — — — — — — 4
2) Miner forming triangular tent in later larval instars—*Gracilaria vacciniella* Ely Miner not a tent-former — — — — — — 3
3) Mine with larval exit hole, pupation in trash on ground—*Ornix preciosella* Dietz Mine without larval exit hole, pupation within mine—*Lithocolletis diversella*
4) Mine on upper surface along edge, causing leaf to curl—the apple trumpet miner, *Tischeria malifoliella* Clem.

Life Cycle. This insect is both a leafminer and a leafroller. The larvae spend the first two instars in mines in the lower surface of the leaf, but in the later stages they leave the mine and draw the tip of the leaf to a point along the edge of the leaf, fastening it with silk along the edges to form a surprisingly symmetrical triangular tent. They spend the rest of the larval period in this tent, not as a miner but as a surface feeder. Larvae leave the tents to make tough white paperlike cocoons in trash on the ground and hibernate as larvae. In New Jersey pupae are formed in April and moths emerge through the month of May, with the peak in mid-May. Another flight of moths occurs in July and August, and second-generation larvae are seen from late August until leaves drop late in the fall.

Control. Undoubtedly the parathion treatments for fruitworms and the malathion applications for maggots keep the leafminers from becoming more numerous. A brocarnid parasite, *Apanteles ornigis* Weed, is a very effective biological control agent. In 1955 in New Jersey, 80 per cent of the overwintering cocoons were destroyed by parasites.

* Read key by starting at 1; choose appropriate description and proceed to indicated number; continue until description fits specimen in question.

Blueberry Crown Girdler

Nomenclature and Distribution. Leaves on bushes in newly planted fields in New Jersey sometimes turn red prematurely in late summer. This may often be a symptom of root-feeding grubs or the crown girdler, *Cryptorhynchus obliquus* Say. The latter is a weevil larva that feeds only on the crown of the blueberry bush, in contrast to rootworms or scarab grubs, which feed on the finer roots. The girdler is a serious blueberry pest because a single larva can kill a bush up to six years of age by girdling the cambium of the crown. In 1951 it caused severe damage in the Chatsworth area in New Jersey, where it was found in 10 of 11 newly planted fields surveyed for its presence. Two of these fields suffered a loss estimated at 20 to 40 per cent of the plants (Marucci, 1951).

Life Cycle. The crown girdler passes the winter as a nearly full-grown legless larva, $\frac{1}{4}$ to $\frac{3}{8}$ inch in length, deep in the wood at the crown, about 1 to 2 inches below the surface of the ground. Early in the spring the larva resumes feeding, and just before pupation it deepens the channel in the wood. Pupation occurs throughout June and the first adults are noticed after the middle of June. Egg-laying has not been observed, but newly hatched larvae are found in early July.

Control. Field tests in New Jersey have shown that control of infested bushes can only be obtained by the use of strong fumigants that can penetrate the soil and permeate into the burrow of the borer. Ethylene dibromide, ethylene dichloride, and paradichlorobenzene (PDB) all gave complete control. However, the ethylene dibromide and ethylene dichloride were injurious to bushes as well as the larvae, while the plants tolerated the $\frac{1}{2}$ ounce of PDB necessary to give control. Preventive soil treatment of 25 pounds per acre of actual DDT before planting eliminates infestation. Allowing cleared and plowed land to lie fallow for a year also helps in reducing girdler as well as root-feeding insects.

Black Vine Borer

Nomenclature and Distribution. Although several insects attack wild berries in Washington, only one has become serious in the recently planted cultivated blueberry fields (Schwartze and Myhre, 1954b). This is the black vine borer, *Brachyrhynus sulcatus* Fabricius, a snout beetle larva which feeds on the bark of the crown below the soil, greatly weakening and sometimes killing the bush. This insect has been found in a few instances in cutting beds in New Jersey but has never been found there on plants grown in the field. It is a very general feeder but is best known as a strawberry pest.

Life Cycle. In Washington these insects hibernate as legless grubs in the crowns of the bush or as adults in trash around the base of the plants. The borers become adults in late May and through June. The weevils are

⅝ inch long and are black with patches of gold and white on the wings. The wing covers of the beetles are fused so that they are unable to fly. All of the beetles are females and they reproduce parthenogenetically (without mating), beginning egg-laying about two weeks after they emerge. The small, white, spherical eggs are laid in the soil close to the roots and crown and the small grubs start to feed immediately. Bushes infested by these grubs are poor in vigor, and their leaves begin to redden in the summer. If a plant is completely girdled it slowly dies. Plants exhibiting red color should be suspected of harboring black vine borers, which may sometimes be identified by the lumpy callous formation on the stem of the plant below the soil surface.

Control. Commercially prepared baits containing dried fruit and arsenicals, or bran, molasses, and arsenicals, have been used with some success. Recently chlordane and aldrin soil treatments have given better results. Application is made to the feeding areas of the bushes in mid-April at the rate of 10 pounds of actual chlordane and 5 pounds of actual aldrin per acre. Applying these materials before planting is a good preventive treatment that may give protection for three years or more.

Cranberry Rootworm

Nomenclature and Distribution. This insect, *Rhabdopterus picipes* Oliv., feeds as a grub on the roots and as an adult on the leaves of blueberries, cranberries, and related plants. The root injury is serious but the foliage feeding is of minor consequence. When new land is planted, rootworm grubs, feeding on the roots of wild host plants, concentrate on the blueberry plant and often cause serious injury unless they are controlled. Injury may also become serious at any time in established fields that are mulched or allowed to become excessively weedy for several seasons. As indicated by the name, the grub of this insect attacks the roots, eating the bark of the larger roots and completely consuming the fine roots. Infested plants are not fruitful, are pale yellow, and the leaves turn red and drop prematurely in the fall. In extreme cases the plant is killed (Weiss and Beckwith, 1945).

Life Cycle. The six-legged grubs are white and comma-shaped with a light brown head and are only about ¼ inch long when fully grown. The beetle measures about ¼ inch in length and is a shiny mahogany brown. They are active from mid-June until late July, feeding most often on leaves low in the bush and leaving characteristic small, irregular oblong holes in the leaves. Eggs are laid in the soil in July. The grubs hatch late in the summer, do some feeding through the fall, and hibernate as partly grown larvae.

Control. Allowing virgin land to lie fallow for a full year after it is plowed generally eliminates this pest. Twenty-five pounds of actual

DDT or 10 pounds of chlordane per acre broadcast and disced in before planting is an effective preventive measure (Tomlinson, 1948). In established fields rootworms can be controlled by applying 6 ounces of 5 per cent DDT or 3 ounces of 5 per cent chlordane per plant and thoroughly working it into the soil within a 2-foot radius of the base of the plant.

Insects of Minor Importance

The following insects are minor pests that have only rarely become sufficiently numerous to require control operations. They are usually abundant only in fields that do not receive insecticide treatments for the other major insects or that are otherwise poorly cared for.

White (Scarabeid) Root Grubs. Various white (scarabeid) grubs are sometimes found feeding on the roots of blueberry bushes in newly planted or in weedy fields. These include larvae of the Japanese beetle, *Popillia japonica* Newman, which are about $1\frac{1}{2}$ times the size of the cranberry rootworm, and larvae of *Phyllophaga* sp., some of which may get to be more than an inch long, and one or two of which may seriously weaken or kill small bushes up to three years old.

The adults of these insects, especially the Japanese beetle, are sometimes very destructive to berries and foliage. Blueberry fields near pastures or under sod culture may have heavy populations of the Japanese beetle.

The method of controlling these grubs is similar to that used against the cranberry rootworm.

When Japanese beetle infestations are severe, the regular malathion treatments every 10 days for maggot control are not adequate; the interval must be reduced to five or seven days.

Termites. When bushes are weakened by the attack of grubs described above, termites may gain access to dead and weak living tissue in the plant and kill it. In New Jersey termites are always secondary, but in North Carolina some growers report that termites may attack live roots of bushes not previously attacked by grubs. Termites are sometimes a problem in propagating beds in both states, where they attack unrooted cuttings. The control measures described for cranberry rootworm also control termites.

Stem Gall, Hemadas nubilipennis Ashm. These are pithy, kidney-shaped galls $\frac{3}{4}$ inch to $1\frac{1}{4}$ inches in length, always produced on new stem growth (Fig. 62). They are caused by a small, black chalcidoid wasp (Driggers, 1927). As many as 200 galls have been counted on a single bush, and such heavy infestations reduce the fruitfulness of the bush. The galls are eliminated in the normal pruning operation, and unless pruning is neglected no damage will result from this insect.

Figure 62. The kidney-shaped blueberry stem gall *Hemadas nubilipennis*, showing individual larval chambers. (Walter Fort, Pemberton, N.J.)

Blueberry Stem Borer, Oberea myops Hald. Occasionally young stems of the blueberry plant are girdled 3 to 6 inches from the tip during late June or July. Two parallel girdles or rings of punctures are cut around the stem about ½ inch apart, between which an egg is laid under the bark. The grub that hatches from the egg is the stem borer. It tunnels the stem and, if undisturbed, emerges as a large, showy, long-horned beetle after three years. The first year it tunnels but a few inches, the second year it may reach the base or crown of the plant, and the third year it may enter another stem from the crown (Beckwith, 1934). The emission of sawdust from infested canes and the weakened appearance of these canes makes detection easy. Removal of infested canes below the tunnel by pruning controls this insect. In normal pruning very few infested canes are missed.

Fall Webworm. This general forest feeder is sometimes seen in cultivated blueberry fields in both the spring and the fall. The fall webworm, *Hyphantria cunea* Drury, is a pale yellow, black-spotted, hairy caterpillar that lives in large colonies in a large, unsightly tent made of a tightly spun web. The caterpillars feed on the foliage inside the web until it is entirely consumed, then make another web. The prominent webs may be readily destroyed by hand picking. Malathion applied for the blueberry maggot is effective against the smaller but not the larger instars of this caterpillar.

Datana Worm. One of the most obvious leaf-feeding worms is the datana worm, *Datana* sp. It usually feeds in colonies of about 100, a voracious group capable of defoliating an entire bush. When disturbed, these insects all raise their heads and anal segments in concert, as though assuming a united fighting position. These worms can be identified by the yellow stripes along their bodies. Although these are minor pests, they may become alarmingly abundant in fields in the Pine Barrens of New Jersey two or three years after a forest fire. The fires stimulate the growth of very dense stands of the favored host plant of the worms, staggerbush, *Pieris mariana* L., upon which the insect flourishes.

The datana worms have two generations a year, with the adult moths laying more than 200 eggs in each generation. The insects soon become so numerous that they strip all of the wild foliage and march into blueberry fields in veritable armies. In extreme cases growers have had to plow deep furrows around their fields and place DDT or malathion dust at the bottom of the furrow. Regular dusting or spraying with malathion along the edges of the field normally will check the marchers. It is necessary to spray or dust every four or five days, however, instead of the 10-day interval used in maggot control.

Blueberry Tip Borer. In June, before new growth has begun to harden, some blueberry shoots may begin to wilt, arch over, and become discolored, the leaves turning yellowish with red veins and the stems purplish. This injury, which may be mistaken for primary mummy berry infection, is caused by the tip borer, *Hendecaneura shawiana* Kearfott. The newly hatched worm, tiny and pink, enters the soft stem and bores channels that may extend for 8 or 10 inches by autumn and result in the destruction of the stem's fruit-production potential in the following year. Hutchinson (1951) found as many as 87 injured shoots on 20 Cabot bushes and listed the varieties most susceptible to its attack in this order: Cabot, Concord, Rancocas, and Pioneer. The insect is considered potentially of economic importance in blueberry plantings in New York (Schaefers, 1962).

Schaefers' study of the life history of the blueberry tip borer showed that the larva spent its entire life of 10 to 11 months in a single stem, during which it changed from pink to white in the later stages and grew from 1.5 to 10 mm. Pupation occurred in May, emergence of adults in early June, and reinfestation by mid-June.

The tip borer is hardly ever observed in New Jersey in fields receiving insecticide treatments for fruitworms. It is occasionally seen in neglected or untreated fields, and this suggests that the parathion applications for fruitworm control are keeping this insect in check.

Red-Striped Fireworm. This caterpillar, *Gelechia trialbamaculella* Cham., more often a cranberry and huckleberry feeder, also attacks the blueberry. In early spring, reddish worms bore into the tips of young stems, making tunnels 3 to 4 inches long. After emerging from the stem the worms web several leaves together and feed inside this web (Weiss and Beckwith, 1945). There are at least two generations a year in the latitude of New Jersey, with the stem-boring habit noted only in the first generation. This pest is never troublesome in fields receiving dusts for other major blueberry insects.

Sparganothis Fruitworm. Sparganothis fruitworm, *Sparganothis sulfureana* Clemens, is another cranberry pest whose wide taste range occasionally includes blueberry. In Maine it is known as a leafroller (Phipps, 1930), and it is apparently a more abundant pest on lowbush blueberries than on the cultivated highbush species. It has been reported on cultivated blueberries in Michigan and New Jersey.

The life cycle of this insect has been investigated on cranberries in New Jersey (Marucci, 1953b). It is a light caterpillar with a black head in its early stage, turning green with a light yellow head as it grows. The moth is sulfur yellow in color, about $3/8$ inch long, with a conspicuous red X-shaped mark on the wings. There are two generations per year in New Jersey, with one flight of moths in early July and the second one in September. This insect hibernates as a newly hatched larva in a flimsy webbing. It feeds on both berries and leaves, which it fastens together in tube-shaped nests.

Good control of this pest is obtained on cranberry bogs in New Jersey by the use of 60 pounds of 5 per cent DDT per acre. On blueberries it is controlled by the parathion treatments made for cherry and cranberry fruitworms.

Blueberry Budworm. Budworms are major problems in lowbush blueberry areas of Maine and Canada, but they rarely cause appreciable damage to cultivated highbush blueberries. The blueberry budworm may be troublesome in mulched or excessively weedy fields (Weiss and Beckwith, 1945).

One species, *Rhyncagrotis anchocelioides* Grn., has been troublesome in New Jersey. The worms hide in the day in trash under bushes and climb up the stems at night to feed on young, tender buds early in the spring. They continue feeding later in the season on leaves, blossoms, and even green berries.

Clean cultivation eliminates this problem entirely. Where mulches are used, early season applications of DDT may be necessary. To protect pollinating insects, such treatments should be made before blossoms open.

Red-Banded Leafroller. This general fruit pest, *Argyrotaenia velutinana* Walker, may attack blueberry leaves and fruit, which it also webs. In Michigan it has been reported causing some damage to blueberry plantings near apple orchards (Vergeer, unpublished data). This insect became a pest in apple orchards after the use of DDT, which reduced its natural enemies. The increased use of DDT in blueberry fields should make blueberry growers vigilant.

Dasystoma salicellum *Hbn*. This leafroller has become troublesome on cultivated blueberries in British Columbia (Raine, 1965). The insect feeds on many of the native plants typical of the peat bogs on which blueberries are grown. It is of chief concern because large numbers of the large whitish worms are dislodged into the picking crates when vibrator machines are used in harvesting. The early larvae may also destroy flower buds and invade or scar the berries. After harvest the mature larvae often cause severe leaf damage, occasionally defoliating some bushes.

This leafroller has only one generation a year. It hibernates as a pupa in the litter under the bushes. Flightless females emerge in mid-March and lay eggs among scales of the new flower buds, in the axils of leaves, and under loose bark on old canes. The eggs hatch within a two-week period starting about mid-May. Larvae develop slowly all summer, pupate within their shelters in October and drop to the ground with the leaves.

Malathion is recommended for control. The rate of application is 2 quarts of 50 per cent emulsible concentrate in 40 gallons per acre; time of spraying is about June 12 or later, well after pollination. Thirty pounds per acre of 5 per cent malathion dust is also effective.

Apple Bud Moth. This species, *Platynota flavedana* Clem., also does a little damage to blueberries near apple orchards in Michigan (Vergeer, unpublished data). In undusted New Jersey fields it has been found both as a leaf feeder and as a fruitworm. It is never a pest in dusted fields.

Dogwood Borer. This new pest of blueberries, *Thamnosphecia scitula* Harris, was discovered in the summer of 1961 in Michigan. Howitt (1964) considers it of economic importance. It is closely related to the lesser peach tree borer, the moth of which it resembles.

The clear-winged moth lays its eggs singly, probably within 8 inches of the ground. Entries are made close to the ground, often in pruning scars or mechanical injuries. Another common point of entry is at the junction of wood two and three years old on an old cane. The larva bores into the wood, killing small stems and weakening larger canes. Larvae are half grown by August and fully grown by October. Pupation occurs in May and moths emerge in about 15 days (Howitt, 1961).

The insect has shown an apparent preference for certain varieties, being found most frequently in Berkeley, Weymouth, Jersey, Pemberton, Coville, Burlington, and Stanley (Howitt, 1964). Study of chemical control methods is in progress.

Currant Fruit Weevil. This insect, *Pseudenthonomor validus* Dietz, is a serious problem on the highbush blueberry only in Massachusetts. It is discussed under lowbush blueberries.

Miscellaneous Insects. Other insects that have been reported attacking the cultivated highbush blueberry but that only occasionally cause minor damage are:

Blueberry spittle bug (*Clastoptera saint-cyri* Prov.)
Snowy tree cricket (*Oecanthus niveus* DeGreer)
Pandemis moth (*Pandemis limitata* Robinson)
Blueberry leafminer (*Gracilaria vacciniella* Ely)
White-marked tussock moth (*Hemerocampa leucostigma* Smith and Abbott)
Southern corn rootworm or spotted cucumber beetle (*Diabrotica duodecimpunctata* Fabricius)
Milkweed bug (*Oncopeltus fasciatus* Dallas)
Tarnished plant bug (*Lygus pratensis* Linné)
Blueberry case bearer (*Chlamys plicata* Fabricius)
Blueberry midge (*Contarinia vaccinii* Felt.)
Chain-spotted geometer (*Cingilia catenaria* Drury)
Blueberry flea beetle (*Altica torquata* Lec.)
Strawberry rootworm (*Paria cannella* Fabricius)
Forest tent caterpillar (*Malacasoma disstria* Hbn.)
Bagworm (*Thyridopteryx ephemeraeformis* Haworth)
Gypsy moth (*Porthetria dispar* Linné)
White fly (*Trialeurodes* sp.)
Red spider (*Tetranychus telarius* Linné)
Spring cankerworm (*Paleacrita vernata* Peck)
Fall cankerworm (*Alsophila pometaria* Harris)
Green fruitworm (*Graptolitha bethunei* Grote and Robinson)
Buffalo tree hopper (*Ceresa bubalus* Fabricius)
Psyllids (*Psylla* spp.)
Green peach aphis (*Myzus persicae* Sulzer)
Rose chafer (*Macrodactylus subspinosus* Fabricius)
Yellow-headed fireworm (*Peronea minuta* Robinson)
Strawberry leafroller (*Ancylus comptana* Froelich)
Blueberry sawfly (*Neopareophara litura* Klug)

Information about the life cycle and control of most of these insects or their close relatives may be found in Metcalf and Flint (1951) or may be obtained from a county agricultural agent.

LOWBUSH INSECTS

The important lowbush blueberry industry of Canada and Maine, well established before the inception of highbush blueberry culture, is beset with insect problems quite unlike those existing in cultivated blueberry fields because of the great differences in the ecological, climatic, and cultural conditions of the lowbush blueberry barrens. As shown in Table XXIV, the highbush and lowbush crops have a few important insect pests in common; these were discussed in the previous section. Some of the more important insects of the lowbush blueberry alone are discussed below. For a more complete discussion of lowbush insects see Phipps (1930) and Woods (1915).

Black Army Cutworm

Nomenclature and Distribution. This destructive noctuid, *Actebia fennica Tausch.*, has occurred in outbreak proportions in both Maine and Canada. A serious infestation occurred in Maine as early as 1884, and in 1925 this pest caused a complete crop failure, costing an estimated $100,000 in Hancock County alone. In 1944 and 1945 it again caused serious destruction in Maine and also in New Brunswick, Canada, where many hundreds of acres of commercially grown blueberries were badly damaged (Maxwell, 1950). Kinsman (1957) has "not found it feeding in alarming proportions" in Nova Scotia. The areas generally most susceptible to the black army cutworm are those displaying fresh, vigorous sprout growth after a recent burn. The fact that the first crop year after a burn is the most productive makes this cutworm's attack more important. Areas in production two or more years after a burn are not frequently invaded by this insect. It has a wide range of host plants, including several that are common around the blueberry barrens; hence, sprouting areas may be repopulated from an ever-present wild abundance of the moths.

Since the larvae are nocturnal feeders and hide in the duff during the day, their presence may not be detected before they have done extensive damage. The browning caused by the worm's feeding on buds has sometimes been mistaken for frost damage. An estimation of the prevalence of the cutworm is, therefore, important. Maxwell (1950), working in New Brunswick, Canada, found that sweeping at night with a heavy cotton insect net 18 inches in diameter gave a good estimation of populations.

According to his data, a level of 25 worms per 50 sweeps indicated that immediate control measures were necessary to avoid economic loss.

Life Cycle. Phipps (1927) gives an excellent account of the life history and habits of this insect. The moth lays as many as 400 eggs on succulent sprouts in August. The eggs do not hatch until early October in Canada, and apparently the young larvae go into the soil to hibernate before the end of November after very little feeding (Maxwell, Wood, and Neilson, 1953). As the buds of the blueberry begin to expand in early April the wintering larvae, about ¼ inch to ½ inch long and velvety black in color, become active and bore through the bud scale to eat out the interior of the bud, later feeding on open blossoms and foliage. Larvae become fully grown in early June and are about 1½ inches long with wide black longitudinal bands. Pupation occurs in the soil, the moth emerging after about a month. Moth flight extends from July to September and there is only one generation a year.

Control. Dusting experiments in New Brunswick, Canada, have shown 3 per cent DDT applied at 40 pounds per acre to be very effective in controlling this insect. Poison bran baits, cryolite, and calcium arsenate did not give good control (Maxwell, 1950).

W-Marked Cutworm

Nomenclature and Distribution. The W-marked cutworm, *Spaelotis clandestina* Harr., obtains its common name from the black markings which form W-shaped figures on the first and eighth abdominal segments. Phipps (1930) reported that this worm, in company with two other cutworms, *Actebia fennia* Tausch. and *Polia purpurissata* Grt., completely stripped 50 acres in Maine. It is one of the most destructive cutworms in both Maine and Canada.

Life Cycle. Wood, Nielson, Maxwell, et al. (1954) carefully worked out the complete life history of this insect in Charlotte County, New Brunswick. Larvae are commonly found feeding on blueberries from May through June, with pupation occurring in the soil in late May or early June. Moth flight, sometimes so extremely heavy as to be a nuisance in the homes of Maine blueberry growers (Phipps, 1930), occurs from mid-July to late September. Oviposition occurs throughout September, and after an unusually long incubation period of from 27 to 57 days, larvae hatch and feed until they reach their second instar stage in Canada, and probably much longer in Maine, before hibernating in the soil. Feeding resumes in April on swelling buds and later on foliage. Phipps (1930) believed that there might be a partial second generation in Maine but that there was only one in New Brunswick.

Control. Control of this and the following cutworm is the same as that used against the black army cutworm.

Polia purpurissata Grt.

Nomenclature and Distribution. Polia purpurissata Grt. is another voracious and destructive blueberry-infesting cutworm, prevalent in Maine and Canada. It is a light brown worm with broad white stripes in its early instars, but a mottled light brown without stripes in the later stages, attaining a length of almost 2 inches.

Life Cycle. The life cycle of this noctuid is described in detail by Wood, Nielson, Maxwell, et al. (1954). It is somewhat similar to the above cutworm, but its eggs are laid sooner, in August, and the preoviposition and egg incubating periods are much shorter, so the larva does more feeding before entering the soil for hibernation in November.

Chain-Spotted Geometer

Nomenclature and Distribution. The chain-spotted geometer, Cingilia catenaria Drury, is another blueberry insect with a wide range of preferred wild host plants that occasionally invades blueberries and causes extensive damage. Serious outbreaks have occurred in blueberry fields in both Maine and Canada, and it has also occasionally attacked cranberries in Massachusetts and Nova Scotia (Franklin, 1916; Gorham, 1924). Pickett (1943) termed it a periodic pest in eastern Canada, while Kinsman (1957) considered it "generally of minor importance" in Nova Scotia. Sweet fern, Myrica asplenifolia L., bayberry, Myrica carolinensis Mill, and lambkill, Kalmia angustifolia L., are favored hosts that are quite common in the blueberry areas of both regions.

Life Cycle. Phipps (1930) has outlined the life cycle of this spanworm. It hibernates in the egg stage on fallen leaves of sweet fern and many other plants. Hatching occurs early in June and the voracious defoliating larvae mature in August, at which time they are from $1\frac{1}{2}$ to 2 inches long and yellow with conspicuous black spots on the sides. The pupa, white with black dots and about $\frac{3}{4}$ inch in length, is formed on the bush in a loosely spun silken cocoon. Moths begin to emerge in September. Oviposition occurs over a long period; a maximum of 368 eggs and an average of 140 per female has been reported.

Control. Calcium arsenate, used to control fruitflies, adversely affects this insect. In Canada an effective control is achieved with 30 pounds of 3 per cent DDT or 20 pounds of 5 per cent dust per acre (Maxwell and Wood, 1961).

Blueberry Thrips

Nomenclature and Distribution. Two species of thrips, Frankliniella vaccinii Morgan and Iaeniothrips vaccinophilus Hood, are economically important on lowbush blueberries. Thrips are found in Maine but are not as troublesome there as in Canada. According to Kinsman (1957) it is an annual pest in some fields in Nova Scotia and has recently increased in number in several fields. The fact that Pickett (1943) did not discuss it in his presentation of blueberry insects may indicate that it has become important in Nova Scotia only in the past few years. In the Atlantic Provinces infestations of thrips are spotty in most fields, but in some they may reduce the crop by more than 50 per cent (Maxwell and Wood, 1961). As with the black army cutworm, recently burned fields are more susceptible to attack, and a constant population reservoir for reinfestation is maintained on wild plants.

Thrips injury, which becomes evident early in June, may be recognized by tight curling of leaves about the stem, accompanied by reddening and malformation of the leaves. In sprout fields all the foliage of infested plants may be wrapped around the stem, while in crop fields the injury is confined to a few leaves. Affected plants are much reduced in vigor and productivity and may be more prone to winter injury. The two species of thrips produce the same type of damage.

Life Cycle. The two species of thrips have similar life histories. Adult thrips, yellowish-grey in color and only $\frac{1}{16}$ inch in length, appear in early June, soon laying eggs on foliage. The very small nymphs spend their entire feeding period of about three weeks in the curled leaves, and by the first week in July practically all of the thrips in Maine fields are in the adult stage. Another generation of nymphs appears in about two weeks, matures in late August, and drops to the ground to hibernate (Phipps, 1930).

Control. In New Brunswick, Canada, 5 per cent DDT dust, as well as parathion, malathion, and DDT sprays, applied late in May before curling of leaves resulted in good control of thrips. Applications made after leaves had curled gave little or no control (Maxwell, Wood, and Neilson, 1953). Kinsman (1957) and Maxwell and Wood (1961) recommend 20 pounds of $2\frac{1}{2}$ per cent dieldrin dust when the new sprouts are appearing above the soil in the spring following a burn.

Blueberry Flea Beetle

Nomenclature and Distribution. Phipps (1930) found the blueberry flea beetle, *Altica sylvia* May, to be the most abundant and destructive beetle on lowbush blueberries in Maine. But its importance has undoubtedly declined in this state with the increased use of calcium arsenate for the blueberry maggot.

Life Cycle. The adult beetle is coppery bronze in color and $\frac{1}{4}$ inch long. It has the typical flea beetle behavior of suddenly jumping when disturbed, having been called "snapper bug" by some Maine growers. It passes the winter as a minute orange-yellow egg on or near the ground.

Black, slug-like larvae appear in late May and begin feeding on unopened buds and later on leaves and blossoms, growing to a length of about $\frac{3}{8}$ inch. Pupation is completed by late June and adults emerge from the first of July. The beetles also feed extensively on blueberry foliage.

Control. Maxwell, Wood, and Neilson (1953) found both 3 per cent DDT dust and 70-30 gypsum-cryolite dust to be effective at 30 pounds per acre. The first arsenical application for maggot control is very effective for the flea beetle but this is too late for the grubs (Maxwell and Wood, 1961). Burning also combats this insect, since it destroys the eggs (Phipps, 1930).

Currant Fruit Weevil

Nomenclature and Distribution. This fruit infesting grub, *Pseudanthonomus validus* Dietz, is sometimes troublesome on lowbush blueberries in Maine. Pickett (1943) states that in Canada more weevils than blueberry maggots are sometimes found in the fruit. It has also caused damage to cultivated highbush blueberries in Massachusetts, where Tomlinson (1959) rated it as a minor pest. Other blueberry states have not reported the currant fruit weevil as a pest.

Life Cycle. The life cycle of this weevil closely resembles that of the other important curculionid attacking blueberries, the plum curculio. It winters as an adult, reddish-brown and about $\frac{1}{8}$ inch long. In Maine the beetles begin laying eggs in the calyx lobes of the green berry about the middle of June (Woods, 1915). The legless, colorless, brown-headed grubs, which can be found throughout July and most of August, spend their entire life of about 30 days in one berry. The first pupae, which are pure white and are formed in the larval excavation in the berry, are seen at about the time of first ripening in mid-July. Adults of the second generation appear in late July and throughout August, doing some additional damage by feeding on berries. In Maine there is evidence of a partial second generation.

Control. Control operations against the blueberry maggot undoubtedly help against this insect. Observations in Canada indicate that burning makes this insect economically insignificant (Maxwell and Wood, 1961).

Blueberry Leaf Beetle

Nomenclature and Distribution. This defoliator, *Galerucella vaccinii* Fall, reported by Woods (1915) as seriously injurious to blueberries in Maine had become unimportant commercially by 1930, according to Phipps (1930). Canadian entomologists do not consider it a pest of any consequence, although Pickett (1943) showed that both larvae and adults accepted only lowbush and highbush blueberries in tests in which it was offered 83 species of plants.

Life Cycle. This beetle, ⅛ inch long and reddish-brown in color, winters among the debris, becoming active early in June in Maine. Eggs are laid about the middle of June and larvae are abundant by early August. Unlike the blueberry flea beetle, which scallops the leaf margins and later destroys the entire leaf, the greenish-gray larvae of this species skeletonize the leaves as they feed on their undersurfaces. The adults appear again in early September; and there is only one generation per year (Phipps, 1930).

Control. Burning destroys this insect, or it may be easily checked by arsenical sprays.

White-Marked Tussock Moth

Nomenclature and Distribution. In 1956 a widespread and destructive outbreak of this caterpillar, *Hemerocampa leucostigma* Smith and Abbott, occurred in blueberry fields of Nova Scotia (Kinsman, 1957). This was the first known economic damage to blueberries caused by this insect. Defoliation of bushes resulted in a failure of the berries to ripen.

The caterpillar stage of this insect, a general shade-tree and fruit-tree feeder, is striking in appearance. Its head is coral red, with a fringe of white hairs, and yellow and black bands extend the entire length of the abdomen. It is about 1½ inches long when fully grown and bears three pencil-like tufts of long black hairs, one projecting forward from each side of the head and the third projecting backward, like a tail, from the hind end. There are also four tufts of short, erect white hairs on the back.

Life Cycle. The winter is passed in the egg stage in conspicuous masses of from 50 to 100 attached to empty cocoons on tree trunks or branches. Larvae appear in Canada during the latter part of June or early July. The caterpillar matures in seven to eight weeks, after which they spin their cocoons on the bushes. Fully winged male moths emerge in about a week, but it takes two weeks for the female, which has rudimentary stubby wings and is incapable of flight, to emerge. In the latitude of Ohio there are two complete generations a year and as many as three in more southern regions (Houser, 1918).

Control. Control of this insect was secured in Nova Scotia by applying 3 per cent or 5 per cent DDT dust per acre in the third week of July (Kinsman, 1957).

Insects of Minor Importance

Sawflies. Phipps (1930) reported that a sawfly stripped plants on nearly 50 acres in Maine in 1929. Two sawflies, *Neopareophora litura* Klug and *Pristiphora idiota* Nort., are potentially damaging to blueberries in Maine and Canada. They have also, though rarely, been minor pests of cultivated highbush blueberries in mulched fields.

Blueberry Tip Worm. A midge that forms leaf galls, the blueberry tipworm, *Contarinia vaccinii* Felt, is sometimes found on blueberries in New Jersey, Maine, and Canada but has never been economically significant. Injury caused by the tipworm resembles that of thrips; the presence of one or more small maggots identifies it as the work of tipworm (Maxwell and Wood, 1961).

Scarab Beetle. A scarab beetle, *Serica cucullata* Dawson, has been plentiful in some Canadian fields but has not caused significant damage (Maxwell, Wood, and Neilson, 1953).

Cutworms. Besides those already mentioned as important, a number of climbing cutworms, spanworms, and leaf tiers feed on blueberries in both Maine and Canada. Wood (1951) collected 38 species of lepidopterous larvae from lowbush blueberries in Charlotte County, New Brunswick, most of which could probably be very readily found in Nova Scotia and Maine. This array of insects capable of causing economic losses points out the need for constant vigilance for possible outbreaks. The areas of commercial lowbush blueberry production in both Maine and Canada are so extensive that it is necessary for each grower to be alert to the problems and to make constant checks for unusual insect abundance. The departments of agriculture in both regions conduct excellent research and extension services for the blueberry grower and may be called on for assistance in any insect problem that may arise.

CHAPTER 10

Diseases and Their Control

by **Eugene H. Varney**, *Department of Plant Biology, Rutgers University, New Brunswick, New Jersey, and* **Allan W. Stretch**, *Crop Research Division, Agricultural Research Service, United States Department of Agriculture, Rutgers University, New Brunswick, New Jersey*

Although the cultivated blueberry industry is relatively new, it has already been threatened with destruction by stunt, a virus disease, and has been severely hampered in the South by stem canker, a fungus disease. Other serious fungus diseases are mummy berry, botrytis blight, and powdery mildew.

Most of the diseases vary in severity and economic importance from one blueberry growing region to another. Stem canker, for example, is very destructive in the South but not in the North, while mummy berry and botrytis blight are of importance principally in the northern states including the lowbush areas. All the blueberry diseases are probably indigenous.

Disease cycles and control measures, where known, are discussed. Control measures, however, should be treated as suggestions rather than definite recommendations. A fungicide may be very effective in one growing region but ineffective in another. A chemical may cause injury under certain climatic conditions and, therefore, may be safe in one area but injurious in another location. For these and similar reasons no chemical control program should be followed without first consulting local agricultural authorities to determine the proper materials to use, the time they should be applied to be most effective, and restrictions on their use before and during the harvest period.

Diseases and Their Control 237

FUNGUS DISEASES

Mummy Berry

Causal Organism and Distribution. Longyear (1901) identified the fungus that causes mummy berry disease as *Sclerotinia vaccinii* Wor. A similar disease was reported earlier in Europe by Weinmann and subsequently studied in detail by Woronin (1888). Reade in 1908 studied the fungus in New York state and felt it differed from the European species sufficiently to warrant the new name *S. vaccinii-corymbosi*. With the creation of the new genus *Monilinia* by Honey in 1928, the fungus became known as *Monilinia vaccinii-corymbosi* (Reade) Honey (Honey, 1936). More critical studies may show whether our fungus is truly distinct from the European species.

Mummy berry was first described in 1901 by Longyear in Michigan on various wild species of *Vaccinium* and has since been reported as far north as Canada and as far south as Georgia and southern Mississippi. In this country mummy berry is of major importance in the northern blueberry regions. Some years the entire crop may be destroyed on susceptible varieties, and then several years may pass before the disease will again reach epiphytotic proportions. The disease is apparently rare in the Pacific Northwest (Crowley, 1933).

Disease Cycle and Symptoms. The fungus survives the winter as mummies, or pseudosclerotia, on the ground near blueberry bushes. The mummies look like miniature gray pumpkins. In the presence of adequate moisture the mummies break their dormancy in early spring and form apothecia or mummy cups. Although mummies will remain viable for several years, the majority form one to several apothecia the following year. Apothecia produce ascospores which are discharged at about the time blueberry buds begin to swell. According to Woronin, the ascospore germ tubes are able to penetrate only the noncuticularized tips of the new growth. No symptoms are visible for one to two weeks, at which time infected twigs and flower clusters suddenly darken and wilt (Fig. 63). Severely infected bushes look as though they were affected by frost. In the cultivated fields of New Jersey, blight is generally limited to the current season's growth, and any extension of the injury into the older wood is usually associated with other fungi, particularly *Botrytis*. Powdery masses of conidia are produced in a few days along the main axis of blighted shoots or the petiole and midrib of affected leaves. As the conidia mature a sweetish, aromatic odor attractive to insects is evident. Conidia are carried by air currents and possibly insects to open flowers. The conidia germinate and the fungus makes its way into the ovaries, where it soon fills the ovarial cavities. The blueberry seeds abort, but

Figure 63. Mummy berry. *Above.* A branch with all but one of the young twigs blighted, and small, wrinkled, mummified berries *(left and right)*. *Below.* Apothecia, or mummy cups of *Monilinia vaccinii-corymbosi*. (Eugene H. Varney, Rutgers University, New Brunswick, N.J.; USDA, Beltsville, Md.)

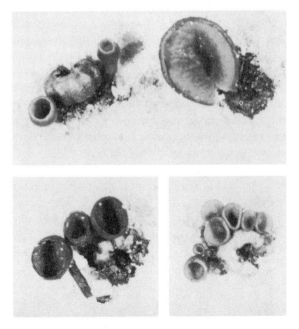

infected berries continue to develop normally until they near maturity, at which time they turn tan to salmon in color rather than normal blue. Affected berries shatter easily. About the time they drop to the ground, the flesh is replaced by fungus tissue. The skin and other remaining host tissue eventually slough off, leaving only the fungus tissue or mummy.

Varietal Susceptibility. Highbush blueberry varieties show various degrees of susceptibility to the blight and mummy phases of the disease. A variety may be very susceptible to one phase and resistant to the other phase. Burlington, Rubel, and Jersey, for example, are relatively resistant to blight but are very susceptible to the mummy phase of the disease. In tests in New Jersey with plants two to four years old, Jersey, Rubel, Burlington, Pemberton, and Dixi fell in the group most resistant to blight; Rancocas, Stanley, Weymouth, Berkeley, Bluecrop, Herbert, Coville, and Scammell in an intermediate group; and Atlantic, Ivanhoe, Earliblue, June, and Blueray in the most susceptible group of the varieties tested. It was also found that seedlings from a cross between two parents susceptible to blight showed more blight and fewer disease-free plants than did seedlings from crosses where one or both parents were relatively resistant to the blight phase.

Control. Control measures include mechanical or chemical destruction of the mummies and apothecia and direct protection of the bushes with fungicides. Early and thorough cultivation has been the most widely practiced control measure. Mummies between rows can be covered with soil by discing or shallow plowing in the spring before apothecia are mature. Mummies between bushes should be raked into the aisles before cultivation. If the field is too wet to work in the spring, cultivation should be done in the fall. Hoeing and raking around bushes at weekly intervals during the period of apothecia formation is important. This will destroy many apothecia before they mature and discharge spores. Mummies lodged in plant crowns should be removed by hand.

Monilinia twig and blossom blight of lowbush blueberry is largely controlled by the burning that is normally practiced every two or three years.

Chemical ground treatments have been tested in the major blueberry areas of Michigan and New Jersey. Dinitro-o-phenol and dinitro-o-cresol give good control if the ground is thoroughly drenched about the time the apothecia are maturing in early spring. Dust and granular preparations of calcium cyanamide have given very good control and have the advantage of being easier to apply in fields too wet for heavy loads of a dinitro spray. Calcium cyanamide dust applied by airplane in Michigan gave as good control as dust applied with ground equipment and with no apparent injury to the plants (Fulton, 1958a).

All chemicals used in ground treatments are caustic, however, and must be applied before buds show any green. It is important that the grower get details from his county agent or agricultural experiment station before using chemicals to kill and prevent the formation of apothecia.

Fungicides applied as protective sprays give partial control of mummy berry blight. Bailey and Sproston (1946) got about 50 per cent control of blight with ferbam in Massachusetts. Johnston (1949) obtained partial control in Michigan with lime sulfur, Bordeaux mixture, ferbam, wettable sulfur, and zineb. Studies in New Jersey indicate that ferbam, ziram, maneb, dichlone, and wettable sulfur will reduce blight 50 to 70 per cent (Varney, 1956). Fungicides are applied as soon as apothecia are ready to discharge spores and blueberry buds begin to break. Two to four applications at weekly intervals may be needed. Fungicides in general have given better control of flower cluster blight than twig blight. In New Jersey, for example, dichlone alone or in combination with ferbam reduced flower blight on the June variety by approximately 80 per cent and twig blight by about 50 per cent on the same bushes. Fungicides are of considerable value on susceptible varieties, such as June, where a high percentage of flowers are lost to blight and in situations where ground treatments alone have not given adequate control. Ferbam and zineb dusts are used in lowbush plantations where mechanical or chemical ground treatments cannot be practiced (Lockhart, 1961b). Mercury fungicides severely injure both lowbush and highbush blueberries and should never be used. Copper injury has been observed on some varieties sprayed with Bordeaux mixture. In New Jersey a gray, corky spotting was observed on plants of the Weymouth variety sprayed with zinc fungicides (ziram and zineb).

Stem Canker

Causal Organism and Distribution. Stem canker caused by the fungus *Botryosphaeria corticis* (Demaree and Wilcox) Arx and Muller was first observed in a cultivated field in North Carolina in 1938. It was subsequently found that most fields of the highbush blueberry in North Carolina and of rabbiteye blueberry in Alabama, Mississippi, Florida, and Georgia were more or less infected. The disease was found on wild plants throughout the southern blueberry sections, indicating that the fungus causing the disease was indigenous. Stem canker was not known in the North until 1951 when Goheen reported it in New Jersey on plants that had originated from southern nurseries. Climatic conditions, however, have not favored the rapid spread of the disease in New Jersey. In the South the disease is a serious problem and has made certain varieties worthless. Losses to stem canker are not so great now, for in most plantings very susceptible varieties have been replaced with more resistant ones.

Disease Cycle and Symptoms. The pathogen probably enters the plant through lenticels on the current season's growth (Demaree and Wilcox, 1942). Reddish, broadly conical swellings develop in late summer or early fall at the infection points (Fig. 64). Small black fruiting bodies (pycnidia) usually develop on the conical growths. Later symptoms vary with the susceptibility of the variety. The second year on very susceptible varieties such as Cabot and Pioneer the conical swellings lose their reddish color, enlarge somewhat, and become fissured. The fungus invades the cortex and cambium surrounding the original injury and the newly attacked tissue swells, becomes blisterlike in appearance, and turns light gray. The wood is slightly attacked and may show some discoloration. Each year the cankers are extended, become deeply fissured, and finally girdle the shoot or branch. Girdled branches are weakened and finally die. Of interest is the fact that cankers appear first on the side of the shoot exposed to the sun's rays. Cankers on less susceptible varieties, such as Rancocas, are not conspicuously swollen but may be extensive, roughened, and cracked. On more resistant varieties, cankers are scattered and only slowly increase in size, or they may occur only on some of the smaller branches. On very resistant varieties, the fungus can establish itself in small shoots but dies out about the second year as the stem tissues mature. Black, conical to subglobose fruiting bodies (both pycnidia and perithecia) are produced in abundance on cankers of all ages. Spores are expelled during wet periods and are carried by the wind and splashing rains to the current season's growth, where new infections take place. Taylor (1958) exposed susceptible Weymouth plants for one-month intervals from March through November to natural inocula. Plants exposed during the period extending from June 15 until August 10 were most severely affected, but infection occurred from April through early September.

Varietal Susceptibility. The great differences in susceptibility that blueberries show to stem canker have been discussed at length by Demaree and Morrow (1951). Atlantic, Jersey, and Scammell are the only highly resistant varieties among the several northern blueberry varieties that do well in the South. Pemberton, Rancocas, and Rubel are moderately resistant. Rancocas, widely grown in North Carolina, is productive as well as tolerant of the destructive stunt disease and will probably not be immediately discarded on the basis of canker susceptibility alone. Cabot, Concord, Dixi, June, Pioneer, Stanley, and Weymouth are so susceptible that they are no longer suitable for growing in the South.

Since northern blueberry varieties are a poor source of canker resistance, plant breeders have sought resistance in several wild North Carolina

Figure 64. Left. Typical stem canker symptoms showing a gradation in severity from small conical swellings to a completely girdled, swollen, deeply fissured stem (*left to right*). Note the black pycnidia. *Right.* Fusicoccum cankers on the highbush blueberry with conspicuous black pycnidia. (USDA, Beltsville, Md.; E. H. Barnes, Michigan State University, East Lansing.)

highbush clones known as the Crabbe selections. Good commercial varieties with adequate canker resistance have been obtained by crossing Crabbe selections with northern varieties possessing desirable fruit qualities. Murphy, Wolcott, Angola, and Croatan, named in the early 1950's, and Morrow, named in 1964, all possess canker resistance. The resistance of Wolcott, the most extensively grown variety in North Carolina, may be breaking down. In greenhouse inoculation tests with three *Botryosphaeria corticis* isolates, Clayton and Fox (1963) demonstrated that isolates do exist that readily infect Wolcott. This variety was rated as highly resistant when it was introduced. The possibility of the existence of more pathogenic fungus races is suggested by this work.

Canker is at present no problem in the rabbiteye blueberry plantings. All of the approximately dozen named varieties are either immune or highly resistant to canker. Fortunately, those who made the original selections from the wild took only canker-free bushes, even though they were unaware at the time that abnormally rough bark was a symptom of a serious disease.

Control. The use of resistant varieties is the only practical means of control. Fungicides have given inadequate control. Roguing, the removal of infected plants, and pruning out all visible cankers are of little value in the South. In the North, however, where the disease has been introduced, all affected plants should be promptly removed and burned. Only disease-free plants and propagating wood should be accepted from southern canker areas.

Fusicoccum Canker

Causal Organism and Distribution. Shear (1917) was the first to identify *Fusicoccum putrefaciens* isolated from cranberries with end rot. Later Shear and Bain (1929) considered *F. putrefaciens* was the conidial stage of *Godronia cassandrae* Peck, which had been described in 1886 by Peck from dead branches of *Chamaedaphne calyculata* (L.) Moench. Shear and Bain (1929), Creelman (1958), and Groves, as noted by Creelman, all found minor variations between the fungus from *Vaccinium* and the original description of the fungus on *Chamaedaphne*. McKeen (1958) found a difference in pathogenicity and morphology between cranberry and blueberry isolates of *Fusicoccum*, and suggested that further cultural and infection studies are needed. Seaver (1945) considered *G. cassandrae* a synonym for *G. urceolus* (Alb. and Schw.) Karst. According to Creelman, Groves believes the fungus from *Vaccinium* is probably undescribed and should be referred to some other genus. Dennis recently referred the blueberry fungus to *Crumenula urceolus* (Fries) de Notaris (Gray and Everett, 1961). Until more detailed studies are available, *G. cassandrae* and *G. urceolus* should probably be considered synonyms for *C. urceolus*.

The fungus was first reported causing cankers on highbush blueberry in Quebec in 1931 (Conners, 1932). Since then it has been reported in British Columbia, Nova Scotia, Massachusetts, New Brunswick, Maine, Finland, England, and Holland (Barnes and Tweedie, 1964). Creelman (1958) proposed the name fusicoccum canker to identify this malady. He noted that this disease was the most serious one in Nova Scotia, making it very difficult to establish new plantings. McKeen (1958) also noted this difficulty in British Columbia. The importance of the disease in other areas has not been estimated.

Disease Cycle and Symptoms. The fungus overwinters as mycelia in cankers on living stems and in crowns of affected plants. In Massachusetts, spores are released from pycnidia from the first of March through mid-July and artificial inoculations were successful in March and April (Zuckerman, 1960b). In Michigan, Barnes and Tweedie (1964) trapped spores typical of *F. putrefaciens* from July through September. Pycnidia are reported to appear in late July through early September in Nova Scotia, and infection apparently occurs after June in British Columbia, since lesions are found on plant parts formed in July and August (McKeen, 1958). Artificial inoculations by McKeen (1958) on the Jersey variety were unsuccessful in June and July but successful in November and December. The wide difference noted by the various workers in time of spore release and susceptibility of the blueberry plant is probably a function of the climate and the varieties used. Differences may also exist in fungus isolates. Ascospores of the perfect stage do not appear to be important in the disease cycle (McKeen, 1958).

The most striking symptom of fusicoccum canker is the sudden wilting and death of girdled stems during periods of warm, dry weather. Cankers appear first as small, reddish discolorations on the stems and many times are centered about a leaf scar. These infection spots frequently occur at the ground line, but many occur higher on the stem. As the cankers enlarge the margins remain reddish, while the bark over the canker turns gray, then brown, and finally the tissue dies (Fig. 64). The cortex, cambium, and to a limited extent the xylem are penetrated by the fungus. Zuckerman (1960b) and Barnes and Tweedie (1964) described a bull's-eye pattern around the locus of infection. Creelman (1958) and McKeen (1958) did not describe this particular symptom. In Nova Scotia (Creelman, 1958), pycnidia, the fruit of the fungus, appear in dead bark in cankers for about five weeks beginning in late July, whereas in Massachusetts pycnidia were found from March through mid-July. McKeen (1958) reported that in British Columbia pycnidial initials form in early March.

Varietal Susceptibility. McKeen (1958) observed a marked variation in varietal susceptibility, Jersey being the most susceptible and Rubel and Rancocas the most resistant. The disease was found on Rubel, Jersey, Pioneer, and Cabot in Massachusetts (Zuckerman, 1959b) and on Earliblue, Jersey, Pemberton, Rubel, and Stanley in Michigan (Barnes and Tweedie, 1964). Berkeley appeared resistant in Michigan. Atlantic, Jersey, and Pemberton seem to be particularly susceptible in coastal British Columbia (Conners and Savile, 1953). Heavy infections were noted in Nova Scotia on Burlington and Jersey, but Rancocas and Stanley were only lightly affected (Lockhart, 1956).

Diseases and Their Control

Control. Control efforts with fungicides have been notably unsuccessful. Creelman (1958), McKeen (1958), and Zuckerman (1960b) all reported failure with spring applications of eradicant and protectant fungicides. Nelson (1964, personal communication) obtained some promising results in Michigan with applications of phenyl mercury starting at the beginning of leaf drop in the fall and continuing until most of the leaves had dropped. Some control is afforded by removing and burning diseased canes.

Coryneum Canker

Causal Organism and Distribution. Coryneum canker, caused by the fungus *Coryneum microstictum* Berk. and Br., was first described by Zuckerman (1959a). He found the disease occurring in 21 of 25 plantings examined in Massachusetts. Zuckerman isolated the organism from bushes of the Rubel, Cabot, Pioneer, Berkeley, Burlington, Earliblue, Jersey, and Pemberton varieties, but he was of the opinion that most varieties are susceptible (Zuckerman, 1960b). The disease has not been reported in other blueberry growing areas.

Symptoms. No definitive symptoms are produced by the fungus. Acervuli (saucer-shaped fruiting bodies) are commonly associated with sun-scalded areas of blueberry stems, and small sunken areas in the bark containing numerous acervuli have been noted (Zuckerman, 1960). The cankers enlarge on weakened bushes and cause dieback, but infections on vigorous bushes generally remain localized.

A comparison of wounding and nonwounding techniques of artificial inoculation of blueberry plants with pure cultures of *C. microstictum* indicated that wounds are apparently necessary for infection to take place.

Control. Satisfactory control of this weak pathogen can probably be achieved with good cultural practices to maintain vigorous bushes. Zuckerman (1960) reported no success with a limited chemical control test.

Phomopsis Twig and Cane Blight

Causal Organism and Distribution. Stevens (1924) had difficulty convincing himself that twig blight was something other than winter injury. After repeatedly isolating a species of *Phomopsis* from affected twigs and reproducing the disease by inoculating healthy twigs, however, he concluded that he was dealing "with a genuine fungus disease capable of attacking somewhat weakened blueberry bushes." He noted that the blueberry *Phomopsis* was very similar to, if not identical with, the *Phomopsis* associated with a rot of harvested cranberries. M. S. Wilcox in 1939 reported identical results in inoculation experiments with *Phomopsis* cultures obtained from either blueberry or cranberry and found no morphological or cultural differences. She concluded the *Phomopsis* from blueberry was identical with *P. vaccinii*, the cause of a rot of cranberries. In 1940, M. S. Wilcox showed that the ascigerous or perfect stage of the blueberry fungus was *Diaporthe vaccinii*.

Twig blight was one of the first diseases observed in cultivated blueberry fields. Stevens in 1924 reported that it was serious in East Wareham, Massachusetts, and that it was found occasionally in New Jersey. *Phomopsis* is a common blueberry fungus but causes a disease primarily on weak bushes and is seldom a problem in well-drained and well-tended fields. Phomopsis twig blight and diebacks caused by winter or cold injury, drought, poor drainage, and boron deficiency are so similar in appearance and development that it is difficult to diagnose the primary cause of a dieback disease without knowing something about the recent history of the planting in question.

Symptoms. Symptoms of phomopsis blight appear first on the smaller twigs. Injury may extend into the larger branches and eventually into the crown. Once the crown is invaded older canes are frequently girdled. Foliage on a girdled cane may suddenly wilt and turn brown during hot weather. Succulent twigs that are affected often become crooked. Pith in diseased twigs becomes discolored, but this is not a diagnostic feature, for pith damaged by cold also becomes brown or discolored. A late frost may kill not only the tender tip of the current season's growth but the pith for several inches beyond the point where there is external evidence of frost damage. Superficially, such injury resembles phomopsis twig blight, and one might erroneously conclude that a fungus had killed the tips and was progressing rapidly down the pith of the stem.

Studies by M. S. Wilcox (1939) indicate that only young, succulent twigs are blighted and that only localized lesions develop when healthy woody tissue is inoculated. Leaf symptoms occurred as reddish spots that increased in size to about 10 mm in diameter. R. B. Wilcox (1940) observed phomopsis leaf spots in the field and in some cases the fungus moved from the leaf spots into the petioles and twigs, producing typical twig blight symptoms.

Control. Since the disease is closely associated with weakened bushes, it is important to maintain plants in a vigorous condition. Horticultural practices that minimize winter injury will also decrease twig blight. Excessive water, drought, overfertilization, and starvation are to be avoided. All weak wood should be removed at the time of pruning. If the pith is discolored, pruning should continue until healthy pith is reached. R. B. Wilcox (1946) reported that the disease was almost completely eradicated from a seriously affected field by pruning severely to healthy wood, improving growing conditions, using adequate fertilizer, and applying fungicides. Since all treatments were applied to the entire field there is no way to know which treatment was the most effective. The spray program consisted of a delayed dormant application of 1:9 lime sulfur followed by an application of 8-8-100 Bordeaux to the foliage. No information is available on the control of phomopsis blight with the newer fungicides.

Botrytis Twig and Blossom Blight

Causal Organism and Distribution. The *Botrytis* commonly associated with blueberries is of the *cinerea* type. So far as known, no one has definitely established a connection between the blueberry *Botrytis* and *Botryotinia fuckeliana* (deBary) Whetzel, the perfect or sexual stage of *B. cinerea* (Pelletier and Hilborn, 1954). Strains of *Botrytis*, as evidenced by different cultural characteristics, have been reported in Washington (Eglitis, Johnson, and Crowley, 1952).

The common gray mold fungus, *Botrytis cinerea*, was one of the first fungi that Stevens (1924) observed on cultivated blueberries in New Jersey. The same year Anderson reported that botrytis blight was perhaps the most destructive disease on many plants, including wild blueberries, in southwest Alaska, where the climate is cool and wet. Ten years later Heald (1934) recorded botrytis twig and fruit blight of cultivated blueberries as a new disease of increasing importance in Washington. By 1950 botrytis blight was considered the most important fungus disease on blueberries in the Pacific Northwest (Darrow, Demaree and Tomlinson, 1951). Markin (1931) was the first to report botrytis blight in Maine on lowbush blueberries, but the disease was not considered to be of great importance until recent years. The disease is of minor importance most years in New Jersey. According to Boller (1951), the disease is often negligible in Oregon and perhaps beneficial where it helps thin an excessive bloom.

Disease Cycle and Symptoms. The fungus overwinters as mycelia and conidiospores in and on affected stems, blossoms, and fruits. *Botrytis* is capable of living as a saprophyte on a wide range of organic matter. Dead blueberry material collected in Maine in mid-March was covered with sporulating *Botrytis* within 24 hours after being placed in a moist chamber held at room temperature (Pelletier and Hilborn, 1954). Spores, produced in vast numbers, are carried by wind to susceptible blueberry tissue. Relatively cool temperatures and high humidity for several days are essential for *Botrytis* to do serious damage. Pelletier and Hilborn (1954) found that three to four days of high humidity are necessary for infection of blossoms of lowbush blueberries and that six to nine days are needed for infection of dormant buds. These conditions are frequently met in the coastal areas of Washington, Maine, and the Canadian Maritime Provinces, where the disease is frequently of major importance.

Botrytis attacks blueberry twigs, blossoms, leaves, and fruit. Twigs are usually attacked at the tips and the fungus may extend downward from less than an inch to the entire length of succulent twigs of the current season. Affected twigs are at first brown to black and finally become gray with weathering. Twigs injured by frosts, winter temperatures, sunscald, and salt injury are similar in appearance to twig blight. *Botrytis*, which can frequently be isolated from such twigs, is present only as a saprophyte and is not responsible for the injury.

Corollas of blighted blossoms turn brown and often resemble frost injury. Botrytis blight is readily distinguished from frost injury by the presence of abundant mycelia. The effect of lack of pollination also can be mistaken for blossom blight (Pelletier and Hilborn, 1954). Blighted flowers show abundant mycelia, and corollas adhere to the calyx for a long period instead of falling off as they do from unpollinated flowers. The fungus attacks leaves that have been injured by various agents and those that are in direct contact with blighted flowers. Infected corollas falling from pollinated flowers often stick to leaves. *Botrytis*, which seldom attacks healthy leaves, is thus able to move from the infected corolla into the injured leaf tissue, causing leaf spots that are readily recognized by their association with the adhering corollas. Unlike blighted leaves, frosted leaves are usually completely killed. *Botrytis* is found in the dead terminal leaves that frequently remain attached to succulent shoots all winter, and these dead leaves may be a source of further infection the following spring (Eglitis, Johnson, and Crowley, 1952).

The spoilage of mature fruit on plants before harvest is usually of minor importance in New Jersey. After harvest, however, *Botrytis* is one of the principal fungi associated with spoilage in storage (Goheen, 1950b). Anderson in 1924 stated that *Botrytis* infrequently attacked the fruit of wild blueberries in southwest Alaska, but in Washington it has been associated with an extensive fruit rotting of cultivated and wild blueberries (Goheen, 1950a).

Varietal Susceptibility. Although there is evidence that varieties of highbush blueberries differ in susceptibility to *Botrytis* (Darrow, Demaree and Tomlinson, 1951; Eglitis, Johnson, and Crowley, 1952), no data are available on their relative susceptibility. Atlantic is one of the most susceptible varieties. Clones of the lowbush blueberries *Vaccinium angustifolium* Ait. and *V. myrtilloides* Michx. differ markedly in their susceptibility to the disease, varying from no infection to almost 100 per cent (Pelletier and Hilborn, 1954).

Control. This disease appears to be most severe on heavily fertilized plants and on plants growing on peat soil with a high water level where rapid and succulent growth is characteristic. Rapid forcing of plants should be avoided, particularly late in the growing season. In Maine, Pelletier and Hilborn (1954) found that more infection was evident among lowbush blueberries fertilized early than among those fertilized late. They also found more infection (38.5 per cent) on bushes with a peat mulch than on those with a sawdust mulch (3.5 per cent). It was suggested that peat, with its greater water-holding capacity, made conditions more favorable for infection.

Ferbam or zineb dusts have successfully controlled blossom and twig blight in Canada and in Maine (Lockhart, 1961a; Hilborn, 1955). If mummy berry is also present, applications should begin at the pre-bloom stage (before flowering) and should be repeated at seven-day to 10-day intervals until the end of the blossom period, or petal-fall. Since infection depends on high humidity, all the applications may not be necessary in the years when rain is infrequent during the blossoming period.

Blueberry Stem Blight

Causal Organism and Distribution. This disease was first reported by Taylor (1958) as caused by an apparently new species of *Botryosphaeria*. Witcher and Clayton (1963) identified the causal organism as *Botryosphaeria dothidea* (Moug. ex Fr.) Ces. and deNot. Surveys by Witcher and Clayton (1963) have indicated widespread occurrence of the stem blight in the blueberry growing areas of North Carolina. They found no correlation between occurrence of stem blight and blueberry variety since infected plants were scattered over the fields except that in some instances they were localized in poorly drained areas. Greenhouse inoculations confirmed field observations in that all 16 inoculated blueberry varieties became infected. Raniere (1961) also reported a similar fungus in New Jersey associated with a blighting of winter-damaged branches.

Symptoms. The most conspicuous symptom of blueberry stem blight is one or more dead branches close to living branches with normal green leaves. In an early stage of infection, leaves on an affected branch may be yellow or reddish. The woody stem tissue of affected branches is a pecan brown and this color frequently occurs only on one side of an affected stem, extending from a few inches to the entire length of the stem. The disease has possibly been confused with winter injury, drowning, and certain phases of stem canker caused by *B. corticis*. Because of confusing symptoms, isolation from infected stems is sometimes necessary to prove a diagnosis (Witcher and Clayton, 1963).

Control. It is difficult to recommend cultural practices which may aid in control since the disease occurs in both apparently weak and strong plants growing in both poorly drained and well-drained fields. In general, though, a thriving bush will better withstand disease than a poor one. Pruning and burning of infected canes as soon as the first symptoms appear may give some measure of control. Fungicide control may be possible, but it has not yet been developed (Taylor, 1958).

Powdery Mildew

Causal Organism and Distribution. A fungus, *Microsphaera penicillata* var. *vaccinii* (Schw.) Cooke, is the cause of powdery mildew (Cooke, 1952). *M. alni* var. *vaccinii* is an older name for the same fungus. Mildew is common wherever blueberries are grown and seems to occur on all species of *Vaccinium*. Its effect on production is difficult to evaluate, but it seems likely that productivity and vigor of very susceptible clones and varieties are reduced, particularly if plants are defoliated earlier than normal each year.

Symptoms. Symptoms are conspicuous on very susceptible varieties. Leaf surfaces are often white with fungus mycelium and spores. The upper surface, the lower side or both may be affected, depending on the variety. Chlorotic spots with reddish borders are common on the upper surface of leaves and may at times be mistaken for symptoms caused by the red ringspot virus. Dendritic, water-soaked areas on the lower surface opposite the chlorotic areas readily distinguish mildew-incited rings from those caused by virus (Fig. 69). On lowbush blueberries, spots or blotches with irregular margins are common. Severely affected lowbush and highbush blueberries lose their leaves earlier than normal. Small black fruiting bodies (cleistothecia) of the fungus are conspicuous on affected leaves of some varieties.

Varietal Susceptibility. There is a striking difference in susceptibility to mildew in clones of lowbush and varieties of highbush blueberries. According to Bergman (1939), who rated some of the older varieties, Pioneer, Cabot, and Wareham were the most susceptible; Concord, Jersey, and Rubel were intermediate; and Rancocas, Stanley, Harding, and Katharine were the most resistant. Moore, Bowen, and Scott (1962) observed differences in varietal susceptibility among the newer varieties as well as the older ones. Berkeley was the most resistant variety observed. Earliblue and Ivanhoe were also very resistant but Bluecrop, Rancocas, Weymouth, Pemberton, Coville, and Dixi were moderately susceptible. Susceptible varieties included Collins, Stanley, Rubel, Blueray, Burlington, Herbert, Jersey, and Atlantic. Climatic factors may affect the susceptibility of a variety. Bergman, for example, rated Stanley as resistant, while Moore, Bowen, and Scott included it in their susceptible group.

Control. Since the most conspicuous symptoms appear after harvest, most growers do not attempt to control the disease on cultivated blueberries. In the lowbush areas of Maine, however, a 20–20–60 dust (calcium arsenate, monohydrated copper sulfate, and hydrated lime) is used to control various leaf diseases. One application is made when most of the blossoms have dropped and a second 10 to 14 days later. Experiments in Maine (Hilborn, 1950) indicate that 10 per cent ferbam dust is equal to the 20–20–60 dust for the general control of foliage diseases. Research work in Michigan (Huguelet, Fulton, and Veenstra, 1961) has shown that sulfur dust applied by air was the most practical and effective treatment. It is important that at least one application be made just after bloom. Monthly applications in June, July, and August help prevent secondary buildup. The sulfur can be added as a 20 per cent dust as part of the carrier in the insecticide dusting program (Nelson, 1963, personal communication). When the sulfur was applied as a 30 per cent dust, some injury was noted on Pioneer, Burlington, Berkeley, and Coville because of dry weather and high temperatures.

Red Leaf Disease

Causal Organism and Distribution. Red leaf is a systemic disease caused by the fungus *Exobasidium vaccinii* Wor. In North America and Europe it is of considerable economic importance in the lowbush blueberry fields. Highbush blueberry is a host (Savile, 1959), but the disease is unknown or rare in commercial plantings. Lockhart (1958) recently studied red leaf disease in considerable detail and Savile (1959) has written a monograph on the genus *Exobasidium*.

Disease Cycle and Symptoms. *E. vaccinii* may be systemic and cause shoot galls or witches'-broom or it may be localized and cause bud galls, leaf galls, or unhypertrophied leaf spots (Savile, 1959). Symptoms will depend upon the species of *Vaccinium* affected. Hypertrophies seldom occur on the lowbush blueberries harvested in northern New England and Canada. Typical symptoms of red leaf are reddening of the foliage and a felt-like layer of mycelia and spores on the lower surface. By midsummer the affected leaves drop, and the crop on these plants is usually very small. Hilborn and Hyland (1956) found that infection occurs by direct penetration of the cuticle of the blueberry stem. The fungus spreads through the cortex and then into the pith.

Lockhart (1958) isolated *E. vaccinii* and studied the effect of temperature on growth of the fungus and germination of the spores. He found that nonsporulating flecks appeared on the foliage two to three weeks after the lowbush blueberries were inoculated and that typical red leaf symptoms developed only after a period of three months, and then on leaves of new side branches that had developed near the base of the main stem. After seven months, typical symptoms were produced in new shoots arising from the rhizome of the originally inoculated plants.

Data taken by Lockhart over a two-year period showed that infection in spray and dust plots ranged from 11.7 to 31.1 per cent of the shoots the first-crop year and from 16.7 to 37.8 per cent of the shoots the second-crop year. Of these, from 4.5 to 28.4 per cent of the shoots died between the first and second crop years.

Control. So far all attempts to prevent the spread of red leaf in the field by the use of fungicides and antibiotics have failed. Eradication of infected plants with a mixture of 2,4-D and 2,4,5-T in water is the only means of control at present and is economical in fields not severely infected.

Septoria Leaf Spot

Causal Organism and Distribution. Septoria albopunctata Cke. was described by Cooke in 1883 from leaves of *Vaccinium arboreum* Marsh collected in Florida. Septoria leaf spot, a problem in North Carolina and

Figure 65. Spots caused by *Septoria* on leaves and a twig of the cultivated blueberry variety June. (USDA, Beltsville, Md.)

other southern states, has been described in detail by Demaree and Wilcox (1947).

A species of *Septoria* causes a minor disease of lowbush blueberries in Canada (Lockhart, 1961a), but Demaree and Wilcox (1947) have questioned reports of this leaf spot on lowbush blueberries in Maine.

Symptoms. The fungus causes small, circular white lesions with a purple border (Fig. 65). Sometimes the lesions are tan to russet with a peripheral zone of brown, or the spots may be entirely brown. Lesions on young shoots of the current year are 5 to 6 mm in diameter, circular to irregular, tan to gray, slightly sunken, and surrounded by a zone of reddish brown. A single pycnidium is usually present within the lesion. Young plants may be so severely spotted that defoliation may occur.

The leaf spots found on the lowbush blueberry are irregular in outline and are brown with small raised pycnidia scattered over their surface (Fig. 67). Fungicides used to control double spot will also control this disease.

Double Spot

Causal Organism and Distribution. *Dothichiza caroliniana* was described as a new fungus in 1947 by Demaree and Wilcox. It is known only on highbush blueberries in North Carolina, where it may cause serious leaf spotting when rains are frequent in May and June.

Symptoms. Initially, spots are 2 to 3 mm in diameter, circular, with brown centers, or sometimes light brown, rust, or gray, surrounded by a dark-brown ring. Apparently the leaf tissues build up a protective layer around the infected cells, and the fungus is at first confined to a small area. Later, after midsummer, the fungus breaks through the protective zone and invades a much larger area, causing a cinnamon-brown spot several millimeters in diameter that surrounds the original small spot. This spot around a spot suggested the phrase "double spot" as a common designation for the disease.

Varietal Susceptibility. Apparently all varieties of the highbush blueberries are susceptible. Cabot, Dixi, Pioneer, and Rancocas are the most susceptible varieties; Adams, Concord, Jersey, and Weymouth are moderately susceptible; and Grover, Harding, June, and Sam are considerably resistant.

Control. Ferbam and Dyrene (2,4-dichloro-6-o-chloranilino-s-triazine) at 2 pounds per 100 gallons of water per acre are recommended for the control of all leaf spots in North Carolina. The materials are applied after blossoming and at two-week intervals until early September. Local authorities should be consulted for the latest restrictions on the use of these fungicides prior to harvest.

Witches'-Broom

Causal Organism and Distribution. Witches'-broom is caused by a rust fungus, *Pucciniastrum goeppertianum* (Kühn) Kleb. The disease has been reported from Maine south to Pennsylvania and west to California and Washington (Demaree and Wilcox, 1947). Markin (1931) found that cultivated highbush blueberries were more susceptible to witches'-broom than were native Maine species.

Disease Cycle and Symptoms. *Pucciniastrum goeppertianum* completes part of its life cycle on species of fir (*Abies*) and part on various *Vaccinium* species. Aeciospores, which are produced only in pustules on fir needles, are carried by the wind to blueberries. The spores germinate and the fungus penetrates the blueberry, where it stimulates the production of an excessive number of lateral buds. This gives rise to a broomlike mass of swollen shoots (Fig. 66). Spores produced on the swollen shoots

Figure 66. Witches'-broom, caused by a rust fungus. An old blueberry bush with several large brooms and an infected branch showing proliferation and swollen and cracked growth (*left and right*). (Maine Agricultural Experiment Station, Orono.)

Diseases and Their Control

are carried back to fir trees, where they infect the needles of the current year. This rust fungus cannot spread from one fir tree to another or from one blueberry plant to another, but must go from balsam to blueberry and back again. Witches'-broom is, therefore, common on blueberry plants only within several hundred yards of woods containing fir trees.

Control. Further increase of the disease in blueberry fields can be controlled by removing the balsam fir trees from nearby woods (Hilborn, 1955; Lockhart, 1961a). Since the fungus is perennial in both low and highbush blueberries, witches'-brooms will develop as long as the affected plants live. Spot spraying with weed killers will destroy diseased lowbush blueberries, giving room for the growth of healthy plants and preventing the spread of the rust back to balsam fir to complete the cycle. Burning has little effect since the fungus usually becomes established in the crowns of diseased plants.

Leaf Rust

Causal Organism and Distribution. Part of the life cycle of the leaf rust fungus, *Pucciniastrum myrtilli* (Schum.) Arth., is completed on hemlock, *Tsuga* sp., and part on all *Vaccinium* species. It is commonly found within range of hemlock trees and in some years may reach epiphytotic proportions. Leaf rust plays an important part in the leaf spotting complex of lowbush blueberries in Maine (Hilborn, 1955) and in some years infected leaves drop in September, which is early enough to have an effect on subsequent crops. In Canada, leaf rust causes little damage because of the late appearance of the disease (Lockhart, 1961a). The fungus sporadically reaches epiphytotic proportions in cultivated highbush plantings in North Carolina, Georgia, Alabama, and Florida. The disease is rare in Maryland and New Jersey, the last outbreak being recorded in 1946. Leaf rust occurs in the far West but is probably caused by another species of *Pucciniastrum*.

In the northern states, leaf rust is most serious near hemlock woods. In New Jersey and the southern states there may be sporadic outbreaks of leaf rust in areas far removed from hemlocks. Unlike the rust which causes witches'-broom, the leaf rust spores produced on blueberry leaves are able to infect other blueberries. These spores may overwinter on evergreen species of *Vaccinium* or cultivated deciduous blueberries during mild winters when some leaves remain green all winter (Demaree and Wilcox, 1947). Under such conditions, the fungus could increase to epiphytotic proportions over a period of two or three years of recurrent mild winters. The rare outbreaks in areas such as New Jersey may originate with spores carried by the winds from the south. In 1946, for example, leaf rust was more severe in Maryland than it was in New Jersey, while in North Carolina, Florida, and Alabama, the disease reached epiphytotic proportions. Spores from infected hemlocks could also be blown long distances.

Symptoms. Infected leaves have light green to reddish areas on the upper surface, and on the lower surface opposite these spots the spores are formed in rust-colored pustules. Infected leaves drop earlier than normal.

Varietal Susceptibility. Cultivated blueberries vary in their susceptibility to leaf rust. According to Meckstroth, as cited by Demaree and Wilcox (1947), Harding and Grover were very heavily infected in a North Carolina planting; Adams, Concord, Jersey, and Rubel were heavily infected; Stanley and Sam were moderately infected; Cabot was slightly infected; and Dixi, June, Pioneer, Rancocas, and Weymouth were free of rust. Owens, Myers, and Black Giant rabbiteye blueberries were only slightly infected.

Control. Because of the sporadic nature of leaf rust, little is done to control the disease in highbush plantings. A 7 per cent ferbam dust applied at petal fall and 10 to 14 days later is recommended for the control of leaf rust on lowbush blueberries (Hilborn, 1955).

Blueberry Anthracnose

Causal Organism and Distribution. The imperfect stage of the fungus *Glomerella cingulata* Spaulding and von Schrenk was first associated with the highbush blueberry by White (1935) in New Jersey. He found *Gloeosporium fructigenum* infecting shoots, blossoms, flower buds, fruit spurs, and pedicels of young plants. Demaree and Wilcox (1947) described a pathogenic fungus on leaves, twigs, and green fruit of highbush blueberry as the imperfect stage of *G. cingulata*. In recent storage tests in New Jersey the fungus was noted to cause fruit rot (Stretch, 1959).

The importance of this disease varies according to variety, season, and growing area. No estimate of importance is available except for New Jersey, where it has caused severe losses (25 to 50 per cent) of marketable fruit in two fields of Coville and one field of Jersey. Even though isolated fields are extensively diseased, this problem is minor to the industry.

Disease Cycle and Symptoms. The fungus overwinters in dead terminal twigs, fruit spurs, and cankers. Moist-chamber incubation of all dead twigs from a single Coville bush, growing in a severely diseased field, revealed 66 per cent with actively sporulating acervuli of *G. cingulata*. Examination of diseased dead leaves in the spring did not disclose the perfect stage of the fungus which might provide inoculum for the primary cycle.

The conidia are disseminated by rain or dew and wind. When the conidium germinates on the fruit surface, it probably penetrates directly, although histological evidence is lacking. However, appressoria have been observed, and direct penetration by this fungus has been demonstrated on other crops. The fungus remains latent until maturation of the fruit or until the fruit sustains such damage as might occur in a frost. Spores liberated from infected fruit and other infection loci provide ample inoculum to infect dying or wounded plant parts and insure a source of inoculum the following season.

Anthracnose is difficult to distinguish from other problems on leaves, stems, and terminals, since the blueberry plant reacts similarly to various stimuli. Stem cankers rarely occur in the field and are usually small, ⅛ inch, with raised purple margins and gray to brown centers. Young succulent stems have been girdled by the cankers when artificially inoculated, causing a dieback. Leaf infections vary from small spots (2 mm) to rather large, poorly defined necrotic areas. The disease is most easily recognized by the salmon-colored spore masses exuded on the surface of the mature fruit. Acervuli are produced in large numbers on the fruit, usually within five days after inoculation at prevailing summer temperatures. Green fruit also will decay if damaged, allowing *G. cingulata* to grow.

Varietal Susceptibility. Anthracnose has been observed on the varieties Rancocas, Bluecrop, Blueray, Coville, Concord, and Jersey in New Jersey. Although it probably occurs on most varieties, it has escaped detection. Anthracnose also occurs in North Carolina and Illinois. In Long Island, Demaree and Wilcox identified the fungus on blueberry.

Control. No proven commercial control measure is available. Research has indicated that dormant or delayed-dormant eradicant spray applications are ineffective by themselves, but in combination with three maneb sprays applied at varying times during the season considerable improvement in control was achieved. The combination of maneb and the delayed-dormant application of phenyl mercury acetate was significantly better than maneb alone when applied on the same dates. However, for maximum control at least seven applications of an effective fungicide beginning at midbloom and continued on a 10-day to 14-day schedule are necessary. Fungicides such as Dyrene, maneb or Dithane M-45 have proven effective but are not approved for use on blueberries. Ferbam and ziram have been given clearance and are also effective. Captan is also approved, but it does not give the degree of control possible with the other materials mentioned.

Phytophtora Root Rot

Causal Organism and Symptoms. Raniere (1961) reported that a species of *Phytophthora* was associated with leaf yellowing and reddening and with early defoliation of rooted Coville blueberry cuttings in New

Jersey. A circular area of dwarfed and dead plants in the corner of a young Earliblue planting was also found to be infested with *Phytophthora*. These plants were affected in varying degrees ranging from slight reduction in new growth to winter killing of previously weakened plants. Roots showed some necrosis, and the crowns and main stems showed extensive vascular discoloration.

Royle and Hickman (1963) identified samples of the fungus provided by Raniere as *Phytophthora cinnamoni* Rands. They successfully inoculated Dixi, Jersey, and Pemberton plants, and after about three months extensive destruction of fibrous roots had occurred, accompanied by brown discoloration and dieback on many of the main roots.

Clayton and Haasis (1964) found the same fungus was responsible for the decline and poor growth of Weymouth and Wolcott plants in North Carolina.

Control. Since root rot is usually associated with inadequately drained areas, better drainage will help alleviate the problem in the field. Measures recommended by Zentmeyer, Paulus, and Burns (1962) for the control of phytophthora root rot of avocado should be of value to blueberry growers. Steam or methyl bromide can be used to sterilize cutting beds. They report that Dexon (p-dimethylaminobenzenediazo sodium sulfonate) looks promising in preventing disease development on healthy trees and in arresting early cases of root rot as well as in permitting resumption of root growth. So far as known, Dexon has not been tested on blueberries.

Leaf Spots of Lowbush Blueberry

Leaf spotting of lowbush blueberries (Fig. 67) is still not clearly understood. Some spots are obviously symptoms of fungus diseases such as leaf rust, *Septoria* and *Gloeosporium* leaf spots, and powdery mildew. These fungus-incited leaf spots usually are reduced by the use of fungicides, as recommended for leaf rust control.

The cause of much of the leaf spotting during midsummer, however, is unknown. Such spotting is more severe in the crop years and has appeared in the sprout year during prolonged dry weather or where blueberries are grown on light sandy or gravelly soils (Lockhart, 1961a). Symptoms, as described by Lockhart, include brown blotches or irregular grayish-brown flecks which sometimes form a faint ring pattern with a green area in the center of the spot or a ringspot, or a reddish to dark brown pinpoint spot. Two or more types of spotting may occur on a single leaf, and unshaded leaves show more spotting than those in the shade.

Hilborn and Bonde (1956) investigated the possibility that certain of the leaf spots found on wild lowbush blueberries were caused by viruses.

Figure 67. Leaf spotting on lowbush blueberry. *Above.* Brown spot caused by a species of *Septoria*. *Below, left.* Grayish-brown flecking or faint ring patterns. *Below, right.* Ring spot. (C. L. Lockhart, Canada Department of Agriculture, Research Station, Kentville, Nova Scotia.)

Attempts to transmit a virus by grafting or by juice inoculations to herbaceous indicator plants failed. In 1955 they observed three types of symptoms on *Datura stramonium* L. infected by dodder connected to spotted blueberry plants. Inoculations from a yellowish leaf spot of blueberry caused severe leaf blotching involving the entire tip, those from plants with a brown leaf spot caused circular necrotic spots, and those from blueberry plants with dark ring-like spots caused a necrosis along the midvein. This work has not been confirmed.

Minor Fungus Diseases

Phyllostictina Leaf and Berry Spot. *Phyllostictina vaccinii* was described as a new species in 1947 by Demaree and Wilcox. It causes a foliage disease on various *Vaccinium* species in the eastern and southern United States. Leaf symptoms consist of small circular spots, grayish in the center and surrounded by a zone of brown host tissue. Affected bushes may be partly defoliated near the end of the growing season. Fruit of the rabbiteye blueberry is also attacked. Fruit lesions appear during the preripening period as a hard, dry rot, localized in spots 5 to 6 mm in diameter, grayish and sunken, with numerous black pycnidia in the central region. This fungus is potentially important as a cause of fruit rot.

Gloeocercospora Leaf Spot. *Gloeocercospora inconspicua* was also described as a new species by Demaree and Wilcox (1947). It has been collected from Maryland and North Carolina on highbush and rabbiteye blueberries. The brownish leaf lesions are circular to angular and do not differ greatly in appearance from spots caused by *Dothichiza*. Sporodochia are formed more frequently on the upper than on the lower epidermis.

Stem and Leaf Fleck. Taylor (1958) reported that *Gloeosporium minus* Shear was commonly present in small red lesions on succulent shoots of blueberry in North Carolina. Infections on blueberry stems are indistinguishable from early symptoms caused by *Botryosphaeria corticis*. Leaf symptoms consist of small red flecks. These may not develop further, but young leaves become malformed and puckered. Young leaves and stems are susceptible, but no necrosis is caused the first season. In the second season, however, the bark of shoots heavily infected the preceding year appears rough. *Gloeosporium minus* readily invades dead tissue killed by other agents and sporulates profusely in leaf spots caused by other fungi.

Thelephora. *Thelephora terrestris* Fr. was found in 1946 on year-old nursery plants in New Jersey (Demaree, 1947). The fungus formed irregular 1-inch to 2-inch fruiting structures around the plants at and immediately below the ground level. The structures were loosely attached to the plants and could be broken off easily. The only apparent injury was to leaves in contact with the soil, but the fungus has been reported to cause injury to forest tree seedlings in propagating beds.

Armillaria Root Rot. Blueberries infected with *Armillaria mellea* (Bahl) Quel. have been observed in Michigan (Fulton, 1958b) and Long Island, New York (Varney, unpublished data). In both instances the affected bushes had been planted on recently cleared land. The plants were low in vigor and in various stages of decline, and the characteristic black shoestrings or rhizomorphs were evident on the surface and beneath the bark of larger roots. Fulton inoculated young Jersey plants, and several months later found black rhizomorphs scattered over the root surface and white rhizomorphs in the cortical areas.

Since this fungus disease is rarely encountered, no chemical control measures have been developed. Recently cleared forest land should be avoided.

Others. Several other fungi have been reported as occurring on blueberries. Demaree and Wilcox (1947) collected *Ramularia effusa* Pk. on *V. vacillans* in Maryland and *V. lamarckii* Camp in Maine; *Rhytisma vaccinii* (Schw.) Fr. on *V. atrococcum* Heller in Maryland; and *Cercosporella* sp. on *V. australe* in New Jersey. Wilcox (1936) listed the following as occurring on blueberries: *Helminthosporium inaequale* Shear, *Melanospora destruens* Shear, *Pestalozzia quepini vaccinii* Shear, *Sphaeropsis malorum* Pk., *Dothiorella ribis* Grov. and Dugg., and *Alternaria* sp. According to Demaree and Wilcox (1947), *Septoria difformis* Cke. and Pk. described by Cooke and Peck in 1875 on *V. pennsylvanicum* was actually *Ramularia vaccinii* Pk.

Several fungi not previously recorded from living stems of highbush blueberry were recently found in Massachusetts by Zuckerman (1960a). These were *Rhabdospora oxycocci* Shear, *Epicoccum* sp., *Pezizella lythri* (Desm.) Shear & Dodge, *Macrophoma* sp., and *Karschia lignyota* (Fr.) Sacc. Tentatively identified species were *Septoria vaccinii* E. and E. and *Alternaria tenuissima* (Fr.) Witts. He also collected *Pleospora obtusa* (Fckl.) Hoshn. and two species of *Cladosporium* from dead stems and *Strasseria oxycocci* Shear from a recently killed twig tip.

Silky thread blight caused by *Rhizoctonia ramicola* W. and R. was found on *V. irrigatum* Ait. in Florida (Roberts and Weber, 1952).

During the spring of 1954, *Sclerotinia sclerotiorum* (Lib.) d By. caused a blight of twigs and flower clusters in New Jersey (Varney, unpublished data). Symptoms were similar to those caused by the mummy berry fungus, *Monilinia vaccinii-corymbosi*. This disease has been observed only rarely since 1954.

VIRUS DISEASES

Blueberry Stunt

Distribution. Stunt, the first virus disease known in the cultivated blueberry, seriously threatened the rapidly expanding blueberry industries in New Jersey and North Carolina during the 1940's. Stunt was described in 1942 by Wilcox, who showed that the causal agent was graft-transmissible and presumably was a virus. According to Tomlinson, Marucci, and Doehlert (1950), the disease was first observed in 1928. A possible earlier report of stunt was made by Stevens in 1926 when he described viruslike symptoms in a wild blueberry plant in New Jersey.

Stunt is known from North Carolina to eastern Canada (Goheen, 1953) and west to Michigan. The disease spreads rapidly in North Carolina and New Jersey, slowly in the northern states, and very little, if at all, in Michigan. In a field of 10 acres in North Carolina, stunt increased in one block from 4.7 per cent in 1942 to 47 per cent in 1945 (Demaree, 1946). Neither stunt nor the vector has been reported from the blueberry growing areas of the Pacific Northwest (Boller 1951).

It has been suggested that the stunt virus was introduced at Whitesbog, New Jersey, in blueberry seedlings grown in greenhouses at Washington, D.C. Survey data obtained during the years 1952 to 1956 by Hutchinson, Goheen, and Varney (1960), however, support the view that stunt is indigenous to New Jersey on wild *Vaccinium corymbosum* L., *V. vacillans* Torrey, *V. atrococcum* Heller, and *V. stamineum* L. Surveys in 1953 and 1954 throughout New England, eastern New York, Pennsylvania, West Virginia, and Virginia gave further support to the view that stunt is indigenous to the Atlantic coastal states and was not introduced from outside the country (Hutchinson and Varney, unpublished data). Infected plants of *V. vacillans*, an excellent indicator for stunt, were found as far north as southern New Hampshire as well as in the areas surveyed west and south of New Jersey. Infected *V. myrtilloides* Michx. and the insect vector were found in an alpine meadow in the mountains of New Hampshire near Pinkham Notch. Meader, McCrum, and Rich (1964) concluded that wild blueberries were the original source of infection of a cultivated blueberry in a New Hampshire planting.

Symptoms. Symptoms are variable, differing with variety, time of year, stage of growth, and age of infection. A characteristic symptom on cultivated and wild *V. corymbosum* is a yellowing of the leaves along the margins and between lateral veins (Fig. 68). Normal green is generally retained along the midrib and lateral veins, and affected leaves are often cupped and reduced in size. Chlorotic areas generally turn a brilliant red

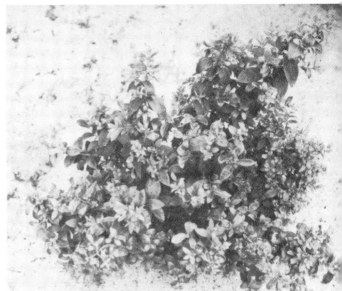

Figure 68. Typical symptoms of stunt disease. *Left.* Twig and leaves of the blueberry variety Atlantic showing chlorosis along leaf margins and between lateral veins, and cupping of leaves. *Right.* Typical growth after a stunt-infected blueberry variety Concord has been cut back. A few leaves are normal in size, and the rest are much smaller and cupped. Note extreme proliferation of lateral shoots on right side of bush. (Eugene H. Varney and M. T. Hutchinson, Rutgers University, New Brunswick, N.J.)

in late summer and early fall, and this symptom is distinctive until the development of normal fall coloration.

Diseased plants live for many years but are much reduced in vigor. Branches may continue to set some fruit, but the berries remain small and of poor quality. If a diseased plant is cut back to the ground the new growth is weak and stunted. Branches are twiggy in appearance as a result of shortened internodes and the growth of normally dormant buds. So far as is known all cultivated varieties are susceptible, but some are more severely affected than others. The Rancocas variety, for example, seldom shows definite symptoms and produces apparently normal yields.

Symptoms on wild species of *Vaccinium* are, in general, like those for highbush blueberry. In addition to the species already mentioned, symptoms have been observed on graft-inoculated *V. amoenum* Ait., *V. altomontanum* Ashe, *V. elliottii* Chap., and one plant of *V. darrowi* Camp infected by means of dodder (Hutchinson, unpublished data). *V. tenellum* Ait., *V. angustifolium* Ait., and *V. caesariense* Mack. appear to be symptomless.

Dodder has been used to infect periwinkle, *Vinca rosea* L. In *Vinca* the virus causes symptoms distinct from those caused by other yellows-type viruses. The leaves are pale green and marked with patches or transverse bands of dark green. Leaf size is not markedly reduced. Flowers of infected plants are smaller and fewer than those of healthy plants, but virescence, or green petals, does not occur.

Transmission. Stunt is readily transmitted by grafting. Wilcox (1942) made bud grafts in September, and the inoculated plants showed symptoms the following spring. Plants grafted at or prior to budbreak may show symptoms the same year, often within two to three months. There is no evidence that the virus can be transmitted by tools used in pruning or in cultivating.

L. O. Kunkel in 1947 (Tomlinson, Marucci, and Doehlert, 1950) was the first to use dodder, probably *Cuscuta campestris* Yuncker, to transmit the virus to *Vinca*. He concluded that stunt was caused by a yellows-type virus and that a leafhopper was the probable vector.

C. A. Doehlert in 1944 used a mixed colony of leafhoppers to transmit stunt from one diseased to four healthy plants (Marucci, Tomlinson, and Doehlert, 1947). In 1950, Tomlinson, Marucci, and Doehlert reported that a complex of *Scaphytopius magdalensis* Prov. and *S. verecundus* Van Duz. was responsible for the transmission of stunt disease. Hutchinson (1955) and Maramorosch (1955) later reported that *S. magdalensis* was a vector and that *S. verecundus* was not. Many other insects have been tested and so far *S. magdalensis*, the sharp-nosed leafhopper, is the only known vector. It occurs along the Atlantic seaboard from Florida to Canada. A report by Suit (1945) of transmission by aphids, presumably *Amphorophora vaccinii*, was never substantiated.

Control. Certification programs, roguing, and vector control have greatly reduced the losses caused by the stunt virus. Boyd (1956) compared the results of 11 years of inspection and certification from 1945 through 1955 and found that there was no significant reduction in stunt after the first two years when there were 4.7 and 5.7 infected bushes per acre. From 1947 through 1955 the level ranged from 1.4 to 2.4 bushes per acre. The certification program initiated in New Jersey in 1945 did succeed in holding the disease in check. Doehlert (1951) also showed that persistent and efficient roguing of an 8-acre field kept stunt under control. In 1943, 13 plants per acre, or about 1 per cent of the plants, were infected. After the fifth year and for four consecutive years thereafter the rate of new infections was held to 0.06 per cent.

Isolation from old plantings with a high incidence of stunt is not enough if wild blueberries are present. It is advisable either to destroy wild blueberries along the edges of commercial fields or to spray them with insecticides. Roguing should be done promptly but only after an insecticide has been used to prevent the dispersal of infected leafhoppers to neighboring plants. Doehlert (1946) concluded that the best method of roguing plants was to chop out the crown with a sharp grubbing hoe and to fill the hole with soil to prevent new shoots from coming up. He tested the weed killers 2,4-D and Ammate and found that the former was too slow in killing treated plants and the latter was too toxic to adjacent plants.

In summary, stunt can be controlled by planting virus-free plants, by conscientious and persistent roguing, and by using on schedule the insecticides recommended by local authorities.

Red Ringspot

Distribution. Hutchinson first described red ringspot disease (RRS) in 1950 but without determining the causal agent. Four years later Hutchinson and Varney (1954) demonstrated that RRS was virus induced. Further work on this disease showed that it was spreading, but efforts to determine the vector have failed (Moore and Stretch, 1963; Raniere, 1959; Stretch, 1963).

The presence of the virus in the wild blueberry has been well established only for New Jersey. The virus is probably indigenous and has been spread from the wild to commercial plantings. Red ringspot has not been reported in other blueberry growing areas except Michigan, where RRS was found in plants whose origin could be traced to New Jersey. No reports of RRS in the wild or in most commercial plantings in Michigan, North Carolina, Massachusetts, Washington, and Oregon may mean that the vector is not present in those areas.

Red ringspot has become a serious problem to the blueberry industry in New Jersey. This problem is concerned primarily with the evaluation of virus-free seedlings and selections and with the production of virus-free nursery stock. The effect of RRS on fruit production has not been determined as yet.

Symptoms. The symptoms of this disease are usually distinct and not easily confused with other diseases. Characteristic red spots and rings, and oak-leaf patterns, generally begin to show on the leaves in late June and July in New Jersey (Fig. 69). The leaf symptoms develop first on the older leaves. The decrease in severity of symptoms from the older to younger leaves is very useful in distinguishing red ringspot. Leaf symptoms can be confused with powdery mildew, but mildew occurs on both leaf surfaces and RRS symptoms occur only on the upper surface. Mildew also does not produce well-defined patterns, as does RRS. The virus does not consistently produce the stem symptoms, red blotches or rings. This is not evidence of absence of disease, however. The red blotches and rings on stems may become obscured by the normal fall coloring of red-stemmed varieties but remain visible all winter on the stems of light-colored varieties. Red ringspot symptoms in the early stages may occur on only one branch.

Varietal Susceptibility. Red ringspot was first described on the Cabot and Pioneer varieties (Hutchinson and Varney, 1954). Burlington, the next variety in which the virus was found, is one of the most susceptible. Raniere (1960) inoculated 13 varieties by means of chip-bud grafts and found that infected Blueray, Collins, Coville, Earliblue, Herbert, Pemberton, Rancocas, and Weymouth exhibited typical ringspot symptoms. Bluecrop showed chlorotic rings instead of the red rings and Berkeley, Jersey, and Stanley did not develop symptoms. Contrary to Raniere's findings,

Figure 69. Typical red ringspot symptoms showing spots, rings, decreasing symptom severity on younger leaves, oak leaf patterns on the leaves (*single leaf at left*), and red blotching of the stem. (Eugene H. Varney, Rutgers University, New Brunswick, N.J.)

Moore and Stretch (1963) reported that RRS occurs on Stanley. Further observations have revealed symptoms of the disease in Dixi, Atlantic, Concord, Rubel, Berkeley, Darrow and in the numbered selections G-3, G-72, and G-80. Berkeley was not infected in Raniere's test. Jersey, Weymouth, and Bluecrop have shown resistance in the field. Jersey appears to be symptomless, but Weymouth and Bluecrop have shown symptoms when graft-inoculated.

Transmission. The mode of transmission of RRS has not been determined. Surveys in connection with vector studies have indicated rapid spread, as noted by Hutchinson and Varney (1954) in Cabot. Since 1957 a survey of an interplanted Burlington-Jersey block has also shown that an active vector is spreading the virus in the Burlington variety. Out of 1,039 plants inspected in the 1957–1962 period and in 1964, 2.5, 5.7, 14.5, 24.0, 28.3, 34.9 and 51.6 per cent respectively were diseased. Further surveys of seedling blocks in New Jersey have shown random spread across all progenies (Moore and Stretch, 1963).

Control. Heat treatment of dormant rooted cuttings at 100 F for varying intervals has not proved effective in killing the virus before killing the plants. Dipping of blueberry cutting wood in hot water at 115 to 140 F at 5 degree intervals for periods of up to one hour did not eliminate the virus before the wood was killed. Attempts to grow excised shoot tips in sterile culture to obtain plant material free of virus failed. This technique works well on herbaceous plants but is difficult with woody plants.

At present the only control that can be recommended is planting of virus-free plants and roguing diseased bushes. Production of virus-free nursery stock has been hampered by lack of knowledge about the vector. A special program to provide virus-free propagating stock is needed. When the vector is determined, it may be possible to reduce or prevent spread through vector control.

Necrotic Ringspot

Distribution. Blueberry necrotic ringspot, formerly known as the Pemberton disease, was first noted by Varney in 1955 on the Pemberton variety on a farm near Taunton Lakes, New Jersey (Varney, 1957b; Raniere, 1960). The viral nature of the observed abnormality was proved in 1956 by Varney by mechanical transmission to tobacco and by grafting to three common blueberry varieties.

Necrotic ringspot has been identified in Connecticut, New York, Michigan, Illinois, and New Jersey. In New Jersey the disease is much more prevalent on heavier and less acid soils than in areas of the outer coastal plain where blueberry and related plants naturally flourish. The heavier, less acid soils usually had a history of multicropping (Raniere, 1964).

Symptoms. This virus typically causes stunting, extensive twig dieback, and eventual unproductivity (Varney and Raniere, 1960). Chlorotic spots, rings, and line patterns show on new leaves, and the chlorotic spots may become necrotic and drop out, giving the leaves a tattered appearance (Fig. 70). When leaf infections are severe, affected leaves are often reduced in size and may be wavy and distorted.

Raniere (1960) graft-inoculated 13 varieties and found that they varied considerably in their reaction to necrotic ringspot virus. Symptoms, which appeared in about one month, were typical of those previously described. A shock symptom characterized by a dieback of the young terminals was also observed. Berkeley, Bluecrop, Blueray, and Rancocas showed only the shock symptoms; Pemberton, Earliblue, and Stanley showed only typical symptoms; and Burlington, Collins, Coville, Herbert, Jersey, and Weymouth showed both types. All the plants were indexed the following year on healthy Pemberton plants and, surprisingly, the varieties that had recovered from all symptoms were also free of virus. These varieties were Berkeley, Bluecrop, Blueray, Herbert, Jersey, and Rancocas.

Transmission. Blueberry necrotic ringspot virus can be readily transmitted mechanically from newly emerging blueberry leaves to a number of herbaceous hosts. In attempts to determine the means of natural transmission the nematode, *Xiphinema americanum* Cobb, was consistently observed in soil samples from several locations in monthly samplings by Raniere (1964). Griffin, Huguelet, and Nelson (1963) transmitted the virus to cucumber seedlings with *X. americanum* collected from roots of infected blueberry plants, but they were unable to get blueberry-to-blueberry transmission.

Investigations by Raniere (1964) indicated that the host range and symptoms exhibited by blueberry necrotic ringspot were similar or identical to those described by Smith (1957) for a strain of tobacco ringspot virus.

Characteristics of the Virus. Isolates of the virus from Illinois, Michigan, and New Jersey differed considerably in their virulence. Of thirty different plant species or varieties tested, the following were found to be most useful as indicators: *Chenopodium amaranticolor* L., *Callistephus chinensis* (L.) Nees, *Cucumis sativus* L., *Phaseolus vulgaris* L. var. Pinto, *Vigna sinensis* (Torner) Savi., *Datura stramonium* L., *Nicotiana rustica* L., *N. tabacum* L. var. Samsun and White Burley, and *Petunia hybrida* Vilm. var. Fire Chief.

Lister, Raniere, and Varney (1963) reported necrotic ringspot virus was inactivated between 62 and 65 C except for one blueberry isolate which was inactivated at 55 to 57 C. Dilution end points have been reported as 1:1,000 to 1:10,000 in Scotland and 1:500 to 1:1,000 in New Jersey. Such differences are probably explained by variations in technique, plants used, and environment. When aged in vitro, the virus remained infective between 18 and 24 hours after extraction in New Jersey.

Figure 70. Stunted growth and dieback on a branch of the Pemberton blueberry infected with necrotic ringspot virus. Note ring and line patterns on the leaves at right. (Eugene H. Varney, Rutgers University, New Brunswick, N.J.)

Lister, however, was able to maintain infectivity between 10 and 14 days at room temperature in Scotland.

Cross protective tests have shown that the various isolates of blueberry necrotic ringspot and known isolates of tobacco ringspot are very closely related (Lister, Raniere, and Varney, 1963). These same isolates are not related to arabis mosaic, tomato blackring, raspberry ringspot, tomato ringspot, and tobacco rattle viruses. Serological tests by Lister confirmed that blueberry necrotic ringspot virus is the same as tobacco ringspot virus.

The purification technique developed by Steere (1956) for tobacco ringspot virus has proved effective in the purification of blueberry necrotic ringspot virus. Raniere (1964) and Lister, Raniere, and Varney (1963) both used this technique to obtain virus material for electron microscopy. They determined that the virus particles were polyhedrons with a maximum diameter of approximately 25 mμ.

Mosaic

Distribution. Mosaic was referred to as "variegation" for several years (Hutchinson, 1950) before Varney (1957a) showed that the symptoms were caused by a virus. Mosaic is common in North Carolina and New Jersey plantings and is likely to occur wherever blueberries, especially the older varieties, are grown. Mosaic has been observed on wild highbush blueberries in areas far removed from commercial plantings in the eastern United States and is presumed to be indigenous.

Symptoms. Leaves of mosaic-affected blueberry are strikingly mottled with yellow and yellow-green areas (Fig. 71). Some leaves are yellow only along the margins of the major veins. The degree of mottling varies with leaf position and blueberry variety. Lower leaves on a twig may be predominantly yellow but the upper leaves may show no symptoms. Mottle on leaves of the Cabot variety is conspicuous at first but frequently fades as summer progresses. In contrast, mottle on leaves of the Stanley variety becomes more brilliant and colorful as yellow areas turn various shades of red and orange-red. Observations suggest that a bush with one or two affected branches has been infected recently. Usually all branches show symptoms in another two to three years, depending on the size of the bush.

Transmission. Mosaic is readily transmitted by grafting. A few leaves on plants inoculated at budbreak by means of chip buds may show typical symptoms in only two to three weeks (Varney, 1957a).

Mosaic is known to spread in the field in North Carolina and New Jersey. In one field in New Jersey the number of mosaic-infected plants doubled in one year (Raniere, 1959). The bud mite, *Aceria vaccinii* Kief., is suspected as a vector because of such circumstantial evidence as

Figure 71. Leaves infected with mosaic virus showing mosaic patterns and symptoms associated with the veins. (Eugene H. Varney, Rutgers University, New Brunswick, N.J.; USDA, Beltsville, Md.)

the association of a high incidence of mosaic with heavy mite infestations. Aphids, which usually do not infest blueberries heavily, are also leading suspects.

Control. All infected bushes should be promptly removed and only certified plants should be used in new plantings.

Shoestring

Distribution. Shoestring is another abnormality that has been known in New Jersey (Hutchinson, 1950) for an undetermined number of years. Varney (1957a) showed that shoestring is also caused by a virus.

This virus disease is common in New Jersey and Michigan and probably occurs wherever cultivated blueberries have been introduced. June, Jersey, Burlington, and Rancocas are varieties on which symptoms are commonly found. Shoestring has also been observed on wild highbush blueberries and is probably indigenous to the United States. Lockhart and Hall (1962) reported shoestring symptoms in Canada on one seedling of the lowbush blueberry, V. *angustifolium*. All other plants of V. *angustifolium* which they tested were symptomless carriers.

Symptoms. Blueberry plants infected with shoestring virus show both twig and leaf symptoms. Red streaks of varying length occur in the spring on new twigs, especially on the surface exposed to direct light. As the summer progresses the streaks become progressively less evident and are finally masked as the wood matures. Red blotches that are difficult to distinguish from those caused by the red ringspot virus may be present throughout the year on some of the branches or canes. Diseased twigs may be scattered among apparently normal branches, and a symptomless twig may occur on a severely diseased branch.

Severely affected leaves are narrow and pointed or straplike and are frequently found at the base of bushes and on shoots from dormant buds on stubs left during pruning (Fig. 72). Moderately affected leaves are wavy and distorted as a result of the abnormal development of the leaf tissue. Affected leaves may be green or dull red. Red vein-banding, a common symptom on both normally shaped and malformed leaves, may be limited to a portion of a vein or may be so extensive that most of the leaf is red. Vein-banding is often limited to the midvein and to the bases of the lateral veins, resulting in a conspicuous oak-leaf type of pattern.

Figure 72. Two naturally infected twigs of the Burlington blueberry showing leaf malformations caused by the shoestring virus. Twig at right is healthy. (Eugene H. Varney, Rutgers University, New Brunswick, N.J.)

In addition to the characteristic leaf and stem symptoms, Fulton (1958b) pointed out that immature fruit develop a premature red to purple cast on the surface exposed to light. This coloring covers part or all of the upper half of the berry, in contrast to the green fruit on healthy bushes. Flowers on affected branches may be deformed or streaked with pink.

Transmission. The disease is readily transmitted by grafting. Plants grafted at budbreak may show symptoms the first year, while plants grafted much later usually show symptoms the following season. Red streaks on the upper surface of new twigs and red vein-banding are the first symptoms to appear. Lockhart and Hall (1962) reported transmission to beans by sap inoculations from the lowbush blueberry showing shoestring symptoms. Natural spread occurs in New Jersey and Michigan, but so far vectors are undetermined.

Control. The only available control measures are roguing and planting of virus-free stock.

Witches'-Broom Virus

In Europe a witches'-broom disease of *Vaccinium myrtillus* has been described by Blattny (1955), Bos (1960), and Uschdraweit (1961). The disease is transmitted by grafting and by a leafhopper, *Idiodonus cruentatus* Panz (Blattny, 1963). The relationship of this virus to blueberry stunt is uncertain.

Virus-like Abnormalities

Abnormal Leaf Shapes. Blueberry seedlings and water shoots of named varieties occasionally are affected by a disorder that superficially resembles shoestring. Leaves are narrow, usually rugose, and frequently marked with areas lacking chlorophyll. This disorder has not been proved to be caused by an infectious agent.

Variegation. Genetic variegations that resemble virus-induced mosaic occur frequently among blueberry seedlings. The difference between mosaic and a true genetic variegation is difficult to describe and is best learned through experience. Genetic variegation is often associated with abnormal leaf shapes. As a rule, seedlings with viruslike symptoms are genetic freaks. So far there is no evidence that the blueberry transmits the shoestring and mosaic viruses through the seed.

Iron Deficiency. Terminal leaves are a pale green, pale or bright yellow, or almost white. The veins remain green and stand out prominently against the lighter background (Hutchinson, 1950). The prominent green veins in iron deficiency provide a marked distinction from stunt.

Magnesium Deficiency. Symptoms are very much like both the spring and fall symptoms of stunt. A diagnostic difference is the occurrence of magnesium deficiency symptoms on the basal leaves of a shoot while stunt symptoms are most pronounced on the terminal leaves. See Chapter VII for a more detailed discussion of iron and magnesium deficiency in the blueberry.

BACTERIAL DISEASES

Crown Gall

Causal Organism. Galls have been known for many years on branches and stems of blueberry plants. As early as 1914, F. V. Coville photographed galls on plants grown in the USDA greenhouses in Washington, D.C.

Gall formation has been attributed to a fungus and a bacterium. Brown (1938) reported that the most common type of blueberry gall was caused by the fungus *Phomopsis*, which she frequently isolated from gall tissue. She was unable to isolate a bacterium pathogenic to blueberry and was also unable to infect blueberries artificially with strains of crown gall bacterium from peach, raspberry, hop, and dahlia. Blueberries inoculated with *Phomopsis* developed "outgrowths" at the point of inoculation, but the symptoms illustrated were more suggestive of cankers than galls. Demaree (Demaree and Smith, 1952a) isolated fungi and bacteria from galls which appeared to be identical with so-called phomopsis galls. None of the fungi produced galls but two bacterial isolates did.

Subsequent morphological and physiological studies of eastern and western strains of the pathogen showed that the nutritional requirements of the former resembled those of *Agrobacterium rubi* (Hildebrand) Starr and Weiss, the cause of cane gall of brambles, while those of the western strain were like *A. tumefaciens* (Smith and Town.) Conn., the cause of crown gall on various plants. Demaree and Smith did not attempt to separate the blueberry pathogens into two species on the basis of a difference in nutritional requirement. They concluded that the blueberry pathogen was merely a new strain or variety of the crown gall organism, *A. tumefaciens*. Like Brown, they were unable to infect blueberries with strains from other hosts (apple and peach). They found no evidence that blueberry gall disease was caused by the fungus *Phomopsis*.

Although the disease is occasionally a problem in nurseries, it is of minor importance in cultivated fields, possibly because of the acidity of the soil in which blueberries are usually grown. Crown gall is more often a problem in relatively alkaline (pH 6.8) than acid (pH 5.0) soils (Siegler, 1938).

Symptoms. Galls occur on branches and small twigs and sometimes at the base of canes near the ground line. They are usually black or dark

Figure 73. Above. Galls on the Cabot blueberry. The root galls (*left*) and the branch galls found at the base of lateral twigs (*center*) are caused by an unknown agent. Crown galls (*three twigs at right*) are larger and are not always associated with buds. *Below.* Bud-proliferating galls caused by *Nocardia vaccinii* occur at or immediately below the ground line. (B. M. Zuckerman, University of Massachusetts, East Wareham; USDA, Beltsville, Md.)

brown, rough, and hard (Fig. 73). Galls vary from globose pea-size structures to large globose or elongated structures that result from smaller galls growing together. Demaree and Smith (1952a) observed that galls from Washington and British Columbia were more often elongated than those from the East. New galls sometimes break through the cortex either above or below old galls. The same investigators found that galls formed only when rapidly growing shoots, young leaves, and buds were injured and then inoculated with the pathogen. No bud proliferation has been observed which distinguishes this disease from tumors caused by *Nocardia* (Demaree and Smith, 1952b).

Control. Sanitation and production of clean nursery stock are the principal means of control. Infected stock should be destroyed promptly. If the presence of crown gall is suspected, pruning equipment should be sterilized with an antiseptic in order to prevent spreading the disease from plant to plant. Zuckerman (1957) tested streptomycin, actidione, and neomycin with negative results.

Root-Gall Disease

In 1956, Zuckerman and Bailey described a new gall disease characterized by gall formation on the roots and stems and by basal stem cankers. The cause of the disease is still unknown. Crown gall is similar to the root-gall disease, and the pattern of spread along the row is about the same for the two diseases. Galls have not been observed, however, on the root systems of plants infected with crown gall (Demaree and Smith, 1952a). The galls produced by the crown gall bacterium are also larger than those that occur on bushes infected with root gall. Branch galls associated with the root-gall disease were always found at the base of a fruiting shoot, whereas crown galls frequently appeared to break directly through the cortex and were not always associated with buds.

The disease has been observed in one field in eastern Massachusetts. Pioneer, Cabot, and Wareham were very susceptible and Jersey, Rubel, and Dixi were very resistant. Symptoms occurred on all woody portions of affected bushes. Galls were found on both main roots and fibrous roots. The young galls were white and coriaceous (leathery) and the old galls were dark brown, woody, and covered with bark. Basal cankers were associated with the galls on about a third of the affected bushes. Small branch galls were frequently associated with the fruiting twigs. These galls occurred at the base of the fruiting twig and appeared to arise from the hypertrophy or hyperplasia of elements associated with the fruiting bud.

Bud-Proliferating Galls

Causal Organism and Distribution. Demaree and Smith (1952b) studied the causal organism in detail and proposed the new name *Nocardia* *vaccinii*. The bacterium resembles *Nocardia minima* (Jensen) Waksman and Henrici but differs in certain cultural characteristics and in its pathogenicity. So far as known this was the first report of a plant tumor caused by a member of the family *Actinomycetaceae*.

This unusual gall disease was first reported in 1947 by Demaree. It was observed at Beltsville, Maryland, as early as 1944 on blueberry seedlings growing in soil composed largely of unsterilized leaf mold taken from a wooded area where huckleberry, *Gaylussacia baccata* (Wang.) K. Koch, was affected by a witches'-broom. It seemed probable that the latter and the bud-proliferating galls were caused by the same organism. The disease is unknown in commercial blueberry plantings.

Symptoms. Galls occur on plants at or immediately below the ground line. No galls have been observed above ground and have been seen only rarely on roots. Demaree and Smith (1952b) were unable to infect stems, branches, or buds. Galls range from $\frac{1}{2}$ to 2 inches in diameter and are first white and soft and later brown to black and hard. Bud proliferation is common. Most of the abnormal buds abort but some grow into weak shoots, forming a witches'-broom effect at the base of the plants (Fig. 73). The characteristic bud proliferation distinguishes this gall disease from crown gall ("phomopsis gall"). The latter is further characterized by galls on stems and branches.

Bacterial Canker

Distribution. Stace-Smith, Wooley, and Vaughan reported in 1953 that a serious stem canker caused by a species of *Pseudomonas* had been observed for the past five years in Oregon. Boller referred to it in 1951 as "blueberry X disease" and said it was also known in Washington. Vaughan (1956) further identified the bacterium as a strain of *Pseudomonas syringae* Van Hall. It would infect *Syringa* species but not native *Vaccinium* species, other ericaceous plants, or either sweet or sour cherry. The disease has not been recorded from other blueberry growing areas.

Symptoms. Symptoms appear on year-old stems, usually in January or early February in Oregon, as water-soaked lesions which rapidly become reddish-brown to black cankers. Cankers may extend from a few millimeters to the length of a cane. The disease can kill young plants and destroy half the year-old stems of older plants. All buds within the canker area are killed. Stems are sometimes girdled, but in other instances vascular tissue is not damaged and buds outside the canker area grow normally.

Varietal Susceptibility. Varieties vary in susceptibility to the disease. Jersey, Atlantic, Scammell, Coville, Evelyn, and N51G are severely affected; Rubel, Pioneer, and Burlington are seldom affected; and Weymouth, June, Rancocas, and *V. australe* × *V. lamarckii* hybrids are highly resistant.

Control. Vaughan and Boller (1954) significantly reduced the severity of the disease with 8–8–100 Bordeaux mixture applied in October and November. Indications were that other copper fungicides are also effective. They found that bases of buds on plants sprayed with copper fungicides turned dark brown to black but that such buds developed normally.

NEMATODES

The nematodes associated with blueberry roots have been investigated only in recent years. Goheen and Braun (1955) were the first to list several parasitic nematodes collected from blueberry roots. The only endoparasite found was the root-knot nematode *Meloidogyne incognita* Kofoid and White in a North Carolina collection. In 1960, Hutchinson, Reed, and Race reported that *Tetylenchus joctus* Thorne, *Hemicycliophora* sp., and *Trichodorus* sp. occurred with sufficient frequency or in high enough numbers to be the possible cause of economic injury to bushes in New Jersey.

Hemicycliophora similis Thorne was associated with the formation of small, terminal root galls and reduced root growth in laboratory experiments conducted by Zuckerman (1964). The economic importance of this nematode in commercial cutting beds and plantings is unknown.

Tetylenchus joctus Thorne parasitized blueberry seedlings and cuttings but did not cause easily recognized root symptoms or consistent reduction in root growth in pot tests conducted by Zuckerman. *T. joctus* and other parasitic nematodes may be important because their feeding wounds may provide a mode of entry for soil fungi and other soil organisms.

Tetylenchus christiei, the stubby root nematode, was shown to be the cause of serious losses in cutting beds in New Jersey (Hutchinson, Reed, and Race, 1960). Affected cuttings callus normally but any roots growing from the callus are attacked by the nematodes. Parasitized roots fail to grow, resulting in poorly rooted cuttings, cuttings with callus and no roots, and eventually dead cuttings (Fig. 74).

Zuckerman in a detailed study (1962) confirmed that root growth of cuttings inoculated with *T. christiei* at the time of setting was significantly reduced but found that the growth in cuttings inoculated after root growth had started varied greatly. He concluded that *T. christiei* would be unlikely to exert an adverse effect on the vigor and productivity of large bushes unless the root systems were already limited by some other factor, such as poor drainage.

Figure 74. Cuttings from an area in a cutting bed heavily infested with stubby root nematodes compared with cuttings from a nematode-free area (*left and right*). (M. T. Hutchinson, Rutgers University, New Brunswick, N.J.)

Xiphinema americanum Cobb, a dagger nematode, has not been shown to have any direct economic effect on blueberries, but it is important as a vector of tobacco ringspot virus, which is the cause of blueberry necrotic ringspot disease.

Losses in cutting beds can easily be prevented by the use of steam or fumigated soil. Sanitary measures must be enforced to prevent contamination of the beds. When vigorous, pest-free plants are put in a field they can usually withstand limited nematode populations. Field fumigation with ethylene dibromide or the dichloropropanes is recommended if a soil check reveals a potential nematode problem (Jenkins, 1961). If established plants are infected, side dressings of dibromochloropropane may be helpful.

CHAPTER 11

Harvesting, Processing, and Storage

by Warren C. Stiles and Dennis A. Abdalla, Department of Plant and Soil Sciences, University of Maine, Orono

Whether blueberries are marketed fresh or in processed form, appropriate harvesting, processing, and storage techniques are essential to insure high quality.

LOWBUSH BLUEBERRY

Harvesting

Lowbush blueberries are harvested commercially in Maine, New Hampshire, Massachusetts, the Canadian Maritime Provinces and Quebec. Hand rakes have been used to harvest this fruit since the early 1880's, and this is still the principal method, although mechanical harvesters now show promise.

Harvest Dates. Massachusetts, New Hampshire and southwestern Maine usually begin raking fruit about July 20. Up Maine's coastal blueberry belt, which extends from east to west and inland from the Atlantic Ocean some 40 miles, the central area begins harvesting approximately August 1. Washington and Hancock Counties, Maine, where the largest production is found, begin some 10 days later. The Maritime Provinces and Quebec follow with harvests commencing about August 20. The total lowbush harvest season encompasses six to seven weeks.

Harvest in a lowbush field does not take place until almost all of the berries are ripe. The lowbush blueberry has the characteristic of remaining on the plant fully ripe until the greener fruit reaches maturity.

Hand Raking. Hand raking technique and supervision of raking crews contribute more to the quality of the finished blueberry product than any other factors (Abdalla, 1963). Proper raking and handling help to eliminate clusters, large stems, leaves, and damaged fruit.

The raking crew may consist of local teenagers and adults as well as migratory workers (Fig. 75a). Training sessions prior to harvest or the day harvest begins are recommended for both old and new rakers. This training should be done by the field foreman with the crew he will supervise during harvest. One foreman should not attempt to supervise more than 40 rakers.

Prior to harvest, blueberry fields are lined off with string in lanes 15 to 20 feet wide, which facilitates raking and supervision. The raker then begins to harvest in a corner of the lane and works across and forward. He utilizes a metal rake to detach the fruit (Fig. 75b). Rakes are available in various sizes from 8 to $15\frac{3}{4}$ inches in width. The $10\frac{1}{2}$-inch (40 teeth) rake is the most popular. The rake is held flat on the ground and moved forward into the blueberry plants. It is then rolled upward and backward to remove the berries, using the bottom edge of the rake as a fulcrum. After the rake is filled, the fruit is poured into a half-bushel field basket. The filled basket is carried to a field winnowing machine and tabulated by a field foreman. This is done by recording the total weight from a scale or punching a ticket to indicate the number of baskets brought in. Payment is then made on a pound basis or by the basket. Presently the average price paid to the raker is 4 or 5 cents per pound or between 90 cents and $1 per basket.

The field winnower (cleaner) can be operated by the rakers themselves, or preferably by one experienced operator (Fig. 75c). As the berries are poured into the winnower a large amount of foreign material is forced out by air through the top of the machine. Acceptable fruit falls downward through a trough and into a square, wooden shipping box with a capacity of 20 pounds. The blueberries are then ready for their journey to the processing plant.

Mechanical Harvesting. The lowbush blueberry is difficult to harvest mechanically because of the woody weed growth, rocks, and generally uneven ground. However, through the use of land-leveling techniques and the development of maneuverable mechanical harvesters, researchers and growers are optimistic.

Two harvesters are now in use. One, designed by the Maine Agricultural Experiment Station, utilizes rotating raking teeth with a vacuum air flow as a raking mechanism (Fig. 76). Blueberries are detached by the rotating raking head, carried by the air flow into a separation chamber to remove foreign material, and then routed onto a conveyer belt. The berries travel on the conveyer to another winnowing machine and then into field lugs. This harvester rakes the plants relatively clean, but problems are encountered in removing foreign material, especially sand.

Figure 75. Harvesting lowbush blueberries. a. A typical lowbush blueberry raking crew in operation. b. Harvesting with a hand rake. The rake is moved forward into the plants, then tilted upward and backward to detach berries. c. Cleaning lowbush blueberries in a field winnowing machine. Green fruit, leaves, stems, and other debris are blown out through the opening directly behind the large pulley. Marketable fruit

is routed into 20-pound shipping lugs at the lower left. d. An experimental mechanical harvester for lowbush blueberries. After berries are detached they are carried upward and emptied into a field box. The unit has nine aluminum rakes attached to an endless chain and is side mounted on a small tractor. (Maine Extension Service, Orono.)

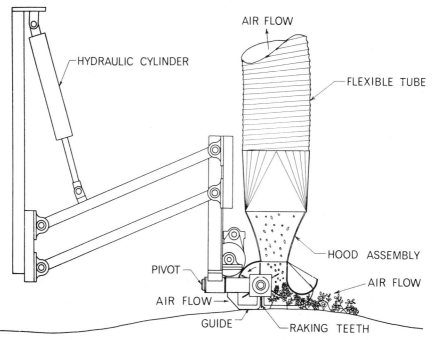

Figure 76. Vacuum mechanical harvesting unit for lowbush blueberries. After berries are detached they enter a separation chamber where leaves, stems, green fruit, and other debris are removed. (R. Rhoades, Maine Agricultural Engineering Department, Orono.)

The other harvester has been designed by a blueberry grower to simulate the hand rake. This harvester employs nine aluminum rakes, each 2 feet wide, attached to an endless chain driven by a gasoline engine (Fig. 75d). The entire unit is side-mounted on a small tractor. As the rakes are drawn through the lowbush plants, berries are detached. Immediately after making a complete pass the rake abruptly makes a 90 degree turn upward. As the rake travels on its overhead ride it inverts and the fruit falls into a lug box.

Both harvesters are promising and could be in general use in the near future.

Processing

Freezing. Ninety-five per cent of lowbush blueberry production is processed, and of this percentage 70 per cent is processed by freezing methods. Freezing is usually done in ultramodern tunnel and flo-freeze type Individually Quick-Frozen (IQF) units (Fig. 77). These units are

Figure 77. A typical tunnel of an IQF (Individually Quick-Frozen) freezer in Maine which can process 5,000 pounds of blueberries per hour. The tunnel is approximately 40 feet long; defrosting doors are located under the tunnel. (Maine Extension Service, Orono.)

Figure 78. Freezing lowbush blueberries by the flo-freeze method. a. Unloading blueberries into a cleaning machine. b. In the air blast room cold air is blown upward through the 6-inch layer of blueberries and "bubbles" the fruit. The aluminum frame and connecting arm agitate

the berries to help keep them from sticking together. c. Pick-over belt for removal of foreign material after freezing. d. Filling a 30-pound paper carton with frozen blueberries. Note polyethylene liners. (Maine Extension Service, Orono.)

capable of freezing 4,000 to 7,000 pounds of fruit per hour. At the height of the season an operating schedule of 22 hours per day is not unusual.

As the berries are brought from the field to the plant, they are weighed and winnowed (Fig. 78a). In a typical tunnel-type IQF freezer the fruit is washed, shake-dried, and conveyed to a three-tier mesh freezing belt. The belt and tunnel are approximately 40 feet long. The temperature in the tunnel is below -25 F and the blueberries are instantly and individually quick-frozen. The berries remain in the tunnel for approximately 17 minutes, after which they enter the recleaning and packaging room at a berry temperature of from -8 to -17 F.

The flo-freeze IQF freezer operates essentially on the same principle as the tunnel-type IQF freezer except that an air blast room about 15 feet long replaces the tunnel (Fig. 78b). After shake-drying, the blueberries enter the room where air blasts at -35 to -40 F "bubble" up through a 6-inch layer of fruit. The berries are agitated slowly and remain in the room approximately 15 minutes, being carried slowly by flotation to an exit trough.

Immediately after freezing, the blueberries enter a "squirrel cage." This rotating metal cylinder is perforated, and as it rotates small and split berries are ejected. The fruit is then picked over for foreign material on a horizontal moving belt (Fig. 78c). This operation exposes the berries for a period of only about 45 seconds after it leaves the freezer. The blueberries are then ready for packaging, storage, and shipping (Fig. 78d).

In addition to IQF freezing, blueberries can be handled for freezing by hand-bulking or machine-bulking. In hand-bulking, field blueberries are emptied onto sorting screens where inferior berries and debris are removed. Sorted berries are then transferred into freezing containers and weighed. With the machine-bulking process, blueberries are first winnowed, then passed onto a horizontal moving belt where workers sort out foreign material and poor fruit. The berries are then routed into the freezing containers.

Three standard wholesale containers are used for packaging frozen blueberries: a 20-pound enameled tin, a 30-pound paper carton with polyethylene liner, and a 60-pound multiwalled paper bag with polyethylene liner (Fig. 79). Almost all the frozen fruit is packaged in 20-pound tins. Upon filling, weighing, and sealing, the containers are immediately transported into a cold storage holding room or onto refrigerated trucks for prompt shipment.

Canning. Blueberries were first canned in the United States in Cherryfield, Maine, in 1866. In 1938 over 80 per cent of Maine's blueberry crop was processed in No. 10 cans (Highlands, 1950). Michigan and New Jersey also can blueberries, but the process is declining in popularity be-

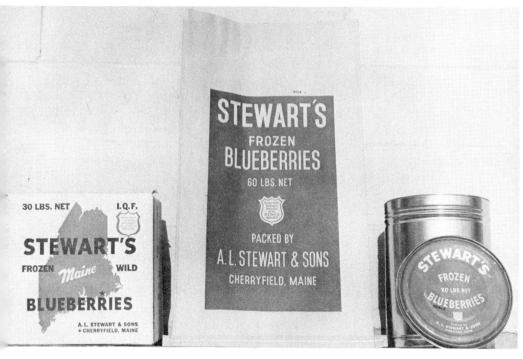

Figure 79. Containers used for packaging frozen blueberries. *From left to right:* 30-pound paper carton with polyethylene liner, 60-pound multi-wall paper bag, and 20-pound enameled fruit tin. (Maine Extension Service, Orono.)

cause of the higher quality and faster packaging provided by freezing. Presently the No. 10 can and the No. 108 muffin-tin water pack are utilized most frequently. Other canned blueberry products include the No. 300 can (pie size) syrup, water, and dietetic pack and the No. 2 can pie-filling mix. These items take their place on the canned fruit shelf and are excellent for making pies and other products any time of the year.

For canning, field blueberries are winnowed and passed onto a de-stemming machine (not necessary for highbush blueberries). This machine breaks up fruit clusters and removes stems. The fruit is then flumed in water to moving inspection belts where foreign material is removed, as in the freezing process. The blueberries then move into the filling machine where either a 22-degree Brix sucrose syrup or water is added. The cans are given the normal exhaust box, capping, cooking, cooling, and labeling process. Twenty-four No. 300 cans, 24 No. 2 cans, 120 No. 108 tins, and 6 No. 10 cans are packed in shipping cartons and sealed.

Freeze-Drying. The blueberry is well adapted to freeze-drying, a recent food-processing development. In freeze-drying, blueberries are first frozen, then placed under vacuum. Sublimation of moisture occurs with the addition of controlled heat in the vacuum. The process produces a finished product containing only about 1 per cent moisture. "Soakback" (incorporation of the blueberries into a product) restores the fruit to 80 per cent of its original weight. This rehydration is very rapid, taking place within a minute. Freeze-drying of blueberries is expensive at present. Estimated costs of a pound of freeze-dried blueberries is $3.50 without overhead and $6 to $7 per pound including overhead costs. One pound of freeze-dried fruit is equivalent to about 8.5 pounds of fresh blueberries. Unless the blueberries are processed under a very high vacuum or in a nitrogen atmosphere, off flavors can develop after six months in storage. With further refinements and reduced production costs, freeze-drying could become a major processing method for blueberries.

Dehydration. A recent development by the Eastern Regional Processing Laboratories called "puff-drying" shows promise for processing blueberries. Blueberries are dried in a controlled atmosphere to a moisture content of 55 per cent. They are then placed in a cereal puffing gun, where the temperature is increased and the gun opened quickly. The steam present "explodes" the fruit. The blueberries are then finish-dried in controlled atmospheres, producing a product containing 5 to 6 per cent moisture. When rehydrated, however, the appearance of the berries is sometimes poor. This new development probably will replace earlier oven dehydration methods because of its speed.

Storage

Cold storage and standard warehouse storage is used for keeping processed blueberries until ready for market. If fruit quality is high and processing and packaging is done properly, the blueberry can be kept for several years in proper storage with little or no loss in quality.

Cold Storage. Frozen blueberries are usually held in cold storage at a temperature of 0, ± 5 F. Unless control mechanisms are extremely accurate, it is hard to keep the temperature at exactly 0 F, at which blueberries usually store very well. The optimum storage temperature is -10 F, but a partial vacuum equivalent to 2 to 3 inches of mercury is required, which is rather expensive to maintain.

The most important factor in successfully storing frozen blueberries is proper sealing of the packaging container. This eliminates air contact with the fruit and also prevents "freezer burn," or dehydration. If sealing is improperly done, moisture can be pulled from fruit by the evaporator coils in the storage room.

Care must also be taken in stacking containers in cold storage rooms. The 20-pound tin, except the locking type, should not be stacked more than five high unless a cardboard separator is used between every fifth tin. If the locking type is used the tins can be stacked up to 14 high. The 30-pound paper box can usually be stacked seven high before crushing

Harvesting, Processing, and Storage

and sagging results. Some problems can occur when stacking the 60-pound multiwall paper bag. Normally they are stacked 10 high, but they can shift as much as 4 feet to one side upon settling.

The 60-pound bag may show evidence of "leakage" inside the polyethylene liner in storage, and so occasionally will the 30-pound box. The cause of this leakage is unknown, but research is under way. This problem is the exception rather than the rule.

Refrigerated trucks used for transporting frozen blueberries are also kept as close to 0 F as possible.

Warehouse Storage. Canned blueberries can be stored up to several months in warehouse temperatures of 65 to 70 F. Long-term storage, up to 18 months, can be best accomplished at temperatures of 45 to 50 F. At 35 F "sweating" of the cans may occur and corrosion may be evident. After 18 months of storage pinholes may appear in the cans, depending on the type of cans used.

HIGHBUSH AND RABBITEYE BLUEBERRIES

Harvesting

Cultivated highbush blueberry plantings generally reach full production within six to 10 years after planting (Beckwith, Coville, and Doehlert, 1937); however, a reasonable commercial crop may be harvested four or five years after planting. In Michigan, Johnston (1959) reported that on the average 400 to 800 pints per acre may be harvested the third season and 1,400 to 2,000 pints the fourth summer. Average yields in the eighth year should range from 4,000 to 7,000 pints per acre. In New Jersey, Beckwith, Coville, and Doehlert (1937) reported that the average yield per acre was about 2,500 pints, although yields of 5,000 pints had been obtained. The average yield from a mature bush is between 4 and 8 pints. Moderate pruning and proper cultural practices can increase the yield to as much as 25 pints per bush.

Cultivated highbush blueberries are harvested by hand picking and with mechanical harvesting aids. The trend toward the use of mechanical aids is increasing rapidly, mainly because of a shortage of labor and increasing labor costs.

Harvest Dates. The harvest season of cultivated blueberries depends on a number of factors. The most important are the variety and the climate of a particular production area. The harvest season begins about May 20 in North Carolina, June 20 in New Jersey, July 1 in Oregon and Washington, and July 10 in Michigan. The use of early, midseason, and late-maturing varieties in most areas, along with specific cultural practices, provides the market with fresh berries over a period of approximately three months. In Michigan, for example, bushes are pruned lightly and produce smaller berries because this practice helps to delay the harvest and avoid competition with the eastern crop.

Prices received for the later Michigan crop are not affected greatly by the smaller berry size. A harvest season normally lasts from six to eight weeks at any one location, so that some overlap between the different areas does occur (Table XIV).

Rabbiteye blueberries are harvested in Florida from late May or early June for 10 to 12 weeks (Mowry and Camp, 1928).

Hand Picking. Hand picking is done mainly by migratory workers and some locally hired pickers, often women and children (Fig. 81a). The quantity of berries harvested by a picker depends partly on skill but to a greater extent on berry size. The average picker can harvest 60 to 80 pints of berries per eight-hour day. Highly efficient pickers working on well-loaded bushes obviously harvest more berries. Figures presented by Baker and Butterfield (1951) indicate that about two pickers per acre are needed early in the season, eight at the harvest peak, and about two during the latter part of the season.

The harvesting operation must be well organized for efficient supervision. The owner or foreman may supervise the entire operation. The harvesting operation may be broken down further to provide for pickers, row bosses, field checkers, and head checkers. In this type of organization the pickers harvest the berries and assemble them at a central collecting point in the field. Row bosses are responsible for maintaining quality and for regulating the number of pickers in the crew.

Baker and Butterfield (1951) have outlined the organization of a typical bushberry harvesting operation in California involving the various job assignments mentioned (Fig. 80). Pickers are usually paid on the basis of

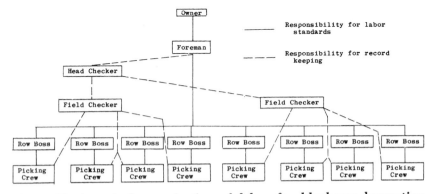

Figure 80. A typical organization of labor for blueberry harvesting. (Baker and Butterfield, 1951.)

the quantity of berries picked. This requires the keeping of records to protect both the employer and the pickers. In most areas, field checkers are assigned to perform this task. The field checker may maintain a card record that shows how many berries each laborer has picked, or he may distribute tickets to the laborers as they deliver the full picking crates to the collecting center. These tickets are turned in daily or weekly, then counted and recorded (Fig. 81b). Records also may be kept by using printed tags bearing the grower's name and address. As the picker delivers berries, his card number is punched with the grower's individual punch, the design of which may be changed from day to day.

Sometimes the checking operation extends beyond the field and a head checker is given the responsibility of tabulating field records and checking them against the quantity of berries being shipped. A head checker may not be necessary in smaller operations.

Cultivated highbush blueberries do not ripen evenly throughout the cluster and must be picked several times during the harvest period. In most commercial operations it is considered necessary to pick over the fields about once a week. Depending on the variety and climate, the number of pickings in any one year will vary from three to seven. Blueberries, unlike some other small fruits, can remain on the bush for a considerable time after ripening.

The size of blueberries at harvest seems to be governed mainly by moisture conditions if nutrition is adequate. The larger berries are obtained from the first two pickings. In general, there seems to be no definite pattern of ripening within the clusters. The Jersey variety has more mature fruit in the middle of the clusters at the first picking, but ripe berries subsequently are more evenly distributed (Shoemaker, 1955). The fruit in late-maturing clusters tends to mature more evenly and to require fewer pickings.

Under normal moisture conditions, berries left on the bushes for periods of up to 10 days after ripening do not decrease in size. The blueberries should be picked only when dry, or the bloom on the surface may be lost and molds encouraged to develop.

There are various types of picking equipment and procedures. One general method involves picking directly into the market container, which has the advantage of minimum handling and better preservation of natural bloom. In this method the picker may carry the market box in a small container, such as a metal can strapped to his waist. Filled boxes are placed into a field box which usually holds 12 pints. These field boxes may be taken to the central pickup area by the picker, or they may be picked up by workers assigned to this task alone, leaving the pickers to continue working. Field boxes are then loaded onto trucks containing pallets for easy unloading at the packing shed (Fig. 81c). Pallets are

Figure 81. Harvesting highbush blueberries in New Jersey. a. A picking crew in the field. b. Tickets being distributed to pickers for tabulation of the harvest. c. Field boxes filled with 12 pints of blueberries are

transported on pallets to the packing shed. d. Processing for the fresh market. (Norman F. Childers, Rutgers University, New Brunswick, N.J.)

unloaded by hand-operated fork lifts and transferred to the capping and crating lines. If desired, mechanical grading precedes capping (Fig. 81d).

A second picking method is to place the fruit into a small pail fastened to the picker's waist. The pail's capacity should not be over 5 quarts. Although more bloom may be lost by this method, undesirable berries and debris can be removed from the harvested fruit as it is poured into the market containers. Pails or boxes strapped to the picker's waist enables a picker to use both hands. In picking, the hands are cupped under the cluster, and ripe berries are removed with slight pressure of the thumbs moving over the cluster. Pickers are cautioned to harvest only ripe berries, avoiding immature fruit, leaves, and stems. When berries are picked directly into market boxes most of the grading responsibility rests on the harvester. Under these conditions only competent, dependable pickers should be hired, and they should be supervised closely to maintain quality.

Mechanical Harvesting. Mechanical harvesting aids now provide a faster, cheaper, and more complete harvest of cultivated highbush blueberries. Thousands of pounds were, and still are, lost yearly because of labor shortages in hand picking. Thus, the mechanical harvester has come into common use in many areas. In Michigan approximately 65 per cent or more of the crop is harvested mechanically. Adoption of mechanical aids in New Jersey and North Carolina has been somewhat slower.

The value of mechanical harvesting aids has been demonstrated by Heddon, Gaston, and Leven (1959) in Michigan. Various methods of detaching fruit were tested (Table XXVII). These included picking by hand and immediately dropping into a large collecting unit, striking stems with a rubber hose to knock berries into a collecting unit, and detaching fruit with a hand-held mechanical vibrator.

The mechanical vibrator has come into common use (Fig. 83). The vibrating unit is an electric hoe converted into a picking head. Two types of vibrators are available; one requires only one operator, the other a crew of four. The crew of four is teamed in pairs between two rows, each pair harvesting one side of each row. The central unit of this harvester is a self-propelled air compressor which is connected by long hose lines with four picking units. Each of the four picking units has four metal "fingers" about three inches long vibrated by air pressure. The fingers are placed on the branch, generally above the fruit cluster, and held firmly in such a way that the vibration is transmitted to the clusters. The ripe blueberries are shaken off and drop into large canvas catchers held above the ground. From the catching frames, fruit is transferred to field lugs and taken to the cleaning shed.

Several larger row types of units are now under development in Michi-

Table XXVII. Comparison of Harvesting Results Using Various Combinations of Equipment in a Commercial Planting of Highbush Blueberries in Michigan

Picking Method and Equipment			Berries Picked by an Adult Worker in 1 Hr [a] (lb)	Berries Picked by an Adult Worker in 8 Hr (lb)	Time Required to Pick 1,000 Lb in 8 Hr (man days)	Total Labor Cost per Lb [b] (cents)	Total Equipment Cost per Lb [c] (cents)	Theoretical Cost per Lb of Picking [d] (cents)
Separation	Collection	Cleaning						
By Hand	4½-pt pail	tip board	9	72	13.9	8.0	0.00	8.00
By Hand	flair pail	tip board	12	96	10.4	6.0	0.02	6.02
Pick and Drop	one-bush unit	tip board	16	128	7.8	4.5	0.23	4.73
Rubber Hose	two-bush unit	air blast	23	184	5.1	3.1	0.43	3.53
Mechanical Vibrator	two-bush unit	air blast	28	224	4.6	2.5	1.00	3.50

Source: Heddon, Gaston, and Leven, 1959.

[a] The fruit was weighed after it had been cleaned.
[b] The amount of money made by an average picker ($5.76) divided by pounds picked in 8 hours.
[c] Annual cost of equipment divided by number of pounds harvested in 30 days, plus 10% (4% interest, 4% taxes, and 2% insurance).
[d] These figures include both labor and equipment costs. Labor costs are calculated at 72 cents per hour—the average amount earned by those who picked by hand. Equipment costs include depreciation, maintenance, interest, taxes, and insurance.

Figure 82. Harvesting highbush blueberries at the plantation of Elizabeth C. White at Whitesbog, New Jersey, in the 1930's. (Whitesbog, Inc., Whitesbog, N.J.)

gan. One typical harvester utilizes two hydraulically operated rods approaching the blueberry bush from each side of the row (Fig. 83). These rods are 18 inches above the soil surface and clamp onto the central trunk of the bush. The rods are vibrated to detach the blueberries, which in turn fall into a canvas catcher and then are conveyed by belt into a field box.

Most growers prefer hand picking for their first and second picking since the mechanical unit tends to remove green berries that can ripen into marketable fruit. On later pickings, however, the mechanical picker is much more efficient. A four-man crew with the mechanical picker is able to match the output of 30 or 40 hand pickers (Reeves, 1962). Experienced hand pickers earn about eight cents per pound. The mechanical unit is operated at 3.5 cents per pound.

Fruit that is harvested mechanically requires additional cleaning before packaging and is best suited for processing. The mechanized cleaning

Figure 83. Above. Hand-held mechanical harvesting aid for highbush blueberries. Vibrating "fingers" are placed on stem to detach berries. Fruit falls into catching frame and is transferred to field boxes. *Below.* Experimental highbush mechanical harvester straddles a blueberry row in Michigan. Hydraulically operated rods reach in toward the plant and detach fruit by vibration. Berries are collected and routed into field boxes as shown on right side of machine. (H. P. Gaston, Michigan State University, East Lansing.)

Figure 84. Field-run Burlington blueberries prepared for storage in field boxes containing 12 pints each. *From left to right:* an open box, a box covered with a perforated polyethylene bag unsealed, and a box covered with a sealed 5-mil polyethylene bag. The 5-mil polyethylene bag gave the best results up to six weeks. (H. W. Hruschka and L. J. Kushman, USDA, Beltsville, Md.)

shed contains a pneumatic winnower (cleaner). The fruit is poured into the winnower's hopper and chaff, leaves, and lighter berries are blown up through the main duct of the cleaner and removed. Quality fruit is routed through a duct at the bottom of the cleaner and onto a horizontal moving belt. Here workers pick over and cull any undesirable berries. The remaining blueberries are delivered to a filling tray, where they drop into containers.

Processing

Although the majority of highbush blueberry production is utilized as fresh fruit, all of the processing methods discussed for the lowbush blueberry are used also for the highbush blueberry. The extent to which the highbush blueberry fruit is processed depends upon the factors of supply and demand for the fresh fruit.

Storage

High quality fresh fruit may be stored up to four weeks at 32 F with some loss in quality (Chandler, 1942; Bünemann, Dewey, and Watson, 1957). At 85 per cent relative humidity and 40 F the acceptable storage time of packaged fruit was reduced to one week, and when the temperature was held at 72 F, quality deteriorated after two days (Hruschka and Kushman, 1963).

In other tests by Hruschka and Kushman (1963) in New Jersey, field-run Burlington blueberries in field boxes containing 12 pints were stored at 32 F and 85 per cent relative humidity. In each box the 12 pints were either open, enclosed in a 1.5-mil polyethylene liner, 32 × 18 inches, perforated with 20⅛-inch holes, or enclosed in a tightly sealed 1.5-mil polyethylene bag with six ½-inch gussets (Fig. 84). Fruit under these conditions were stored for periods of up to seven weeks. The sealed liner was more valuable in checking weight loss and maintaining turgidity, appearance, and flavor for up to six weeks. After that time off-odors and flavors developed.

CHAPTER 12

Economics and Marketing

by **Frederick A. Perkins,** *Department of Agricultural Economics and Marketing, Rutgers University, New Brunswick, New Jersey*

ECONOMICS

National Value of Blueberry Production

The value of production from tree fruits, nuts, grapes, coffee, berries, and various small fruits grown in the United States increased approximately 44 per cent in a decade, from $977,244,708 in 1949 to $1,407,023,301 in 1959 (Table XXVIII). During this period, the value of

Table XXVIII. Value of United States Fruit, Nut, and Berry Production, 1949, 1954, and 1959

	Value of Production		
Product	1949	1954	1959
All Fruits and Nuts [a]	$977,244,708	$1,204,491,196	$1,407,023,301
Berries and Small Fruits			
Blackberries, Dewberries	$ 3,133,503	$ 3,366,188	$ 2,921,219
Blueberries	4,537,883	9,521,720	11,391,371
Cranberries	6,478,008	12,132,366	11,039,135
Raspberries	11,309,762	11,004,016	9,384,112
Strawberries	50,088,534	59,474,074	74,610,140
Other Berries	3,850,504	4,621,387	2,784,667
Total, Berries and Small Fruits	$ 79,398,194	$ 100,119,751	$ 112,130,644

Source: United States Department of Commerce, Bureau of the Census, *United States Census of Agriculture* for 1949, 1954, and 1959. United States Government Printing Office, Washington, D.C.

[a] Includes tree fruits, nuts, grapes, coffee, tame berries, and small fruits.

blueberry production increased some 151 per cent ($4,537,883 to $11,391,371) and far exceeded changes in the value of other fruits, nuts, and berry crops. In 1959, returns from the blueberry crop represented slightly more than 10 per cent of the United States total for berries and small fruits.

Value of Blueberry Production by States

In recent years, the state of New Jersey has ranked first in the value of blueberry production (Table XXIX). In 1959 the New Jersey crop was

Table XXIX. Value of United States Blueberry Production, by States, 1949, 1954, and 1959

State	Value of Production [a]		
	1949	1954	1959
New Jersey	$2,037,149	$3,744,000	$ 4,216,230
Michigan	434,609	2,560,254	2,841,456
Maine	1,510,083	2,066,873	1,971,925
North Carolina	86,978	263,090	973,402
Washington	161,431	179,492	537,323
Massachusetts	28,328	232,037	290,698
New Hampshire	103,856	77,840	129,934
Indiana	22,360	80,861	90,551
Maryland	26,857	65,949	65,355
Oregon	3,571	34,770	63,213
Other States	122,671	216,554	211,284
Total Value, United States Production	$4,537,883	$9,521,720	$11,391,371

Source: United States Department of Commerce, Bureau of the Census, *United States Census of Agriculture* for 1949, 1954, and 1959. United States Government Printing Office, Washington, D.C.

[a] Includes both lowbush and highbush blueberries; processed and fresh.

valued at $4,216,230 and accounted for approximately 37 per cent of the total farm value of United States blueberry production. Production in the second-ranking state, Michigan, was valued at $2,841,456 and represented approximately 25 per cent of the total. Maine was the third-ranking state in 1959 with a crop valued at $1,971,925, which accounted for approximately 17 per cent of the total. The three states of New Jersey, Michigan, and Maine accounted for slightly more than 79 per cent of the $11,391,371 value for all blueberries produced in the United States in 1959. The remaining 21 per cent was distributed among several other states. As can be seen from Table XXIX, the rate of increase in the value of blueberry production varied considerably by states.

Production Trends

Domestic. Blueberry production in the United States has approximately doubled from 16 million quarts to 32 million quarts in the 10 year period from 1949 to 1959 (Table XXX). Maine led in total production with approximately 10 million quarts produced in 1959 followed by New Jersey, Michigan, North Carolina, Washington and Massachusetts. During

Table XXX. Major Blueberry Producing States Showing Acreage, Production, and Yield per Acre, 1949 to 1959 [a]

State [b]	1949	1954	1959
United States			
acreage	30,881	42,812	43,094
quarts	16,049,969	30,643,102	32,681,901
qt/a	520	716	758
Maine			
acreage	22,440	26,500	24,970
quarts	8,843,295	11,482,631	10,112,439
qt/a	394	433	405
New Jersey			
acreage	2,674	4,879	6,030
quarts	4,526,976	8,706,975	8,432,460
qt/a	1,693	1,785	1,398
Michigan			
acreage	1,731	4,167	5,003
quarts	1,086,521	6,737,510	7,892,938
qt/a	628	1,617	1,578
North Carolina			
acreage	568	1,121	1,895
quarts	334,533	1,052,365	2,163,110
qt/a	589	939	1,141
Washington			
acreage	207	434	619
quarts	448,424	543,915	1,885,348
qt/a	2,166	1,253	3,046
Massachusetts			
acreage	155	2,295	1,326
quarts	55,547	580,092	618,507
qt/a	358	253	466

Source: United States Department of Commerce, Bureau of the Census, *United States Census of Agriculture* for 1949, 1954, and 1959. United States Government Printing Office, Washington, D.C.

[a] Highbush blueberries, except for New England states, which include both highbush and lowbush.

[b] No production data are available for the rabbiteye blueberry producing states.

Economics and Marketing

this same period the total blueberry acreage in production in the nation increased from approximately 31 thousand acres to 43 thousand acres. This increase in acreage occurred principally in the highbush blueberry producing states of New Jersey, Michigan and North Carolina.

Current Trends. Recent statistics compiled from various industry sources indicate that national blueberry production has continued to increase since 1959 (Table XXXI). Average production data from the five major producing states in the United States for the years 1960, 1962, and 1964 (50,167,751 quarts) show an increase in blueberry production of nearly 65 per cent over 1959 estimates for the same states as reported by United States Census Bureau information. The value of blueberry production grown on 42,020 acres in 1964 for the five states is estimated at approximately $16,319,868.

Table XXXI. Production, Acreage, and Value of Production in Major Blueberry Producing States, 1960, 1962, and 1964

State	Production (quarts)			Acreage 1964	Value of Production 1964
	1960	1962	1964		
Maine	14,224,281	20,188,078	14,575,162	21,800	$ 2,888,068
New Jersey	16,940,000	12,584,000	13,068,000	8,100	5,257,000
Michigan	14,284,103	12,196,297	14,866,666	8,000	5,200,000
North Carolina	3,263,333	3,996,667	4,400,000	3,500	2,311,800
Washington	1,916,666	1,766,667	2,233,333	620	663,000
5-State Total	50,628,383	50,731,709	49,143,161	42,020	$16,319,868

Source: Computed from various industry releases. Pounds converted to quarts by dividing by 1.5.

Imports. Large quantities of lowbush blueberries are grown in Canada and are marketed in the United States. Canada supplied more than 90 per cent of the 10,205,000 pounds of blueberries imported by the United States in 1962. Other sources of imports include Poland, France, and Yugoslavia. Almost all of the blueberries imported have been either processed or are destined to be processed.

Utilization of United States Blueberry Crop

Fresh Market. During the period 1959 to 1962, an estimated total of 35 per cent of the United States blueberry crop (approximately 15,800,000 quarts per year) was utilized in the fresh form and nearly 65 per cent (approximately 29,000,000 quarts per year) was processed. Of the portion of the crop processed, about 52 per cent was frozen, 28 per cent was canned, and 20 per cent was used for canned pie filling (Isidro, 1964). In Maine, approximately 99 per cent of the 1960-to-1962 blueberry crop was processed into canned or frozen forms (Abdalla, 1964b). As much as 60 per cent of the Michigan blueberry crop and about 75 per cent of Washington's production have been processed in recent years. In contrast, North Carolina blueberries are marketed largely through fresh market outlets, and in New Jersey approximately 25 per cent are processed and 75 per cent are sold through fresh market channels (New Jersey State Department of Agriculture, Crop Reporting Service, 1963).

Processed. The freezing and canning season for blueberries generally begins somewhat later than the appearance of blueberries on the fresh market. It extends from about mid-June in New Jersey to October in Washington. The Canned Food Pack Statistics issued by the National Canners Association (National Canners Association, 1962) show that in 1962, 570,560 cases of blueberries were canned. The greatest volume canned in one year, however, was in 1951, when 1,155,425 cases were packed. The most popular can sizes were small (No. 300 and 303), followed by the institutional size (No. 10) used by bakeries and institutions. The volume of United States blueberries frozen, as reported in Frozen Pack Statistics issued by the National Association of Frozen Food Packers (National Association of Frozen Food Packers, 1962) has increased substantially over the years, from 1,716,516 pounds in 1942 to 26,452,413 pounds in 1962. Slightly more than 4 per cent of the 1962 frozen pack was in packages weighing 1 pound or under, 2 per cent was in other small units weighing up to 10 pounds, and 94 per cent of the output was in the larger institutional packs.

Consumption

During the 20 years 1942 to 1962, the United States per capita consumption of all fruits, measured on a combined fresh and processed basis, fluctuated around 200 pounds annually. The trend in fruit consumption has been toward increased use of canned and frozen fruit and juices with decreases in the consumption of most fresh and dried forms. The total consumption of blueberries in the United States from domestic and foreign sources has increased substantially over the years and in 1962 was estimated at 85.6 million pounds (Isidro, 1964). The average blueberry consumption per person in the United States is 0.46 pound a year. This is only slightly less than the per capita consumption of cherries, nectarines, and pineapples in recent years.

A study of blueberry consumption using a Michigan State University consumer panel as a source of information, indicated that blueberry consumption tends to be highest in population areas near large producing areas (French, 1957). Home consumption of blueberries and blueberry products by the Michigan panel was reported to have averaged about 1.4 pounds per person a year from 1952 to 1956, more than three times the United States average. An average increase in consumption of 0.1 pound per person per year was reported. For the years 1952 and 1953, participating panel members reported consuming 0.82 pound of fresh blueberries, 0.6 pound of commercially processed pie, 3.5 pounds of canned berries, and 0.04 pound of frozen berries.

Information obtained from Michigan panel studies in 1958 (Shaffer, 1960) showed an average annual per capita expenditure on blueberries of 37.8 cents. This expenditure was estimated to represent 1.54 per cent of the total amount spent for all fruits and 0.14 per cent of the annual expenditure for all food purchased by the Michigan panel members.

MARKETING

Grades and Standards

Fresh Fruit. The first official grade standards for fresh-market, cultivated blueberries were issued in 1930 by the New Jersey Department of Agriculture. Little or no use was ever made of these specifications, however. Early New Jersey grade standards were revised in 1959 and new specifications covering both fresh and processed blueberries became effective for Garden State berries that year.* While these specifications were made available for use, they have not been widely used on a continuing basis by New Jersey producers or their marketing organizations (Edmonston, 1962).

Federal standards for fresh cultivated blueberries were proposed by the Department of Agriculture in 1964 for the first time (May 22 issue of the Federal Register). The standards are intended to provide a means of describing quality for all handlers and consumers of fresh blueberries, as well as serving as a reference point for the development of further specifications.

Several of the larger cooperative marketing associations that handle most of the cultivated blueberries sold in the United States use their own grade classifications. Quality, probably the most important factor in respect to market appeal, is frequently designated by the use of private labels or trademarks, such as Crown, Harvest Moon, Green Leaf, Star, Great Lakes, Lake State Brand, and Bluecrest. Size of berries plus factors such as firmness, color, bloom, ripeness, and freedom from foreign materi-

* Copies of blueberry grade specifications are available from the U.S. Department of Agriculture, Agricultural Marketing Service, Washington 25, D.C., and the N.J. Department of Agriculture at Trenton, N.J.

als and from damage by disease, handling, and moisture are important factors considered in labeling packaged berries. The standard used for measuring berry size is the number of berries required to fill the 2-gill, or ½-pint, cup. Sizing symbols vary from one producing area to another. For example, in New Jersey and North Carolina the Tru-Blu Cooperative uses the following grade sizes:

Crown—not more than 80 berries per cup.

Harvest Moon—not more than 130 berries per cup.

Green Leaf—not more than 190 berries per cup.

Star—not more than 250 berries per cup, or marketable berries of Crown, Harvest Moon, or Green Leaf size, but not perfect in quality because of overripeness, rain crack, loss of bloom, or any other cause.

Of these grade sizes, Crown, Harvest Moon and Green Leaf are marketed under a brand name, while the Star grade is marketed without a brand name.

In Michigan, 95 per cent of the crop is marketed through the Michigan Blueberry Growers Association Cooperative. Their grade sizes are:

Great Lakes—60 to 175 berries per cup.

Lake States—175 to 250 berries per cup.

These organizations may also provide additional controls and specifications on how the berries shall be packed or handled. USDA market reporters frequently classify sizes as extra large, large, medium or small. Blueberries not handled by cooperatives may or may not be graded. Frequently, individual growers have their own brand or grade names which they display on their finished packages.

Processed Fruit. Federal grade standards based on appearance and freedom from undesirable materials have been established for canned and frozen blueberries. Grades are specified according to point scores. A total of 100 possible points can be granted to any one particular lot of blueberries. This score is divided into three factors: color, 20 points; absence of defects, including freedom from damage or harmless materials and undeveloped berries, 40 points; and character, including firmness and freshness, 40 points. To be designated U.S. Grade A or U.S. Fancy, the lot must receive a score of 90 or more for the three categories. The score required for U.S. Grade B or U.S. Choice is 80 to 89 points, for U.S. Grade C or U.S. Standard is 70 to 79 points, and for U.S. Grade D or Substandard is 69 points or less.

Canned blueberries must meet specific drained weight requirements in relation to can size.

Packaging

Whether blueberries are harvested by hand or mechanically, careful handling is necessary to protect the berries from bruising and to maintain the powdery bloom on the berries, which adds to their attractiveness. After harvesting, the blueberries are transported from the fields to packing plants where they are cleaned, sorted, and packaged for market. Procedures used to pack the berries vary considerably depending upon the extent to which mechanized equipment is employed. In many plants the berries are handled entirely by hand, in others the operation is mechanized, and in some plants various combinations of the two techniques are used. A few packers utilize newer developments such as mechanical cleaners, sizers, and automatic filling devices to aid them in sorting and packing their berries.

Containers and Capping Films. The universal consumer package for fresh blueberries is the pint-sized container which contains at least 14.5 ounces of fruit (Fig. 85). Wood-veneer and molded pulpboard cups topped with a sheet of cellophane are the two most common types of containers used in the fresh market, although wax chipboard containers are also being used. More recently, plastic mesh and a paperboard box featuring an open-mesh top and side windows have been introduced to the trade (Perkins, 1964). Berries for processing are put in larger boxes or cans.

Four types of pint-sized containers for marketing were tested in North Carolina in 1958 (Hruschka and Kushman, 1963). These included wood veneer, molded pulp, plastic mesh, and waxed chipboard. The conclusion was that where decay is not a problem and weight loss is apt to be severe, the waxed chipboard container with the capping film should give greater protection; otherwise there was little difference in storage quality.

With the advent of prepackaged fruits and vegetables, many new wrapping films are now available. Several of these films can be used to overwrap fresh blueberries to help maintain quality. The new films have

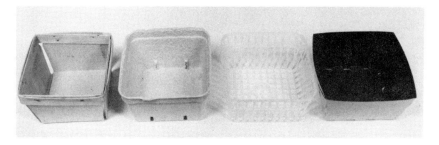

Figure 85. Four types of pint containers used in packaging highbush blueberries. *From left to right:* wood-veneer, moulded pulp, plastic mesh, and waxed chipboard. Boxes are capped with transparent film. (H. W. Hruschka and L. J. Kushman, USDA, Beltsville, Md.)

the characteristic of being partly or fully moisture proof. In tests conducted by Hruschka and Kushman (1963) with Croatan blueberries in North Carolina, it was found that weight loss could be considerably altered by the type of capping film used (Table XXXII). However, with certain films weight loss is kept so low that moisture condensation on the film obstructed the view of the blueberries (Fig. 86). Condensation was slight on the films that allowed the most weight loss. Blueberries packed without capping films showed the greatest weight loss.

Hand Packing. A considerable volume of the cultivated berries destined for the fresh market are packaged by hand. Several new pieces of equipment and improved mechanized techniques for packing blueberries are being developed, however, and show promise of being used more widely in future years. In many areas the pickers in the field are responsible for a part or most of the grading when the packing is done by hand. Most growers, however, perform some additional grading or inspection at the packing shed as the berries are prepared for market. Packers quickly check the boxes for off-grade berries and debris.

Figure 86. A 12-pint retail flat packed for market. Condensation varies according to the type of capping film used. Berries are clearly visible under cellulose acetate caps (*2d and 4th row from left*). This flat was stored at 32 F at a relative humidity of 85 per cent. (H. W. Hruschka and L. J. Kushman, USDA, Beltsville, Md.)

Table XXXII. Weight Loss of Croatan Blueberries Packaged in Wood-Veneer Pint Containers and Moisture Condensation on Various Film Caps during Storage

Type of Film Cap	Per Cent Weight Loss After [a]			Moisture Condensation on Films [b]		
	2 Weeks at 32 F (85% HR [c])	+2 Days at 70 F (50% RH)	Total	After 2 Weeks at 32 F (85% RH)	When Moved to 70 F	After 2 Days at 70 F (50% RH)
none (control)	4.5	2.1	6.6	—	—	—
Cellophane 300 PHD	2.4	1.2	3.6	none	none	none
Cellophane 300 PI	1.7	1.7	3.4	light	none	none
Nylon 42 (0.5 mil)	1.7	1.4	3.1	none	heavy	none
Cellulose Acetate 100-P912	1.9	1.2	3.1	none	moderate	none
Cellophane 300 MS86	1.3	1.5	2.8	moderate	heavy	moderate
Polyethylene (0.8 mil)	1.4	1.3	2.7	heavy	heavy	fairly heavy
Polyester 44 (1 mil)	1.7	0.8	2.5	fairly heavy	heavy	moderate
Cellophane 300 DSB	0.9	1.0	1.9	moderate	heavy	moderate

Source: Hruschka and Kushman, 1963.
[a] Based on three pint packages of blueberries.
[b] Moisture condensed on all fruit in capped or noncapped packages moved from 32 to 70 F.
[c] RH refers to relative humidity.

The standard procedure used in packing highbush blueberries for the fresh market involves "topping" or filling the cups, rounding full, and usually fastening a sheet of cellophane or similar material over them. A square frame having inside dimensions nearly equal to the outside dimensions of the box rim is frequently used as a forming device to draw the cover smoothly down over the box for fastening with a rubber band or sticker tape. A paper label identifying the grower or packer and, if the berries are graded, a label showing the grade plus other information is inserted in each package. In other instances this information is printed on the cellophane used to cover the berries.

When the individual boxes have been capped, they are placed into shipping crates and lidded or closed for shipping. The type of master container used varies somewhat with different regions, but generally the cup boxes are packed in wooden flats holding 12 pint-sized packages. Several types of fibreboard and combination fibreboard-wood shipping containers are used for this purpose, but the all-wooden flat is most widely accepted at the present time. A 12-pint master shipping container holds approximately 11 pounds of blueberries.

Cost of Hand Packing. A study of the cost involved in packing New Jersey blueberries for the fresh market, conducted by Baker (unpublished data) of the Department of Agricultural Economics and Marketing at Rutgers University during the 1964 packing season, revealed a total hand-packing cost for large operations of about 51 cents per 12-pint master container. Of this amount, 37.5 cents was for materials, including pint-sized cups, cellophane, rubber bands, and shipping container; 12.5 cents was for labor costs; and 1 cent was allocated for overhead costs.

Mechanized Packing. As the berries arrive from the field in mechanized packing sheds, they are hand dumped into a hopper where the leaves and other foreign materials are removed by an air blast. From the cleaner the berries are conveyed on a moving belt to an area where workers perform the grading, sorting, and culling operations. The moving belt next delivers the remaining berries to a filling device or station where they are either manually or automatically filled into pint cups (Fig. 87). Berries for processing are similarly handled with mechanical equipment but are placed in larger boxes, tins, or processor lugs rather than pint-sized containers.

Handling for Market

Cooling. Blueberries can be cooled rapidly by cold air blasts (Stretch, 1960). Experiments were conducted in precooling berries in three different types of packages and in layers of varying thickness. Using 18 F air at a velocity of approximately 600 feet per minute for five minutes, berries in cellophane-capped packages cooled hardly at all. After removal

Figure 87. A modern automatic packaging facility for filling pint containers with blueberries. (Frederick A. Perkins, Rutgers University, New Brunswick, N.J.)

of the cellophane, little cooling was achieved in pulp and veneer cups, but in plastic cups berry temperatures in the bottom were reduced 30 F in five minutes. When single layers of berries were cooled, the temperature of the berries was lowered 30 to 40 F in one to two minutes with 18 F air at 600 feet per minute. In the 1-inch layer the berry temperature was lowered 62 F in five minutes. As a result of modifications in 1958, berries in pulp and veneer packages were cooled approximately 22 F in a five-minute exposure to 29 F air moving at 600 feet per minute. Further studies to determine the keeping quality of the berries after cooling are needed before recommendations can be made regarding precooling.

Transportation. In transit, blueberries should be refrigerated at temperatures as close to 31 F as possible with a relative humidity of 85 to 90 per cent. The packed master container should be stacked in well-aligned rows (Fig. 88). Each unit should be stacked squarely on top of the container below it so that its weight will be distributed evenly. The rows should be arranged so as to allow an abundance of cold air to circulate. When the containers are closely stacked they may block the passage of air, permitting unfavorably high temperatures to develop and speeding deterioration of the fruit.

Figure 88. Flats of blueberries ready for shipment to market. (Frederick A. Perkins, Rutgers University, New Brunswick, N.J.)

Fresh Blueberry Sales

Marketing Organizations. The majority of blueberries sold in the United States are handled by cooperative marketing organizations. Examples of these farmer-owned associations in three major states producing blueberries for fresh market are the Tru-Blu Cooperative Association in New Jersey, the Michigan Blueberry Growers Association, and the Tru-Blu Cooperative Association in North Carolina. These cooperatives control the quality of the berries handled by setting standards by which members produce and market their crop. Cooperative marketing associations also arrange for contacts with prospective buyers of fresh blueberries in wholesale terminal markets and retail chain-store outlets in large cities, arrange for shipment, and in most cases receive payment from buyers. Most blueberries for fresh market are packed under cooperative brand names and sold during a relatively short period. Others are sold to large processors and are marketed year-round as canned or frozen berries.

The Michigan Blueberry Growers Association, with approximately 775 members, is one of the larger blueberry marketing organizations. Its sales in 1961 amounted to approximately $3.5 million (Persing, 1964). Returns to members for fresh market berries for this organization are based on a weekly pool-selling plan. Producers receive the average price for the week, less about 4 per cent withheld to cover operating costs and to finance the organization's warehousing facilities. An additional amount of about 1 per cent is levied to carry on advertising and promotional programs. The processing price to be paid Michigan association members is announced in advance of harvest, and it has remained remarkably constant over the years, even when crop yields have fluctuated.

In 1961, the Michigan Blueberry Growers Association spent about $60,000 to advertise and promote blueberry sales. The key advertising program was a six-week campaign directed to 22 Midwestern cities using radio, television, and newspapers in the ratio of 50, 20, and 30 per cent, respectively. A retail display kit, advertising mats, and recipe informational booklets directed to homemakers, food editors, bakeries, and restaurants were also featured.

Roadside stands, blueberry auctions, jobbers, brokers, and other market outlets have also played an important part in getting fresh blueberries from producer to consumer and in building blueberry sales over the years.

Marketing Season. The marketing season for fresh blueberries is normally from early May to late September (Table XXXIII). Approximately two-thirds of all sales occur during July and August. Blueberries sold in fresh form are principally cultivated berries from New Jersey, Michigan, and North Carolina. These three states generally supply well over 90 per cent of all blueberries going to fresh market. Data compiled by the USDA's Agricultural Marketing Service in 1963 indicated that 7 per cent of the annual supply was marketed in May, 24 per cent in June, 40 per cent in July, 26 per cent in August, and 3 per cent in September.

Early blueberries are shipped mainly from North Carolina, midseason berries from New Jersey, and late berries from Michigan. The harvest season in any one locality usually lasts six to seven weeks.

Market Areas. For the most part, fresh blueberries are consumed in a relatively small proportion of the country, the areas near the major areas

Table XXXIII. Monthly Shipments of Fresh Blueberries to 41 United States Cities, by State of Origin, 1963 Season

Origin	Number of Carlots Shipped per Month [a]						Total	Per Cent of Total
	May	June	July	Aug.	Sept.	Oct.		
New Jersey	0	99	435	147	6	0	687	49.7
North Carolina	93	225	8	1	0	0	327	23.7
Michigan	0	0	82	166	26	1	275	19.9
Washington	0	0	10	18	5	0	33	2.4
Massachusetts	0	0	3	14	3	0	20	1.4
South Carolina	2	6	0	0	0	0	8	0.6
Maryland	0	3	4	0	0	0	7	0.5
Other States	1	2	5	7	0	0	15	1.1
Canada	0	0	3	3	2	0	8	0.6
Unknown	0	1	0	0	0	0	1	0.1
Total	96	336	550	356	42	1	1,381	100
Per Cent of Total	7.0	24.3	39.8	25.8	3.0	0.1	100	

Source: United States Department of Agriculture, Agricultural Marketing Service, 1963.

[a] No shipments were reported for other months. Truck shipments converted into carlot equivalents on the basis of 1,400 12-pint trays per car.

of production. During 1963 approximately 70 per cent of the fresh blueberry shipments went to five cities: New York, Boston, Philadelphia, Chicago, and Detroit (Table XXXIV). These and five other cities accounted for more than 85 per cent of all American blueberry unloadings. One hundred and twenty-two carlot-equivalent truckloads of blueberries from United States producing areas were reported delivered to Canada in 1963. These data would tend to indicate that a large part of the nation's population is not regularly exposed to fresh blueberry sales, even during the principal season of fresh marketing.

For maximizing returns, this information emphasizes the need for good management and wise decision-making by marketing agencies to control supplies going to various markets and to expand existing market outlets. Under present marketing procedures, periods of oversupply and depressed prices in terminal cities have been known to exist. In recent years more attention has been devoted to better packaging, grading, quality control, and advertising and promotion programs aimed at enhancing the market appeal of fresh blueberries.

Table XXXIV. Fresh Blueberry Shipments to 41 United States and Five Canadian Cities, by State of Origin, 1963 Season

Destination	Origin of Shipments by States					Total	Per Cent of Total
	N.J.	Mich.	N.C.	Wash.	Other		
	(number of carlots) [a]						
UNITED STATES							
New York City and Newark	322	5	131	—	20	478	34.6
Boston	92	8	48	—	8	156	11.3
Philadelphia	86	1	45	—	—	132	9.6
Chicago	16	67	18	—	1	102	7.4
Detroit	18	68	2	—	8	96	7.0
Milwaukee	8	35	10	—	—	53	3.8
Minneapolis	8	37	6	1	—	52	3.8
Baltimore	31	1	11	—	—	43	3.1
Washington, D.C.	22	—	14	—	—	36	2.6
Cleveland	8	17	6	—	—	31	2.2
Miami	16	—	11	—	—	27	2.0
Los Angeles	—	—	—	18	5	23	1.7
Other Cities [b]	60	36	25	14	17	152	11.0
Total United States Cities	687	275	327	33	59	1,381	100
CANADA							
Montreal	40	1	7	—	18	66	54.1
Toronto	3	4	—	—	40	47	38.5
Other Cities [b]	—	—	—	—	9	9	7.4
Total Canadian Cities	43	5	7	—	67	122	100
Total United States and Canadian Cities	730	280	334	33	126	1,503	100
Per Cent of Total	48.6	18.6	22.2	2.2	8.4		100

Source: United States Department of Agriculture, Agricultural Marketing Service, 1963.

[a] Truck shipments converted into carlot equivalents on the basis of 1,400 12-pint trays per car.

[b] Cities reporting 20 or less truckloads in 1963 as included in 41 city reports.

Blueberry Prices

A compilation of blueberry prices on a national scale is not available, but the following figures are representative of the various sources stated. Prices received for berries sold fresh are normally higher than prices received for bulk sales for processing. Average prices for all sales reflect varying percentages of berries sold fresh and to processing outlets.

Grower Prices. During the five-year period 1959–63, blueberry prices received by New Jersey growers for all sales, both processing and fresh market, ranged from $2.60 per tray (about 21.7 cents per pint) in 1960 to $3.05 per tray (about 25.4 cents per pint) in 1963 and averaged $2.84 per 12-pint tray (about 23.7 cents per pint). In 1963, the average grower price for fresh berries in New Jersey from all sales was $3.17 per tray (about 26.4 cents per pint), while the price for processing berries averaged $2.32 per tray (about 19.3 cents per pint).

The average blueberry price as reported by New Jersey Auctions in 1963 for fresh sales was $3.72 per tray (Persing, 1964). In recent years, however, blueberry auctions as a sales outlet for New Jersey blueberries have been declining in importance. The average farm price for Washington state blueberries in 1962 for fresh market and processing berries was 18.1 cents per pound, or approximately $2 per tray (Weeks and Nakashima, 1964).

Wholesale Prices. Wholesale prices for blueberries vary considerably from one season to another and in various markets, depending on the quantity available, quality and size of the berries to be marketed, source of berries, shipper reputation, and other factors. Information compiled by the New Jersey Crop Reporting Service, highlighted in Table XXXV, shows 1963 wholesale prices for different size classifications of blueberries

Table XXXV. Wholesale Blueberry Prices in Leading Terminal Markets for New Jersey Stock for Selected Periods, 1963 Season

Period	New York City		Philadelphia		Boston		Baltimore
	Large	Medium	Large	Med.-Large	Large	Medium	Med.-Large
(cents/pt)							
July 8–12	32–38	25–32	33–40	27–35	30–40	28–35	27–40
July 22–26	27–33	20–28	28–35	25–30	27–32	26–28	27–37½
Aug. 5–9	32–37	25–30	35–37	30–33	30–32	28	25–35

Source: New Jersey State Department of Agriculture, Crop Reporting Service, 1964.

packaged in pint containers for selected markets. For the time periods reported, wholesale prices during 1963 for large New Jersey blueberries ranged from a low of 27 cents a pint in New York City and Boston to a high of 40 cents a pint in Philadelphia and Boston markets. Prices for medium-sized berries from New Jersey during this same period ranged from 20 cents a pint in New York City to 35 cents a pint in Boston. The lowest prices in 1963 for medium-large berries from New Jersey was 25 cents a pint in Baltimore and Philadelphia, and the highest price reported for the medium-large size was 40 cents a pint for the Baltimore wholesale terminal market.

Representative New York wholesale prices for New Jersey highbush blueberries, as reported in a monthly supply letter by the United Fresh Fruit and Vegetable Association, Washington, D.C., as of July 17 in each of three years, was 27 cents a pint in 1961, 21 cents in 1962, and 33 cents in 1963. Blueberries from North Carolina for this same date were priced on the New York wholesale market at 40 cents a pint in 1961, 28 cents in 1962, and 37 cents in 1963. Michigan berries in pint packages in the Chicago area on July 14 sold for 40 cents in 1963 and 34 cents in 1964. On August 18, 1964, Michigan berries in Chicago wholesale markets were priced at 29 cents a pint, compared with 34 cents in 1963. In many instances these sources of information do not reflect clearly the quality of the berries sold and its influence on establishing prices.

Retail Prices. New York City retail prices for blueberries for the years 1963 and 1964, as reported by the New York City Weekly Market Reports, ranged from a low of 33 cents a pint for poorer quality fruit to 69 cents a pint for premium quality berries (New York City Bureau of Consumer Service and Research, 1964). The most frequent retail prices quoted for fresh blueberries in the New York City market area in recent years generally ranged from 39 cents to 49 cents a pint.

CHAPTER 13

Blueberries in the Home Garden

by **Warren C. Stiles**, *Department of Plant and Soil Sciences, University of Maine, Orono, and* **John S. Bailey**, *Department of Horticultural Science, University of Massachusetts, Amherst*

Blueberries may be grown successfully in gardens in areas where their commercial culture is not profitable. The plant has such specific climatic and edaphic requirements that its successful husbandry depends upon the modifications which the home gardener must make.

SITE

Since frosts after growth starts in the spring may ruin the bloom and crop, low areas that experience a higher incidence of radiation frosts should be avoided. Full sunshine promotes the production of more and better berries, but for ornamental purposes the plants will stand partial shade.

Locating the blueberry plant near trees is undesirable because they compete for water. Areas that are completely surrounded by trees and bushes have poor air circulation, which allows cold air to accumulate, thus increasing the danger of frost damage. This condition also prevents the berries from drying quickly after a rain, and prolonged wetting favors the growth of fungus diseases.

SOILS

Although the blueberry grows best on soils of sandy-peat or sandy-muck nature, it will tolerate some of the heavier soils provided they have good aeration, high organic matter content, an adequate supply of moisture, and good drainage. Many garden soils are of this heavier nature—loamy

Blueberries in the Home Garden

sands, loams or clay loams—and may require a good deal of care and modification to be made suitable for blueberries.

If the soil in the location selected for the blueberries is not sufficiently acid but is otherwise satisfactory, sulfur, aluminum sulfate, or acid peat moss can be used to increase acidity. On sandy soils finely ground sulfur is recommended, used at the rate of ¾ pound per 100 square feet for each full pH unit that the soil tests above pH 4.5. On silt loam soils this rate is increased to 1½ or 2 pounds per 100 square feet (Judkins, 1949). If acid peat moss is used to lower the pH of the soil, it is generally mixed with garden soil, using 1 part soil to 1 part peat by volume. If the soil pH is higher than 6.5 it is not practical to correct this condition without replacing the soil.

At times it may be necessary or advantageous to remove the garden soil and replace it with one which is more suitable for growing blueberries. A soil brought in from an area in which blueberries thrive naturally is excellent, or an artificial soil can be prepared. To prepare for planting a well-drained location is selected. A hole approximately 4 feet in diameter and 8 inches deep is dug and filled with one of the following mixtures (Doehlert, 1950, 1957).

Mixture A

To 6 cubic feet (5 level bushels) of sandy loam, add powdered sulfur (3 ounces if the pH of the sandy loam is about 6.0, or 1½ ounces if the pH is about 5.5) and about 3 quarts of finely chopped composted manure, plant compost, or fertile garden soil. Mix thoroughly. Then mix in 6 cubic feet (well-packed measure) of sawdust or horticultural peat. Failure to mix sulfur and soil thoroughly may cause severe injury or death of the blueberry plants. To insure good mixing, first mix the sulfur with a smaller portion of dry soil. Also remember that a wet soil does not break up well for mixing.

Mixture B

The following materials will usually be harder to assemble, but should produce somewhat better results. Mix together 3 parts sandy soil without sulfur, 3 parts sawdust or horticultural peat (well-packed measure), and 1 part acid muck (pH below 5.0) or cranberry bog peat (or 2 parts acid leaf mold).

An old cement mixer can be used to advantage for mixing the soils. Since the blueberry requires an acid substrate, a soil that has been limed in the last few years is not suitable for use in these mixtures. Weathering in place for at least two months before planting improves the mixtures.

Test the acidity of the soil in the bed every two years. If the pH is above

5.2, scatter 1 ounce of sulfur over each 16 square feet of area and rake it in thoroughly.

Johnston (1953) suggests a method for growing blueberries in tubs of soil specially prepared to meet the requirements of the plants. A 50-gallon metal drum is cut into two tubs and four 2-inch holes are cut in the bottom of each tub to provide drainage. After the tubs are carefully cleaned, they are set in the soil so that about an inch of the top remains above the soil surface and are filled with a good blueberry soil or soil preparation. One plant is set in each tub.

SOURCE OF PLANTS

Blueberry plants for the home garden can be purchased from a reputable nursery or commercial grower more economically than they can be produced at home. However, for those who have the patience and skill, information on how to propagate them is presented in Chapter VI.

Large two-year-old plants have proved to be the most satisfactory to plant. Younger bushes require much additional care and take a longer time to come to bearing (Cain and Slate, 1953). On the other hand, it is possible to move even mature plants while in full leaf if precautions are taken to include enough of their root system, if tops are pruned properly, and if sufficient moisture is provided until they have become established in their new location (Doehlert, 1957).

VARIETIES

Varieties suitable for planting in the home garden vary in different areas. It is advisable to consult a county agent or state experiment station for recommendations. (See Chapter V for variety descriptions.)

Two or more varieties are recommended for a home garden planting. This allows for cross-pollination and promotes better fruit set, production of larger berries, and earlier ripening (Cain and Slate, 1953).

PLANTING

Planting as early in the spring as the soil can be worked without packing is preferred. If the plants arrive before they can be set, they must be protected from drying out. This is done by separating the plants and placing their roots in a shallow trench or furrow and covering with soil. Be sure that the roots are moist when covered. If they are dry, soaking in water for a few hours or watering the soil after they are covered will help to keep the plants in good condition.

As soon as the soil is properly prepared, the plants can be moved into the permanent location. For best results, broken portions of roots are removed before planting, the plants are set so that the uppermost roots are only 1 or 2 inches below the soil surface, and the soil is packed firmly around them. Any dead wood, short weak twigs and fruit buds are removed.

The space required varies with the nature of the soil involved. A spacing of less than 4 feet in a row is seldom desirable. On the better soils, more distance is usually required to allow for maximum growth. Ten feet between rows is a good distance; less than this crowds the bushes too much (Doehlert, 1957).

PRUNING

Pruning has three purposes: (1) to help control disease by removing dead, weak, or diseased parts of bushes, (2) to prevent overbearing, and (3) to encourage new growth which will bear the next year's crop.

The blueberry bush often bears more fruit than it can support to best advantage. This tendency results in the production of a large crop of small berries and an insufficient amount of new, vigorous growth necessary for a good crop the following year.

In late March or early April, the fruit buds can be recognized by their large size in comparison with the smaller leaf buds. Since each fruit bud usually produces five to eight fruits, it is possible to estimate the size of the crop. Pruning can then be used to adjust the size of the potential fruit crop to the bearing capacity of the bush.

To remove excess fruit buds, the less vigorous, twiggy canes are cut back to a strong lateral branch or new shoot, and most of the short twigs bearing fruit buds are removed. This treatment usually reduces the crop sufficiently to allow the bush to develop the remaining berries to large size and to produce a number of strong fruiting canes for the next year's crop.

On strong, vigorous canes, about 50 per cent of the fruit buds can be removed by pruning the weaker new growth. The old canes should not be completely removed, unless diseased or injured, until the bush is several years old and has established a larger root system. At this time new, vigorous canes will arise from the base of the plant and exceed the general height of the bush. These may be headed back to the average height of the bush. This procedure will induce strong lateral growth of fruiting wood. After three or four years, many of the low, spreading branches can be removed as they are replaced by the taller, more vigorous ones (Cain and Slate, 1953).

The following general rules can be applied for pruning. (1) Remove sucker shoots and weak, twiggy branches. (2) Allow one fruit bud for approximately each 3 inches of new shoot growth. (3) Allow a maximum of six to eight canes for old bushes. For a more detailed discussion of pruning, growth and fruiting habit, see Chapter VIII.

MULCHING

Blueberries respond favorably to mulching, which is one of the best and easiest cultural methods for handling them. Therefore mulching should have a special appeal to the home gardener. It is a necessity if bushes are planted in a lawn unless the sod is removed around them. Various materials make good mulches—leaves, sawdust, hay, straw, wood shavings, peat moss, pine needles, spent hops, almost any available organic material.

Sawdust is one of the best mulching materials if it is available. Both softwood sawdust (Schwartze and Myhre, 1954b) and hardwood sawdust have been used. Six to 8 inches of new material will usually be needed to give a 3-inch layer of mulch after settling. The mulch can be applied in a circle around the bushes, in a band along the row, or as a mat covering the entire area of the blueberry bed. It can be put down any time during the year.

Peat moss is especially good where the soil needs to be made more acid. Peat moss is available at most feed stores and garden centers. It is available in several grades of fineness. The coarser grade used for poultry or stable bedding is better for mulching in windy locations.

Baled wood shavings are available at some feed stores. In windy situations they may blow away unless held down by something like evergreen branches or pieces of plastic sheeting.

Straw is less desirable as a mulch. It decomposes slowly, may become a fire hazard, and sometimes brings in weed seeds. Hay, however, makes a good mulch, if it is not too expensive. A good-quality hay will add considerable nitrogen upon decomposition, but too it can bring in many weed seeds and become a fire hazard under very dry conditions.

Pine needles are an excellent mulch, if available. Usually the gardener will need to go into the woods and rake up his own. This may be very time-consuming. Deciduous leaves, however, make only a fair mulch if they can be prevented from drying out and blowing away or becoming a fire hazard.

In a city where there is a brewery, spent hops can usually be obtained for the hauling. They make an excellent mulch, if the odor, which disappears in a couple of weeks, is not too offensive. They are highly fire resistant, do not blow, and decompose slowly.

The new plastic mulches, which are being used extensively for strawberries in some areas, have possibilities as a mulch for blueberries. Since

FERTILIZING

It is generally recommended that blueberries be fertilized in early spring. A single application just before growth starts (Judkins, 1949) or a split application providing half of the fertilizer at this time and half when berries begin to turn blue may be used (Doehlert, 1957). The second application must not be delayed too long or it may stimulate late growth of the bushes and thus increase the hazard of winter injury.

For most home gardeners, a 10–10–10 or similar analysis fertilizer is readily available and will give good results. Ammonium sulfate is an especially good source of nitrogen for the blueberry. The ammonium form of nitrogen is readily taken up and assimilated by the plant, leaving an acid residue which aids in providing the necessary acid soil condition. In case the soil is quite acid (below pH 4.0), sodium nitrate, which leaves an alkaline residue, may be preferred as the nitrogen source.

Fertilizer placed in the planting hole at the time the plants are set is likely to cause serious injury to the roots. Fertilizers spread evenly over an area approximately equal to that covered by the tops of the bushes and raked in lightly give good results (Cain and Slate, 1953). If the blueberries are used in mixed plantings with azaleas or rhododendrons, no special change or precaution which is not used for these other plants is needed. But if the blueberry is interplanted with plants with a soil requirement that is much different, it may be necessary to handle the blueberry plants in one of the special ways already mentioned.

Beginning the year following planting, 1 ounce of ammonium sulfate per bush for each year that the blueberries have been in the garden will supply enough nitrogen; that is, 1 ounce the first year, 2 ounces the second year, and so on to a total of 5 ounces. Thereafter, 5 ounces per bush should provide sufficient nitrogen for mature bushes (Beach, 1948).

If a complete fertilizer is being used, the 10–10–10 fertilizer at a pound per 100 square feet may be spread evenly in one application over the area around the plants (Doehlert, 1957). The application may be split as suggested above.

A newly applied mulch tends to make the nitrogen in fertilizers unavailable for a time. Temporary nitrogen starvation can be avoided by doubling the above fertilizer rates for a year or two.

HARVESTING

Most varieties ripen over a period of three to five weeks and require several pickings. Unless the berries are evenly blue colored to the stem before harvesting, flavor will be impaired. The blue color develops approximately three days before the berries attain maximum flavor. A good picking guide is to wait until one-third of the berries have turned blue before picking (Cain and Slate, 1953). Shriveling or splitting generally indicates that the berries have passed their prime. Yield per bush increases from 1 pint or less for four-year-old plants to 6 to 10 pints for nine-year-olds.

Fresh blueberries keep longer than some other berries. At room temperature they will keep for two or three days, on crushed ice for four days, or in a home refrigerator for 10 to 12 days (Wood and Wright, 1954).

PESTS

Birds are the primary pest in home plantings of blueberries. Several methods have been suggested for controlling them. The best bird protection consists of enclosing the plants in some kind of netting. Numerous kinds have been employed, such as second-hand tobacco cloth, aster cloth, second-hand fish netting (Fig. 90), plastic netting of various sorts, and wire netting, such as hardware cloth and chicken wire. If only a few bushes are involved, single bushes can be covered with a cloth type of netting. If a large number of bushes are involved, it is better to put up a permanent frame over the bushes to support the netting (Fig. 89). To be effective, the netting must be in place all during harvest. Removal of the netting immediately after harvest will make it last longer and increase the amount of sunlight reaching the plants. Green screening blends better with the color of the bushes than other colors and detracts less from the esthetic value of the surroundings. Scarecrows, metal strips and discs, pieces of rubber hose to imitate snakes, and stuffed hawks and owls have been tried with little success (Doehlert, 1957). Exploding firecrackers and other noisemakers do little but annoy the neighbors.

Blueberries are attacked by a number of insects and diseases. These and their control are discussed in Chapters IX and X. If an attack occurs, the home gardener can obtain expert advice on identification and the latest control methods by communicating with the county agricultural agent in his area.

Figure 89. A home garden blueberry planting in a permanent enclosure. (A. F. Sozio, New York City.)

Figure 90. The fish net over this blueberry planting keeps out birds. A fine-mesh netting will also prevent invasions of Japanese and rose chafer beetles. (C. H. Hill.)

ORNAMENTALS

The wide diversity of types of blueberries suggests various possibilities for their use in landscaping. Bush habit varies from very low-spreading to very tall and upright, and from thick and brushy to open and rangy (Table II). Twig colors range from green to yellow, reddish-brown, and brown. The pendant white blossoms are very attractive in the spring. Foliage color is good in summer and excellent in fall, when the leaves take on the reds and yellows characteristic of other plants of the same genus. When fully ripe, the blueberries among the green foliage are very attractive.

The highbush types may be used as single specimen plants, in mass plantings, or as hedges. Lowbush blueberries may be employed as ground covers on steep slopes or where it is necessary to avoid soil erosion.

Bibliography

Aalders, L. E., and Hall, I. V. 1961. Pollen Incompatibility and Fruit Set in Lowbush Blueberries. *Can. J. Genet. and Cytol.* 3:300–307.

Aalders, L. E., and Hall, I. V. 1964. A Comparison of Flower-Bud Development in the Lowbush Blueberry *Vaccinium angustifolium* Ait. under Greenhouse and Field Conditions. *Proc. Amer. Soc. Hort. Sci.* 84:281–284.

Abdalla, D. A. 1963. Raking and Handling Lowbush Blueberries. *Me. Agr. Ext. Serv. Bull.* 497.

Abdalla, D. A. 1964a. Pruning Blueberries with LP-Gas. *Butane Propane News* 26:40–42.

Abdalla, D. A. 1964b. Supply and Use of Blueberries in Maine. *Me. Agr. Exp. Sta. Mimeo. Rep.* 27.

Amling, H. J. 1958. Influence of Nutrient-Element Supply on Leaf Composition and Growth of Highbush Blueberry (*Vaccinium corymbosum* L.) with Special References to Importance of Sampling Date on Leaf and Fruit Composition of Field-Grown Blueberries. Ph.D. dissertation, Michigan State University, East Lansing.

Anderson, J. P. 1924. *Botrytis cinereia* in Alaska. *Phytopathology* 14:152–155.

Bailey, J. S. 1936. A Chlorosis of Cultivated Blueberries. *Proc. Amer. Soc. Hort. Sci.* 34:395–396.

Bailey, J. S. 1944. A Comparison of Manures Applied to Cultivated Blueberries. *Proc. Amer. Soc. Hort. Sci.* 44:299–300.

Bailey, J. S. 1947. Development Time from Bloom to Maturity in Cultivated Blueberries. *Proc. Amer. Soc. Hort. Sci.* 49:193–195.

Bailey, J. S. 1949. Frost Injury to Blueberries. *Fruit Var. and Hort. Dig.* 4:98.

Bailey, J. S. 1958. Blueberry Nutrition—A Progress Report, pp. 3–4. *In Fruit Notes*, Pomology Department, University of Massachusetts, Amherst.

Bailey, J. S. 1960. Weed Control in Cultivated Blueberries with Diuron, pp. 14–19. *In* J. N. Moore and N. F. Childers (ed.), *Blueberry Research—Fifty Years of Progress*, New Jersey Agricultural Experiment Station, New Brunswick.

Bailey, J. S., and Drake, M. 1954. Magnesium Deficiency in Cultivated Blueberries and Its Effect on Leaf Potassium, Calcium and Nitrogen. *Proc. Amer. Soc. Hort. Sci.* 63:95–100.

Bailey, J. S., and Everson, J. N. 1937. Further Observations on a Chlorosis of the Cultivated Bluebery. *Proc. Amer. Soc. Hort. Sci.* 35:495–496.

Bailey, J. S., and Franklin, H. J. 1935. Blueberry Culture in Massachusetts. *Mass. Agr. Exp. Sta. Bull.* 317.

Bailey, J. S., Franklin, H. J., and Kelley, J. L. 1939. Blueberry Culture in Massachusetts. *Mass. Agr. Exp. Sta. Bull.* 358.

Bailey, J. S., Franklin, H. J., and Kelley, J. L. 1950. Blueberry Culture in Massachusetts. *Mass. Agr. Exp. Sta. Bull.* 358 (revised).

Bailey, J. S., and French, A. P. 1946. Identification of Blueberry Varieties by Plant Characters. *Mass. Agr. Exp. Sta. Bull.* 431.

Bailey, J. S., and Kelley, J. L. 1959. Blueberry Growing. *Mass. Ext. Serv. Pub.* 240.

Bailey, J. S., Smith, C. T., and Weatherby, R. T. 1949. The Nutritional Status of the Cultivated Blueberry as Revealed by Leaf Analysis. *Proc. Amer. Soc. Hort. Sci.* 54:205–208.

Bailey, J. S., and Sproston, T. 1946. Fermate for the Control of Mummy Berry of the Cultivated Blueberry. *Proc. Amer. Soc. Hort. Sci.* 47:209–212.

Bailey, L. H. 1919. Vaccinium. *Standard Cyclopedia of Horticulture* 6:3421–3425.

Baker, R. E., and Butterfield, H. M. 1951. Commercial Bushberry Growing in California. *Calif. Agr. Ext. Serv. Circ.* 169.

Ballinger, W. E. 1957. Nutritional Conditions of Michigan Blueberry Plantations. Ph.D. thesis, Michigan State University, East Lansing.

Ballinger, W. E. 1962. Studies of Sulfate and Chloride Ion Effects upon Wolcott Blueberry Growth and Composition. *Proc. Amer. Soc. Hort. Sci.* 80:331–339.

Ballinger, W. E., Bell, H. K., and Kenworthy, A. L. 1958. Soluble Solids in Blueberry Fruit in Relation to Yield and Nitrogen Content of Fruiting Shoot Leaves. *Mich. Agr. Exp. Sta. Quart. Bull.* 40:912–914.

Ballinger, W. E., Kenworthy, A. L., Bell, H. K., Benne, E. J., and Bass, S. T. 1958a. Relation between Nutrient Element Content of Blueberry Foliage and Fruit. *Mich. Agr. Exp. Sta. Quart. Bull.* 40:906–911.

Ballinger, W. E., Kenworthy, A. D., Bell, H. K., Benne, E. J., and Bass, S. T. 1958b. Production in Michigan Blueberry Plantations in Relation to Nutrient-Element Content of the Fruiting-Shoot Leaves and Soil. *Mich. Agr. Exp. Quart. Bull.* 40:896–905.

Ballinger, W. E., Kushman, L. J., and Brooks, J. F. 1963. Influence of Crop Load and Nitrogen Applications upon Yield and Fruit Qualities of Wolcott Blueberries. *Proc. Amer. Soc. Hort. Sci.* 82:264–276.

Barker, W. G., and Collins, W. B. 1963a. Growth and Development of the Lowbush Blueberry: Apical Abortion. *Can. J. Bot.* 41:1319–1324.

Barker, W. G., and Collins, W. B. 1963b. The Blueberry Rhizome: In Vitro Culture. *Can. J. Bot.* 41:1325–1329.

Bibliography

Barker, W. G., and Collins, W. B. 1964. An Effect of Gibberellic Acid upon the Maturing of Seeds in the Lowbush Blueberry, *Vaccinium angustifolium* Ait. *Can. J. Bot.* 42:1102–1103.

Barker, W. G., and Collins, W. B. 1965. Parthenocarpic Fruit Set in the Lowbush Blueberry. *Proc. Amer. Soc. Hort. Sci.* 87:229–233.

Barnes, E. H., and Tweedie, H. C. 1964. Fusicoccum Canker of Highbush Blueberry in Michigan. *Plant Dis. Reptr.* 48:687–689.

Beach, F. H. 1948. The Home Fruit Garden. *Ohio St. Univ. Agr. Ext. Serv. Bull.* 252 (revised).

Beckwith, C. S. 1920. The Effect of Fertilizers on Blueberries. *Soil Sci.* 10:309–314.

Beckwith, C. S. 1930. *Report of the Department of Entomology*, p. 174. New Jersey State Agricultural Experiment Station, New Brunswick.

Beckwith, C. S. 1932. Report of the Cranberry and Blueberry Substation. *Proc. 1st Ann. Blueberry Open House* 1:1–3.

Beckwith, C. S. 1933. Blueberry Fertilizer. *N.J. Agr.* 15:4–5.

Beckwith, C. S. 1934. Blueberry Stem Borer. *N.J. Agr. Exp. Sta. Circ.* 320.

Beckwith, C. S. 1935. Blueberry Progress Report. *Rutgers Univ. Monthly Rep.*, July.

Beckwith, C. S. 1941a. Blueberry Insects. *Proc. 10th Ann. Blueberry Open House* 10:1–4.

Beckwith, C. S. 1941b. Control of Cranberry Fruitworm on Blueberries. *J. Econ. Entomol.* 34:169–171.

Beckwith, C. S. 1943a. Insects Attacking Blueberry Fruit. *N.J. Agr. Exp. Sta. Circ.* 472.

Beckwith, C. S. 1943b. Locating and Preparing Fields for the Cultivated Blueberry. *N.J. Agr. Exp. Sta. Circ.* 473.

Beckwith, C. S., and Coville, S. 1931. Blueberry Culture. *N.J. Agr. Exp. Sta. Circ.* 229.

Beckwith, C. S., Coville, S., and Doehlert, C. A. 1937. Blueberry Culture. *N.J. Agr. Exp. Sta. Circ.* 229 (revised).

Beckwith, C. S., and Doehlert, C. A. 1933. Fertilizer and Tillage for Blueberries. *N.J. Agr. Exp. Sta. Bull.* 558.

Bell, H. K., and Johnston, S. 1962. Hints on Blueberry Growing. *Mich. Agr. Exp. Sta. Ext. Folder* F-119 (revised).

Bell, H. P. 1950. Determinate Growth in the Blueberry. *Can. J. Res.* 28:637–644.

Bell, H. P. 1953. The Growth Cycle of the Blueberry and Some Factors of the Environment. *Can. J. Bot.* 31:1–6.

Bell, H. P. 1957. The Development of the Blueberry Seed. *Can. J. Bot.* 35:139–153.

Bell, H. P., and Burchill, J. 1955. Flower Development in the Lowbush Blueberry. *Can. J. Bot.* 33:251–258.

Bell, H. P., and Giffin, Elspeth C. 1957. The Lowbush Blueberry: The Vascular Anatomy of the Ovary. *Can. J. Bot.* 35:667–673.

Bergman, H. F. 1939. Observations on Powdery Mildew on Cultivated Blueberries in Massachusetts in 1938. *Phytopathology* 29:545–546.

Bessey, O. A. 1938. Vitamin C Methods of Assay and Dietary Sources. *J. Amer. Med. Assoc.* 111:1290–1298.

Blasberg, C. H. 1948. Growing Blueberries in Vermont. *Vt. Agr. Exp. Sta. Pam.* 19.

Blasberg, C. H. 1961. Growing Blueberries in Vermont. *Vt. Agr. Exp. Sta. Pam.* 19 (revised).

Blattny, C. 1955. Einige Ergebnisse der pflanzlichen Virusforschung in der Tschechoslowakischen Republik. *Pflanzenschutz-Kongr.*, Berlin, pp. 47–65.

Blattny, C., Jr. 1963. Die Übertragung der Hexenbesenkrankheit der Heidelbeeren (*Vaccinium myrtillus* L.) durch die Zikade *Idiodonus cruentatus* Panz. *Phytopathol. Zeit.* 49:203–205.

Boller, C. A. 1951. Growing Blueberries in Oregon. *Ore. Agr. Exp. Sta. Bull.* 499.

Boller, C. A. 1956. Growing Blueberries in Oregon. *Ore. Agr. Exp. Sta. Bull.* 499 (revised).

Bos, L. 1960. A Witches' Broom Disease of *Vaccinium myrtillus* in the Netherlands. *T. Plantenziekten* 66:259–263.

Boulanger, L. W. 1965. Studies on Lowbush Blueberry Pollination. Paper presented at the Eastern Branch Meeting of the Entomological Society of America, Nov. 4, 1965.

Boyd, W. M. 1956. Present Status of Blueberry Stunt. *Proc. 24th Ann. Blueberry Open House* 24:5–6.

Brightwell, W. T. 1948. Propagation of the Rabbiteye Blueberry. *Proc. Amer. Soc. Hort. Sci.* 52:289–293.

Brightwell, W. T. 1960. Present Status of Rabbiteye Breeding, pp. 49–50. In J. N. Moore and N. F. Childers (ed.), *Blueberry Research—Fifty Years of Progress*, New Jersey Agricultural Experiment Station, New Brunswick.

Brightwell, W. T. 1962. Rabbiteye Blueberries. *Ga. Agr. Exp. Sta. Mimeo.* N.S. 131.

Brightwell, W. T., and Darrow, G. M. 1950. The Callaway and Coastal Blueberries. *Ga. Coastal Plain Exp. Sta. Mimeo.* 67.

Brightwell, W. T., Darrow, G. M., and Woodard, O. J. 1949. Inheritance in Seedlings of *Vaccinium constablaei* × *Vaccinium ashei* variety Pecan. *Proc. Amer. Soc. Hort. Sci.* 53:239–240.

Brightwell, W. T., and Johnston, S. 1944. Pruning the Highbush Blueberry. *Mich. St. Coll. Tech. Bull.* 192.

Brightwell, W. T., and Woodard, O. J. 1960. Blueberry Breeding in Georgia. *Fruit Var. and Hort. Dig.* 15:39.

Brightwell, W. T., Woodard, O. J., Darrow, G. M., and Scott, D. H. 1955. Observations on Breeding Blueberries in the Southeast. *Proc. Amer. Soc. Hort. Sci.* 65:274–278.

Brown, Nellie A. 1938. Blueberry Galls Produced by the Fungus *Phomopsis. Phytopathology* 28:71–73.

Bünemann, G., Dewey, D. H., and Watson, D. P. 1957. Anatomical Changes in the Fruit of Rubel Blueberry during Storage in Controlled Atmosphere. *Proc. Amer. Soc. Hort. Sci.* 70:156–160.

Bibliography

Cain, J. C. 1952a. A Comparison of Ammonium and Nitrate Nitrogen for Blueberries. *Proc. Amer. Soc. Hort. Sci.* 59:161–166.

Cain, J. C. 1952b. Cultivated Blueberries Must Be Cultivated. *N.Y. Agr. Exp. Sta. Farm Res.* 18:2.

Cain, J. C., and Holley, R. W. 1955. A Comparison of Chlorotic and Green Blueberry Leaf Tissue with Respect to Free Amino Acid and Basic Cation Content. *Proc. Amer. Soc. Hort. Sci.* 65:49–53.

Cain, J. C., and Slate, G. L. 1953. Blueberries in the Home Garden. *Cornell Agr. Exp. Sta. Ext. Bull.* 900.

Camp, W. H. 1940. Necessary Changes in the Status and Nomenclature of Certain Species of *Vaccinium* in Eastern North America. *Amer. Midland Nat.* 23:177.

Camp, W. H. 1942a. A Survey of the American Species of *Vaccinium*, Subgenus *Euvaccinium*. *Brittonia* 4:205–247.

Camp, W. H. 1942b. On the Structure of Populations in the Genus *Vaccinium*. *Brittonia* 4:189–204.

Camp, W. H. 1945. The North American Blueberries with Notes on the Other Groups of *Vacciniaceae*. *Brittonia* 5:203–275.

Carleton, W. M., and Kampe, D. 1954. An Experimental, Hydraulically Manipulated Blueberry Weeder. *Mich. Agr. Exp. Sta. Bull.* 36:426–434.

Chandler, F. B. 1938. The Effect of Lime on the Lowbush Blueberry. *Proc. Amer. Soc. Hort. Sci.* 36:477.

Chandler, F. B. 1941. The Relationship of Different Methods of Expressing Size of Blueberry Fruits. *Proc. Amer. Soc. Hort. Sci.* 39:279–280.

Chandler, F. B. 1942. Blueberry Storage. *Science* 95:104.

Chandler, F. B. 1943. Lowbush Blueberries. *Me. Agr. Exp. Sta. Bull.* 423.

Chandler, F. B. 1944. Composition and Uses of Blueberries. *Me. Agr. Exp. Sta. Bull.* 428.

Chandler, F. B. 1947. Cultivation of Lowbush Blueberries. *Proc. Amer. Soc. Hort. Sci.* 49:205–207.

Chandler, F. B., and Highlands, M. E. 1950. Blueberry Juice. *Food Tech.* 4:285–286.

Chandler, F. B., and Hyland, F. 1941. Botanical and Economic Distribution of *Vaccinium* L. in Maine. *Proc. Amer. Soc. Hort. Sci.* 38:430–433.

Chandler, F. B., and Mason, I. C. 1933. The Effects of Fertilizers on the Native Maine Blueberry. *Proc. Amer. Soc. Hort. Sci.* 30:297–298.

Chandler, F. B., and Mason, I. C. 1935. Blueberry Pollination. *Me. Agr. Exp. Sta. Bull.* 380:215–216.

Chandler, F. B., and Mason, I. C. 1939. Pruning of the Lowbush Blueberry. *Proc. Amer. Soc. Hort. Sci.* 37:609–610.

Chandler, F. B., and Mason, I. C. 1940. The Effect of Growth Substances on the Rooting of Blueberry Cuttings. *Science* 92:35.

Chandler, F. B., and Mason, I. C. 1942. The Effect of Mulch on Soil Moisture, Soil Temperature and Growth of Blueberry Plants. *Proc. Amer. Soc. Hort. Sci.* 40:335–337.

Chandler, F. B., and Mason, I. C. 1943. Pruning of Lowbush Blueberries. *Proc. Amer. Soc. Hort. Sci.* 43:173–174.

Chandler, F. B., and Mason, I. C. 1946. Blueberry Weeds in Maine and Their Control. *Me. Agr. Exp. Sta. Bull.* 443.

Chandler, F. B., Mason, I. C., and Phipps, C. R. 1932. Blueberry Investigations. *Me. Agr. Exp. Sta. Bull.* 363.

Childers, N. F. (ed.). 1966. *Fruit Nutrition—Temperature, Sub-Tropical, Tropical.* 2nd ed. Horticultural Publications, Rutgers University, New Brunswick, N.J. 1100 pp.

Childs, W. H. 1946. Shredded Sphagnum vs. Peat and Sand as a Medium for Transplanted Blueberry Seedlings. *Proc. Amer. Soc. Hort. Sci.* 47:206–208.

Childs, W. H. 1947. Influence of Fineness of Shredding on Value of Sphagnum as a Medium for Transplanted Blueberry Seedlings. *Proc. Amer. Soc. Hort. Sci.* 49:208–210.

Childs, W. H. 1950. Greenhouse Fertilization of Blueberry Hybrids Grown in Shredded Sphagnum. *Proc. Amer. Soc. Hort. Sci.* 56:23–26.

Christopher, E. P., and Caroselli, N. E. 1958. Highbush Blueberries—Culture and Pest Control. *R.I. Agr. Ext. Serv. Bull.* 143 (revised).

Christopher, E. P., and Shutak, V. 1947. Influence of Several Soil Management Practices upon the Yield of Cultivated Blueberries. *Proc. Amer. Soc. Hort. Sci.* 49:211–212.

Clark, J. H. 1936. Blueberries under Mulch. *N.J. Agr.* 18:1–4.

Clark, J. H. 1941. Leaf Characters as a Basis for the Classification of Blueberry Varieties. *Proc. Amer. Soc. Hort. Sci.* 38:441–446.

Clark, J. H., and Gilbert, S. G. 1942. Selection of Criterion Leaves for the Identification of Blueberry Varieties. *Proc. Amer. Soc. Hort. Sci.* 40:347–351.

Clayton, C. N., and Fox, J. A. 1963. Blueberry Canker-Resistant Varieties Affected by Certain Isolates of *Botryosphaeria corticis. Plant Dis. Reptr.* 47:758–761.

Clayton, C. N., and Haasis, F. A. 1964. Blueberry Root-Rot Caused by *Phytophtora cinnamomi* in North Carolina. *Plant. Dis. Reptr.* 48:460–461.

Colgrove, M. S., Jr., and Roberts, A. N. 1956. Growth of Azalea as Influenced by Ammonium and Nitrate Nitrogen. *Proc. Amer. Soc. Hort. Sci.* 68:522–536.

Conners, I. L. (comp.). 1932. *Eleventh Annual Report of the Canadian Plant Disease Survey, 1931,* Canadian Department of Agriculture, Ottawa.

Conners, I. L., and Saville, D. B. O. (comp.). 1953. *Thirty-first Annual Report of the Canadian Plant Disease Survey, 1952,* Canadian Department of Agriculture, Ottawa.

Cooke, W. B. 1952. Nomenclatural Notes on the *Erysiphaceae. Mycologia* 44:570–574.

Coville, F. V. 1910. Experiments in Blueberry Culture. *U.S. Dep. Agr. Bull.* 193.

Coville, F. V. 1921. Directions for Blueberry Culture. *U.S. Dep. Agr. Bull.* 974.

Coville, F. V. 1927. Blueberry Chromosomes. *Science* 66:565–566.

Coville, F. V. 1937. Improving the Wild Blueberry, pp. 559–574. *In 1937 Yearbook of Agriculture,* United States Government Printing Office, Washington, D.C.

Creelman, D. W. 1958. Fusicoccum Canker of the Highbush Blueberry, Especially with Reference to Its Occurrence in Nova Scotia. *Plant Dis. Reptr.* 42:843–845.

Crowley, D. J. 1933. Observations and Experiments with Blueberries in Western Washington. *Wash. Agr. Exp. Sta. Bull.* 276.

Curran, C. H. 1932. New North American *Diptera* with Notes on Others. *Amer. Mus. Novitates* 526.

Darrow, G. M. 1941. Seed Size in Blueberry and Related Species. *Proc. Amer. Soc. Hort. Sci.* 38:438–440.

Darrow, G. M. 1942. Rest Period Requirements of Blueberries. *Proc. Amer. Soc. Hort. Sci.* 41:189–194.

Darrow, G. M. 1947. New Varieties of Blueberry, pp. 300–303. *In Science in Farming—Yearbook of Agriculture, 1943–1947,* United States Government Printing Office, Washington, D.C.

Darrow, G. M. 1949. Polyploidy in Fruit Improvement. *Proc. Amer. Soc. Hort. Sci.* 54:523–532.

Darrow, G. M. 1952. Breeding of Small Fruits in the United States. *Sci. Monthly* 75:288–297.

Darrow, G. M. 1956a. The Big Six Blueberry Varieties for Northern States. *Nat. Hort. Mag.* 35:162–165.

Darrow, G. M. 1956b. Use of Naphthalene Acetamide in Blueberry Breeding. *Proc. Amer. Soc. Hort. Sci.* 67:341–343.

Darrow, G. M. 1957. Blueberry Growing. *U.S. Dep. Agr. Farmers' Bull.* 1951 (revised).

Darrow, G. M. 1958. Seed Number in Blueberry Fruits. *Proc. Amer. Soc. Hort. Sci.* 72:213–215.

Darrow, G. M. 1960. Blueberry Breeding—Past, Present, Future. *Nat. Hort. Mag.* 39:14–33.

Darrow, G. M. 1962. The Blueberry Goes Modern. *Amer. Fruit Grower* 82: 13, 50–52.

Darrow, G. M., and Camp, W. H. 1945. *Vaccinium* Hybrids and the Development of New Horticultural Material. *Torrey Bot. Club Bull.* 72:1–21.

Darrow, G. M., Camp, W. H., Fisher, H. E., and Dermen, H. 1942. Studies on the Cytology of *Vaccinium* Species. *Proc. Amer. Soc. Hort. Sci.* 41: 187–188.

Darrow, G. M., Camp, W. H., Fisher, H. E., and Dermen, H. 1944. Chromosome Numbers in *Vaccinium* and Related Groups. *Torrey Bot. Club Bull.* 71:498–506.

Darrow, G. M., and Clark, J. H. 1940. The Atlantic, Pemberton, and Burlington Blueberries. *U.S. Dep. Agr. Circ.* 589.

Darrow, G. M., Clark, J. H., and Morrow, E. B. 1939. The Inheritance of Certain Characters in the Cultivated Blueberry. *Proc. Amer. Soc. Hort. Sci.* 37:611–616.

Darrow, G. M., Demaree, J. B., and Tomlinson, W. E., Jr. 1951. Blueberry Growing. *U.S. Dep. Agr. Farmers' Bull.* 1951 (revised).

Darrow, G. M., Dermen, H., and Scott, D. H. 1949. A Tetraploid Blueberry. *J. Hered.* 40:304–306.

Darrow, G. M., and Moore, J. N. 1962. Blueberry Growing. *U.S. Dep. Agr. Farmers' Bull.* 1951 (revised).

Darrow, G. M., Morrow, E. B., and Scott, D. H. 1952. An Evaluation of Interspecific Blueberry Crosses. *Proc. Amer. Soc. Hort. Sci.,* 59:277–282.

Darrow, G. M., and Scott, D. H. 1954. Longevity of Blueberry Seed in Cool Storage. *Proc. Amer. Soc. Hort. Sci.* 63:271.

Darrow, G. M., Scott, D. H., and Dermen, H. 1954. Tetraploid Blueberries from Hexaploid × Diploid Species Crosses. *Proc. Amer. Soc. Hort. Sci.* 62:266–270.

Darrow, G. M., Scott, D. H., and Galletta, G. J. 1952. New Blueberry Varieties for New Jersey. *N.J. Agr. Exp. Sta. Bull.* 767.

Darrow, G. M., Scott, D. H., and Gilbert, F. A. 1949. Two New Blueberry Varieties, Coville and Berkeley. *N.J. Agr. Exp. Sta. Bull.* 747.

Darrow, G. M., Scott, D. H., and Hough, L. F. 1956. A New Blueberry Introduced—Blueray. *N.J. Agr. Exp. Sta. Bull.* 783:1–4.

Darrow, G. M., Whitton, L., and Scott, D. H. 1960. The Ashworth Blueberry as a Parent in Breeding for Hardiness and Earliness. *Fruit Var. and Hort. Dig.* 14:43–46.

Darrow, G. M., Wilcox, R. B., and Beckwith, C. S. 1944. Blueberry Growing. *U.S. Dep. Agr. Farmers' Bull.* 1951.

Darrow, G. M., Woodard, O., and Morrow, E. B. 1944. Improvement of the Rabbiteye Blueberry. *Proc. Amer. Soc. Hort. Sci.* 45:275–279.

Demaree, J. B. 1946. Rate of Spread of Blueberry Stunt in North Carolina. *Plant Dis. Reptr.* 30:321–325.

Demaree, J. B. 1947. A Proliferating Gall on Blueberry Plants Caused by an *Actinomyces* (abstract). *Phytopathology* 37:438.

Demaree, J. B., and Morrow, E. B. 1951. Relative Resistance of Some Blueberry Varieties and Selections to Stem Canker in North Carolina. *Plant Dis. Reptr.* 35:136–141.

Demaree, J. B., and Smith, N. R. 1952a. Blueberry Galls Caused by a Strain of *Agrobacterium tumefaciens. Phytopathology* 42:88–90.

Demaree, J. B., and Smith, N. R. 1952b. *Nocardia vaccinii* n. sp. Causing Galls on Blueberry Plants. *Phytopathology* 42:249–252.

Demaree, J. B., and Wilcox, Marguerite S. 1942. Blueberry Cane Canker. *Phytopathology* 32:1068–1075.

Demaree, J. B., and Wilcox, Marguerite S. 1947. Fungi Pathogenic to Blueberries in the Eastern United States. *Phytopathology* 37:487–506.

Doak, K. D. 1928. The Mycorrhizal Fungus of *Vaccinium. Phytopathology* 18:148.

Doehlert, C. A. 1937. Blueberry Tillage Problems and a New Harrow. *N.J. Agr. Exp. Sta. Bull.* 625.

Doehlert, C. A. 1940. Dates for Applying Blueberry Fertilizer. *Proc. Amer. Soc. Hort. Sci.* 38:451–455.

Doehlert, C. A. 1941. Pruning Blueberries. *Proc. 10th Ann. Blueberry Open House* 10:4–9.

Doehlert, C. A. 1944. Fertilizing Commercial Blueberry Fields in New Jersey. *N.J. Agr. Exp. Sta. Circ.* 483.

Doehlert, C. A. 1946. Blueberry Stunt Disease. *Proc. 15th Ann. Blueberry Open House* 15:1–2.

Doehlert, C. A. 1950. Blueberries in the Garden. *N.J. Agr. Exp. Sta. Circ.* 538.

Doehlert, C. A. 1951. Roguing to Control Stunt Disease. *Proc. 20th Ann. Blueberry Open House* 20:11–14.

Doehlert, C. A. 1953a. An Automatic Rotary Hoe for Blueberries. *Hort. News* 34:5.

Doehlert, C. A. 1953b. Facts about Fertilizing Blueberries. *N.J. Agr. Exp. Sta. Circ.* 550.

Doehlert, C. A. 1953c. Propagating Blueberries from Hardwood Cuttings. *N.J. Agr. Exp. Sta. Circ.* 551.

Doehlert, C. A. 1957. Blueberries in the Garden. *N.J. Agr. Exp. Sta. Circ.* 579.

Doehlert, C. A., and Shive, J. W. 1936. Nutrition of the Blueberry (*Vaccinium corymbosum* L.) in Sand Culture. *Soil Sci.* 41:341–350.

Doehlert, C. A., and Tomlinson, W. E., Jr. 1947. Blossom Weevil on Cultivated Blueberries. *N.J. Exp. Sta. Circ.* 504.

Doehlert, C. A., and Tomlinson, W. E., Jr. 1951. Blossom Weevil on Cultivated Blueberries. *N.J. Agr. Exp. Sta. Circ.* 504 (revised).

Doran, W. L. 1941. The Propagation of Some Trees and Shrubs by Cuttings. *Mass. Agr. Exp. Sta. Bull.* 382.

Doran, W. L., and Bailey, J. S. 1943. Propagation of the Highbush Blueberry by Softwood Cuttings. *Mass. Agr. Exp. Sta. Bull.* 410.

Drew, Mrs. A. M. 1953. The Gem. Plant Pat. 1181. April 21. Plant Patent Office, United States Department of Commerce, Washington, D.C.

Driggers, B. F. 1927. Galls on Stems of Cultivated Blueberry Caused by a Chalcidoid, *Hemadas nubilipennis* Ashm. *J. N.Y. Entomol. Soc.* 25:253–259.

Eaton, E. L. 1943. Blueberry Culture. *In The Blueberry, Can. Dep. Agr. Pub.* 754:1–16.

Eaton, E. L. 1949. Highbush Blueberries. *Rep. Dom. Exp. Sta. Kentville, Nova Scotia, 1937–1946* 46:20–25.

Eaton, E. L. 1950. Blueberry Culture. *In The Blueberry, Can. Dep. Agr. Pub.* 754:1–25 (reprinted).

Eaton, E. L. 1951. Propagating Highbush Blueberries by Mounding. *Sci. Agr.* 31:131–132.

Eaton, E. L., and Hall, I. V. 1961. Blueberry Culture and Propagation. *In The Blueberry in the Atlantic Provinces, Can. Dep. Agr. Pub.* 754:5–25 (revised).

Eaton, E. L., and White, R. G. 1960. The Relation between Burning Dates and the Development of Sprouts and Flower Buds in the Lowbush Blueberry. *Proc. Amer. Soc. Hort. Sci.* 69:288–292.

Eck, Paul. 1964. Magnesium Status of New Jersey Blueberry Plantings. *Proc. 32nd Ann. Blueberry Open House* 32:7–9.

Edmonston, M. N. 1962. Blueberry Grades Assure the Consumer of Dependable Quality. *Proc. 30th Ann. Blueberry Open House* 30:15–16.

Eggert, F. P. 1955. Increasing Stand of Blueberry Plants. *In Producing Blueberries in Maine, Me. Agr. Exp. Sta. Bull.* 479:14–16.

Eggert, F. P. 1957. Shoot Emergence and Flowering Habit in the Lowbush Blueberry (*Vaccinium angustifolium*). *Proc. Amer. Soc. Hort. Sci.* 69: 288–292.

Eglitis, M., Johnson, F., and Crowley, D. J. 1952. Strains of Botrytis Pathogenic to Blueberries (abstract). *Phytopathology* 42:513.

Emmett, H. E. G., and Ashley, E. 1934. Some Observations on the Relation Between the Hydrogen-Ion Concentration of the Soil and Plant Distribution. *Ann. Bot.* 48:76.

Entomological Society of America, Committee on Common Names of Insects. 1960, 1961. Common Names of Insects Approved by the Entomological Society of America. *Bull. Entomol. Soc. Amer.* 6:175–211; 7:93.

Filmer, R. S., and Marucci, P. E. 1963. The Importance of Honeybees in Blueberry Pollination. *Proc. 31st Ann. Blueberry Open House* 31:14–21.

Filmer, R. S., and Marucci, P. E. 1964. Honeybees and Blueberry Pollination. *Proc. 32nd Ann. Blueberry Open House* 32:25–27.

Flint, E. M. 1918. Structure of Wood in Blueberry and Huckleberry. *Bot. Gaz.* 65:556–559.

Florida State Department of Agriculture. 1941. Blueberries with Special References to Florida Culture. *Fla. St. Dep. Agr. Bull.* 33.

Franklin, H. J. 1916. Report of the Cranberry Substation for 1915. *Mass. Agr. Exp. Sta. Bull.* 168:1–48.

French, B. C. 1957. Trends in Blueberry Consumption. *Mich. Agr. Exp. Sta. Quart. Bull.* 40:34–43.

Fuleki, T., and Hope, G. W. 1964. Effect of Various Treatments on Yield and Composition of Blueberry Juice. *Food Tech.* 18:166–168.

Fulton, B. B. 1946. Dusting Blueberries to Control the Cranberry Fruitworm. *J. Econ. Entomol.* 39:306–308.

Fulton, R. H. 1958a. Controlling Mummy Berry Disease of Blueberries by Soil Treatment. *Mich. Agr. Exp. Sta. Quart. Bull.* 40:491–497.

Fulton, R. H. 1958b. New or Unusual Small Fruit Diseases and Disease-Like Occurrences in Michigan. *Plant Dis. Reptr.* 42:71–73.

Goheen, A. C. 1950a. Botrytis Twig-Blight of Blueberries in Washington. *Abstr. Northwest Sci.* 24:30.

Goheen, A. C. 1950b. Blueberry Spoilage in Cold Storage. *Proc. 19th Ann. Blueberry Open House* 19:2–5.

Goheen, A. C. 1951. Blueberry Stem Canker in New Jersey. *Proc. 20th Ann. Blueberry Open House* 20:16–19.

Goheen, A. C. 1953. The Cultivated Highbush Blueberry, pp. 784–789. *In Plant Diseases—Yearbook of Agriculture, 1953,* United States Government Printing Office, Washington, D.C.

Goheen, A. C., and Braun, A. J. 1955. Some Parasitic Nematodes Associated with Blueberry Roots. *Plant Dis. Reptr.* 39:908.

Gorham, R. P. 1924. The Chain Dot Moth as an Injurious Insect. *Proc. Acadian Entomol. Soc.* 10:58–59.

Gray, A., and Fernald, M. L. 1950. *Gray's Manual of Botany*. 8th ed. American Book Company, New York.

Gray, Elizabeth G., and Everett, H. F. M. 1961. New or Uncommon Plant Diseases and Pests. *Plant Pathol.* 10:123–124.

Griffin, G. D., Huguelet, J. E., and Nelson, J. W. 1963. Xiphinema americanum as a Vector of Necrotic Ringspot Virus of Blueberry. *Plant Dis. Reptr.* 47:703–704.

Griggs, W. H., and Rollins, H. A. 1947. The Effect of Planting Treatment and Soil Management System on the Production of Cultivated Blueberries. *Proc. Amer. Soc. Hort. Sci.* 49:213–218.

Griggs, W. H., and Rollins, H. A. 1948. Effect of Soil Management on Yields, Growth, and Moisture and Ascorbic Acid Content of the Fruit of Cultivated Blueberries. *Proc. Amer. Soc. Hort. Sci.* 51:304–307.

Hall, I. V. 1955. Floristic Changes Following the Cutting and Burning of a Woodlot for Blueberry Production. *Can. J. Agr. Sci.* 35:143–152.

Hall, I. V. 1957. The Tap Root in Lowbush Blueberry. *Can. J. Bot.* 35:933–934.

Hall, I. V. 1958. Some Effects of Light on Native Lowbush Blueberries. *Proc. Amer. Soc. Hort. Sci.* 72:216–218.

Hall, I. V. 1959. Plant Populations in Blueberry Stands Developed from Abandoned Hayfields and Woodlots. *Ecology* 40:742–743.

Hall, I. V. 1963. Note on the Effect of a Single Intensive Cultivation on the Composition of an Old Blueberry Stand. *Can. J. Plant Sci.* 43:417–419.

Hall, I. V., and Aalders, L. E. 1964. A Comparison of Flower-Bud Development in the Lowbush Blueberry, Vaccinium angustifolium Ait. under Greenhouse and Field Conditions. *Proc. Amer. Soc. Hort. Sci.* 85:281–284.

Hall, I. V., Aalders, L. E., and Townsend, L. R. 1964. The Effects of Soil pH on the Mineral Composition and Growth of the Lowbush Blueberry. *Can. J. Plant Sci.* 44:433–438.

Hall, I. V., Craig, D. L., and Aalders, L. E. 1963. The Effect of Photoperiod on the Growth and Flowering of the Highbush Blueberry (Vaccinium corymbosum L.). *Proc. Amer. Soc. Hort. Sci.* 82:260–263.

Hall, I. V., and Ludwig, R. A. 1961. The Effects of Photoperiod, Temperature, and Light Intensity on the Growth of the Lowbush Blueberry (Vaccinium angustifolium Ait.). *Can. J. Bot.* 39:1733–1739.

Harmer, P. M. 1944. Soil Reaction and Growth of Blueberry. *Soil Sci.* 9:133–141.

Heald, F. D. 1934. Division of Plant Pathology Report. *In Forty-fourth Ann. Report, Washington Agr. Exp. Sta. Bull.* 305:47–51.

Heath, G. H., and Luckwill, L. C. 1938. The Rooting Systems of Heath Plants with a Section on the Underground Organs of Heath Bryophytes. *J. Ecol.* 26:331–352.

Heddon, S., Gaston, H. P., and Leven, J. H. 1959. Harvesting Blueberries Mechanically. *Mich. Agr. Exp. Sta. Quart. Bull.* 42:24–34.

Highlands, M. E. 1950. Producing Blueberries in Maine. *Me. Agr. Exp. Sta. Bull.* 479.

Hilborn, M. T. 1950. Blueberry Disease Research by the Maine Agricultural Experiment Station. *Mimeograph Rep.* 12.

Hilborn, M. T. 1955. Blueberry Diseases and Their Control. In *Producing Blueberries in Maine, Maine Agr. Exp. Sta. Bull.* 479:27–32.

Hilborn, M. T., and Bonde, R. 1956. *Datura* spp. as Indicator Plants for Apple and Blueberry Virus Diseases (abstract). *Phytopathology* 46:241.

Hilborn, M. T., and Hyland, Fay. 1956. The Mode of Infection of Lowbush Blueberry by *Exobasidium vaccinii* (abstract). *Phytopathology* 46:241.

Hill, R. G., Jr. 1956. Iron Chelates Helping Correct Chlorosis in Blueberries. *Ohio Farm and Home Res.* 41:23, 31.

Hill, R. G., Jr. 1958. Berries. *Amer. Fruit Grower* 78:22.

Hindle, R., Jr. 1955. A Study of Some Growth Characteristics of the Highbush Blueberry. M.S. thesis, University of Rhode Island, Kingston.

Hindle, R., Jr., Shutak, V. G., and Christopher, E. P. 1957. Growth Studies of the Highbush Blueberry Fruit. *Proc. Amer. Soc. Hort. Sci.* 69:282–287.

Hitz, C. W. 1949. Increasing Plant Stand in Blueberry Fields. *Me. Agr. Exp. Sta. Bull.* 467.

Hitz, C. W., and Amling, H. J. 1952. Fruit Notes. *Del. Agr. Exp. Sta. Misc. Paper* 158.

Hodges, M. A., and Peterson, W. H. 1931. Manganese, Copper and Iron Content of Serving Portions of Common Foods. *J. Amer. Diet. Assoc.* 7:6.

Hoerner, J. L., and List, G. M. 1952. Controlling Cherry Fruitworm in Colorado. *J. Entomol.* 45:800–805.

Honey, E. E. 1928. The *Monilioid* Species of *Sclerotinia*. *Mycologia* 20:127–157.

Honey, E. E. 1936. North American Species of *Monilinia*. I. Occurrence, Grouping, and Life Histories. *Amer. J. Bot.* 23:100–106.

Houser, J. R. 1918. Destructive Insects Affecting Ohio Shade and Forest Trees. *Ohio Agr. Exp. Sta. Bull.* 332:207–213.

Howitt, A. J. 1961. Entomological Research on Small and Tree Fruits in 1961. *Mich. State Univ. Entomol. Rep.*

Howitt, A. J. 1964. Entomological Research on Small and Tree Fruits in 1964. *Mich. State Univ. Entomol. Rep.*

Hruschka, H. W., and Kushman, L. J. 1963. Storage and Shelf Life of Packaged Blueberries. *U.S. Dep. Agr. Mark. Res. Rep.* 612.

Huguelet, J. E., Fulton, R. H., and Veenstra, H. A. 1961. Control of Powdery Mildew on Blueberry. *Plant Dis. Reptr.* 45:368–372.

Huntington, H. G. 1939. A Statement on Blueberry Conditions in North Carolina. *Proc. 8th Ann. Blueberry Open House* 8:1–7.

Hutchinson, M. T. 1949. Means of Developing Resistance to Blueberry Stunt Disease. *Proc. 18th Ann. Blueberry Open House* 18:9–10.

Hutchinson, M. T. 1950. Can You Recognize the Symptoms of Stunt Disease? *Proc. 19th Ann. Blueberry Open House* 19:9–11.

Hutchinson, M. T. 1951. Experimental Blueberry Insect Control in 1951. *Proc. 20th Ann. Blueberry Open House* 20:2–5.

Hutchinson, M. T. 1954. Control of the Cranberry Fruitworm on Blueberries. *J. Econ. Entomol.* 47:518–520.

Hutchinson, M. T. 1955. An Ecological Study of the Leafhopper Vectors of Blueberry Stunt. *J. Econ. Entomol.* 48:1–8.

Hutchinson, M. T., Goheen, A. C., and Varney, E. H. 1960. Wild Sources of Blueberry Stunt Virus in New Jersey. *Phytopathology* 50:308–312.

Hutchinson, M. T., Reed, J. P., and Race, S. R. 1960. Nematodes Stunt Blueberry Plants. *N.J. Agr.* 42:12–13.

Hutchinson, M. T., Reed, J. P., Streu, H. T., DiEdwardo, A. A., and Schroeder, P. H. 1961. Plant Parasitic Nematodes of New Jersey. *N.J. Agr. Exp. Sta. Bull.* 796.

Hutchinson, M. T., and Varney, E. H. 1954. Ringspot—A Virus Disease of Cultivated Blueberry. *Plant Dis. Reptr.* 38:260–262.

Isidro, Dennis, S. 1964. Trends in Blueberry Utilization and Consumer Preference Tests of Products from Two New Processing Technologies for Blueberries: Dehydrofreezing and Explosive Puffing. M.S. thesis, Michigan State University, East Lansing.

Janifer, C. S., Jr. 1952. A Study of Certain Thermal Features of the Climate of New Jersey. M.S. thesis, Rutgers University, New Brunswick, N.J.

Jenkins, W. R. 1961. Nematodes on Blueberries. *Proc. 29th Ann. Blueberry Open House* 29:6.

Johnston, S. 1928. Investigations in Rooting Blueberry Cuttings. *Proc. Amer. Soc. Hort. Sci.* 25:181–182.

Johnston, S. 1930. The Propagation of the Highbush Blueberry. *Mich. Agr. Exp. Sta. Spec. Bull.* 202.

Johnston, S. 1934. The Cultivation of the Highland Blueberry (*Vaccinium corymbosum*). *Mich. Agr. Exp. Sta. Spec. Bull.* 252.

Johnston, S. 1935. Propagating Low and Highbush Blueberry Plants by Means of Small Side Shoots. *Proc. Amer. Soc. Hort. Sci.* 33:372–375.

Johnston, S. 1937. Influence of Cultivation on the Growth and Yield of Blueberry Plants. *Mich. Agr. Exp. Sta. Quart. Bull.* 19:232–234

Johnston, S. 1939. The Influence of Certain Hormone-Like Substances on the Rooting of Hardwood Blueberry Cuttings. *Mich. Agr. Exp. Sta. Quart. Bull.* 21:255–258.

Johnston, S. 1942a. Observations on the Inheritance of Horticulturally Important Characteristics in the Highbush Blueberry. *Proc. Amer. Soc. Hort. Sci.* 40:352–356.

Johnston, S. 1942b. The Influence of Various Soils on the Growth and Productivity of the Highbush Blueberry. *Mich. Agr. Exp. Sta. Quart. Bull.* 24:307–310.

Johnston, S. 1943a. Essentials of Blueberry Culture. *Mich. Agr. Exp. Sta. Circ. Bull.* 188.

Johnston, S. 1943b. The Influence of Manure on the Yield and Size of Fruit of the Highbush Blueberry. *Mich. Agr. Exp. Sta. Quart. Bull.* 25:374–376.

Johnston, S. 1944. Investigations in Budding the Highbush Blueberry. *Proc. Amer. Soc. Hort. Sci.* 44:301–302.

Johnston, S. 1946. Observations on Hybridizing Lowbush and Highbush Blueberries. *Proc. Amer. Soc. Hort. Sci.* 47:199–200.

Johnston, S. 1947. Essentials of Blueberry Culture. *Mich. Agr. Exp. Sta. Circ. Bull.* 188 (revised).

Johnston, S. 1948. The Behavior of Highbush and Lowbush Blueberry Selections and Their Hybrids Growing on Various Soils Located at Different Levels. *Mich. Agr. Exp. Sta. Tech. Bull.* 205.

Johnston, S. 1949. A Preliminary Report on the Control of Mummy Berry (*Sclerotinia vaccinii*) in Blueberries by Use of a Chemical Weed Killer. *Proc. Amer. Soc. Hort. Sci.* 54:189–191.

Johnston, S. 1951a. Essentials of Blueberry Culture. *Mich. Agr. Exp. Sta. Circ. Bull.* 188 (revised).

Johnston, S. 1951b. Problems Associated with Cultivated Blueberry Production in Northern Michigan. *Mich. Agr. Exp. Sta. Bull.* 33:293–298.

Johnston, S. 1951c. The Keweenaw Blueberry. A Variety for Trial in Northern Michigan. *Mich. Agr. Exp. Sta. Quart. Bull.* 33:299–301.

Johnston, S. 1953. A New Method of Growing Blueberries in the Home Garden. *Mich. Agr. Exp. Sta. Quart. Bull.* 36:226–229.

Johnston, S. 1956. Blueberry Breeding in Michigan. *Fruit Var. and Hort. Dig.* 11:20.

Johnston, S. 1958. Investigations with Plastic Covers for Blueberry Propagating Frames. *Mich. Agr. Exp. Sta. Quart. Bull.* 40:530–534.

Johnston, S. 1959. Essentials of Blueberry Culture. *Mich. Agr. Exp. Sta. Circ. Bull.* 188 (revised).

Judkins, W. P. 1949. Blueberry Culture in Ohio. *Ohio Farm and Home Res.* 34:107–112.

Judkins, W. P. 1951. Sprinklers for Berries. *Amer. Fruit Grower* 71:9, 18, 19.

Keifer, H. H. 1941. Eriophyed Studies—Blueberries. *Calif. Dep. Agr. Bull.* 30:196–216.

Kender, W. J. 1965. Some Factors Affecting the Propagation of Lowbush Blueberries by Softwood Cuttings. *Proc. Amer. Soc. Hort. Sci.* 86:301–306.

Kender, W. J., and Anastasia, F. 1964. Nutrient Deficiency Symptoms of the Lowbush Blueberry. *Proc. Amer. Soc. Hort. Sci.* 84:275–280.

Kender, W. J., Eggert, F. P., and Whitton, L. 1964. Growth and Yield of Lowbush Blueberries as Influenced by Various Pruning Methods. *Proc. Amer. Soc. Hort. Sci.* 84:269–273.

Kenworthy, A. L., Larsen, R. P., and Bell, H. K. 1956. Fertilizers for Fruit Crops. *Mich. Agr. Exp. Sta. Ext. Folder* F-224.

Kenworthy, A. L., Larsen, R. P., and Bell, H. K. 1962. Fertilizers for Fruit Crops. *Mich. Agr. Exp. Sta. Ext. Folder* F-224 (revised).

Kinsman, G. B. 1957. The Lowbush Blueberry in Nova Scotia. *Dep. Agr. and Mark., Halifax, N.S., Pub.* 1036.

Knight, R. J., and Scott, D. H. 1964. Effects of Temperatures on Self- and Cross-Pollination and Fruiting of Four Highbush Blueberry Varieties. *Proc. Amer. Soc. Hort. Sci.* 84:302–306.

Bibliography

Kramer, A., Evinger, E. L., and Schrader, A. L. 1941. Effect of Mulches and Fertilizers on Yield and Survival of the Dryland and Highbush Blueberries. *Proc. Amer. Soc. Hort. Sci.* 38:455–461.

Kramer, A., and Schrader, A. L. 1942. Effect of Nutrients, Media and Growth Substance on the Growth of the Cabot Variety of *Vaccinium corymbosum*. *J. Agr. Res.* 65:313–328.

Kramer, A., and Schrader, A. L. 1945. Significance of the pH of Blueberry Leaves. *Plant Physiol.* 20:30–36.

Krohn, J., and Mahn, F. 1962. Technical Help for Blueberry Growers from the Soil Conservation Service. *Proc. 30th Ann. Blueberry Open House* 30:10–11.

Kushman, L. J., and Ballinger, W. E. 1963. Influence of Season and Harvest Interval upon Quality of Wolcott Blueberries Grown in Eastern North Carolina. *Proc. Amer. Soc. Hort. Sci.* 83:395–405.

Lathrop, F. H. 1945. Ten years of Warfare against the Blueberry Maggot. *J. Econ. Entomol.* 38:330–334.

Lathrop, F. H. 1952. Fighting the Blueberry Fruitfly in Maine. *Me. Agr. Exp. Sta. Bull.* 500.

Lathrop, F. H., and Nickels, C. B. 1931. The Blueberry Maggot from an Ecological Viewpoint. *Ann. Entomol. Soc. Amer.* 24:260–281.

Lathrop, F. H., and Nickels, C. B. 1932. The Biology and Control of the Blueberry Maggot in Washington County, Maine. *U.S. Dep. Agr. Tech. Bull.* 275.

Latimer, L. P., and Smith, W. W. 1938. Improved Blueberries. *N.H. Agr. Exp. Sta. Ext. Circ.* 215.

Lee, W. R. 1958. Pollination Studies in Lowbush Blueberries. *J. Econ. Entomol.* 51:544–545.

Liebster, G. 1960. Hinweise für den Anbau der Kulturheidelbeere. *Der Erwerbsobstbau* 12:236–241.

Lister, R. M., Raniere, L. C., and Varney, E. H. 1963. Relationships of Viruses Associated with Ringspot Disease of Blueberry. *Phytopathology* 53:1031–1035.

Lockhart, C. L. 1956. *Thirty-fourth Annual Report of the Canadian Plant Disease Survey, 1955*, Canadian Department of Agriculture, Ottawa.

Lockhart, C. L. 1958. Studies on Red Leaf Disease of Lowbush Blueberries. *Plant Dis. Reptr.* 42:764–767.

Lockhart, C. L. 1959. Symptoms of Mineral Deficiency in the Lowbush Blueberry. *Plant Dis. Reptr.* 43:102–105.

Lockhart, C. L. 1961a. Diseases and Their Control. In *The Blueberry in the Atlantic Provinces, Can. Dep. Agr. Pub.* 754:31–35.

Lockhart, C. L. 1961b. Monilinia Twig and Blossom Blight of Lowbush Blueberry and Its Controls. *Can. J. Plant Sci.* 41:336–341.

Lockhart, C. L., and Hall, I. V. 1962. Note on an Indication of Shoestring Virus in the Lowbush Blueberry *Vaccinium angustifolium* Ait. *Can. Bot.* 40:1561–1562.

Lockhart, C. L., and Langille, W. M. 1962. The Mineral Content of the Lowbush Blueberry. *Can. Plant Dis. Survey* 42:124–127.

Longley, A. E. 1927. Chromosomes in *Vaccinium*. *Science* 66:566–568.
Longyear, B. O. 1901. A Sclerotium Disease of the Huckleberry. *Ann. Rep. Mich. Acad. Sci. Arts and Lett.* 3:61–62.
Mahlstede, J. P., and Watson, D. P. 1952. An Anatomical Study of Adventitious Root Development in Stems of *Vaccinium corymbosum*. *Bot. Gaz.* 113:279–285.
Maramorosch, K. 1955. Transmission of Blueberry Stunt Virus by *Scaphytopius magdalensis*. *J. Econ. Entomol.* 48:106.
Markin, Florence L. 1931. Progress of Investigations. *Me. Agr. Exp. Sta. Bull.* 360:204–206.
Marucci, P. E. 1947. A Leafhopper Survey in Blueberry Fields. *Proc. 16th Ann. Blueberry Open House* 16:3–5.
Marucci, P. E. 1951. New Soil Grubs Attacking Crowns of Blueberry Bushes. *Proc. 20th Ann. Blueberry Open House* 20:6–8.
Marucci, P. E. 1953a. Life Cycle and Control of Cherry Fruitworm on Blueberries. *Proc. 22nd Ann. Blueberry Open House* 22:10–14.
Marucci, P. E. 1953b. The Sparganothis Fruitworm in New Jersey. *Proc. Ann. Meet. Amer. Cranberry Growers' Assoc.* 83:6–12.
Marucci, P. E. 1958. Malathion Bait Sprays for Control of Blueberry Maggot. *Proc. 26th Ann. Blueberry Open House* 26:19–21.
Marucci, P. E., and Fort, W. Z. 1952. Commentary on Maggot, Cherry Fruitworm and Other Blueberry Insects. *Proc. 21st Ann. Blueberry Open House* 21:11–15.
Marucci, P. E., Tomlinson, W. E., Jr., and Doehlert, C. A. 1947. Cage Tests for Possible Carriers of Blueberry Stunt Disease. *Proc. 16th Ann. Blueberry Open House* 16:9–11.
Mason, I. C. 1950. Use of Fertilizer. *In Producing Blueberries in Maine, Me. Agr. Exp. Sta. Bull.* 479:12–15.
Matzner, F. 1965. Über den Vitamin-C-Gehalt der Kulturheidebeeren. *Erw-Obstb.* 7:105–108.
Matthews, J. R., and Knox, E. M. 1926. The Comparative Morphology of the Stamen in the *Ericacae*. *Trans. and Proc. Bot. Soc. Edinburgh* 29:243–281.
Maxwell, C. W. B. 1950. Field Observations on the Black Army Cutworm, *Actebia fennica* (Tausch), and Its Control in New Brunswick. *Sci. Agr.* 30:132–135.
Maxwell, C. W. B., and Wood, G. W. 1961. Insects and Their Control. *In The Blueberry in the Atlantic Provinces, Can. Dep. Agr. Pub.* 754:26–30 (revised).
Maxwell, C. W. B., Wood, G. W., and Neilson, W. 1953. Insects in Relation to Blueberry Culture. *Progress Report 1949–1953 of Dominion of Canada Blueberry Substation*, Tower Hill, N.B.
McAllister, L. C., Jr. 1933. Results of Dusting Experiments to Control the Blueberry Maggot. *J. Econ. Entomol.* 26:221–227.
McAllister, L. C., Jr., and Anderson, W. H. 1935. Insectary Studies on the Longevity and Preoviposition Period of the Blueberry Maggot and on Cross Breeding with the Apple Maggot. *J. Econ. Entomol.* 28:675–678.

McKeen, W. E. 1958. Blueberry Canker in British Columbia. *Phytopathology* 48:277–280.

Meader, E. M. 1952. Accelerated Increase of Highbush Blueberry by Forced Softwood Cuttings. *Proc. Amer. Soc. Hort. Sci.* 60:97–100.

Meader, E. M., and Darrow, G. M. 1947. Highbush Blueberry Pollination Experiments. *Proc. Amer. Soc. Hort. Sci.* 49:196–204.

Meader, E. M., McCrum, R. C., and Rich, A. E. 1964. Blueberry Stunt and Mosaic Viruses in New Hampshire. *Plant Dis. Reptr.* 48:286–287.

Meader, E. M., Smith, W. W., and Yeager, A. F. 1954. Bush Types and Fruit Colors in Hybrids of Highbush and Lowbush Blueberries. *Proc. Amer. Soc. Hort. Sci.* 63:272–278.

Merriam, O. A., and Fellers, C. R. 1937. Vitamin Content and Other Nutrition Studies with Blueberries. *Mass. Sta. Bull.* 339.

Merrill, T. A. 1936. Pollination of the Highbush Blueberry. *Mich. Agr. Exp. Sta. Bull.* 151.

Merrill, T. A. 1939. Acid Tolerances of the Highbush Blueberry. *Mich. Agr. Exp. Sta. Quart. Bull.* 22:112–116.

Merrill, T. A. 1944. Effects of Soil Treatment on the Growth of the Highbush Blueberry. *J. Agr. Res.* 69:9–20.

Metcalf, C. L., and Flint, W. D. 1951. *Destructive and Useful Insects*. 3rd ed. McGraw-Hill Book Company, New York. 1071 pp.

Mikkelsen, D. S., and Doehlert, C. A. 1950. Magnesium Deficiency in Blueberry. *Proc. Amer. Soc. Hort. Sci.* 55:289–292.

Minton, N. A., Hagler, T. B., and Brightwell, W. T. 1951. Nutrient-Element Deficiency Symptoms of the Rabbiteye Blueberry. *Proc. Amer. Soc. Hort. Sci.* 58:115–119.

Moore, J. N. 1964. Duration of Receptivity to Pollination of Flowers of the Highbush Blueberry and the Cultivated Strawberry. *Proc. Amer. Soc. Hort. Sci.* 85:295–301.

Moore, J. N. 1965. Improving Highbush Blueberries by Breeding and Selection. *Euphytica* 14:39–48.

Moore, J. N., Bowen, H. H., and Scott, D. H. 1962. Response of Highbush Blueberry Varieties, Selections, and Hybrid Progenies to Powdery Mildew. *Proc. Amer. Soc. Hort. Sci.* 81:274–280.

Moore, J. N., and Ink, D. P. 1964. Effect of Rooting Medium, Shading, Type of Cutting, and Cold Storage of Cuttings on the Propagation of Highbush Blueberry Varieties. *Proc. Amer. Soc. Hort. Sci.* 84:285–294.

Moore, J. N., Scott, D. H., and Dermen, H. 1964. Development of a Decaploid Blueberry by Colchicine Treatment. *Proc. Amer. Soc. Hort. Sci.* 84:274–279.

Moore, J. N., and Stretch, A. W. 1963. Incidence of Red Ringspot Virus in Experimental and Commercial Blueberry Plantations in New Jersey. *Plant Dis. Reptr.* 47:294–297.

Morrow, E. B., and Darrow, G. M. 1950. The Murphy and Wolcott Blueberries. *N.C. Agr. Exp. Sta. Spec. Circ.* 10.

Morrow, E. B., Darrow, G. M., and Rigney, J. A. 1949. A Rating System for the Evaluation of Horticultural Material. *Proc. Amer. Soc. Hort. Sci.* 53:276–280.

Morrow, E. B., Darrow, G. M., and Scott, D. H. 1954. A Quick Method for Cleaning Berry Seed for Breeders. *Proc. Amer. Soc. Hort. Sci.* 63:265.

Mowry, H., and Camp, A. F. 1928. Blueberry Culture in Florida. *Fla. Agr. Exp. Sta. Bull.* 194.

National Association of Frozen Food Packers. 1962. *Frozen Food Pack Statistics,* Washington, D.C.

National Canners Association. 1962. *Canned Food Statistics,* Washington, D.C.

Nelson, E. K. 1927. The Non-Volatile Acids of the Pear, Quince, Apple, Loganberry, and Blueberry. *J. Amer. Chem. Soc.* 49:1300–1302.

Newcomber, E. H. 1941. Chromosome Numbers of Some Species and Varieties of *Vaccinium* and Related Genera. *Proc. Amer. Soc. Hort. Sci.* 38:468–470.

New Jersey Agricultural Experiment Station. 1960. New Jersey Fertilizer and Lime Recommendations. *N.J. Agr. Exp. Sta. Circ.* 589:10.

New Jersey State Department of Agriculture. 1963. The Blueberry Industry of New Jersey. *N.J. St. Dep. Agr. Circ.* 425.

New Jersey State Department of Agriculture, Crop Reporting Service. 1962. 1961 New Jersey Agriculture Statistics. *N.J. St. Dep. Agr. Circ.* 422:1–72.

New Jersey State Department of Agriculture, Crop Reporting Service. 1964. 1963 New Jersey Agriculture Statistics. *N.J. St. Dep. Agr. Circ.* 430:1–56.

New York City Bureau of Consumer Service and Research. 1964. Weekly Retail Price Report for New York City.

North Carolina State College Agricultural Experiment Station. 1956. Fertilizer Recommendations. *N.C. Agr. Exp. Sta. Leaflet* 3.

North Carolina State University Agricultural Extension Service. 1964. *Handbook for Agricultural Workers.*

O'Rourke, F. L. 1942. The Influence of Blossom Buds on Rooting of Hardwood Cuttings of Blueberry. *Proc. Amer. Soc. Hort. Sci.* 40:332–334.

O'Rourke, F. L. 1943. The Effect of Indole-Butyric Acid in Talc on Rooting of Softwood Cuttings of Blueberries. *Proc. Amer. Soc. Hort. Sci.* 42:369–370.

O'Rourke, F. L. 1944. Wood Type and Original Position on Shoot with Reference to Rooting in Hardwood Cuttings of Blueberry. *Proc. Amer. Soc. Hort. Sci.* 45:195–197.

O'Rourke, F. L. 1952a. Propagation Studies with the Low Dryland Blueberry, *Vaccinium vacillans* Torr. *Proc. Amer. Soc. Hort. Sci.* 59:150–152.

O'Rourke, F. L. 1952b. The Influence of the Position of the Basal Cut on Rooting of Hardwood Cuttings of Blueberry. *Proc. Amer. Soc. Hort. Sci.* 59:153–154.

Pelletier, E. N., and Hilborn, M. T. 1954. Blossom and Twig Blight of Lowbush Blueberries. *Me. Agr. Exp. Sta. Bull.* 529.

Perkins, F. A. 1964. New Container Boosts Blueberry Sales. *Prod. Mark.* 7:22–23.

Bibliography

Perlmutter, F., and Darrow, G. M. 1942. Effect of Soil Media, Photoperiod, and Nitrogenous Fertilizer on the Growth of Blueberry Seedlings. *Proc. Amer. Soc. Hort. Sci.* 40:341-346.

Persing, D. P. 1964. 1963 New Jersey Fresh Marketing Summary of Blueberries, Strawberries, Peaches. *N.J. St. Dep. Agr. Mimeo. Rep.* pp. 3-11.

Phipps, C. R. 1927. The Black Army Cutworm, a Blueberry Pest. *Me. Agr. Exp. Sta. Bull.* 340.

Phipps, C. R. 1930. Blueberry and Huckleberry Insects. *Me. Agr. Exp. Sta. Bull.* 356:107-232.

Phipps, C. R., Chandler, F. B., and Mason, I. C. 1932. Blueberry Pollination. *Me. Agr. Exp. Sta. Bull.* 363:266.

Pickett, A. D. 1943. Blueberry Insects and Their Control. In *The Blueberry, Can. Dep. Agr. Farmers' Bull.* 120:25-27.

Pickett, A. D., and Neary, M. E. 1940. Further Studies on *Rhagoletis pomonella* Walsh. *Sci. Agr.* 20:551-556.

Popenoe, J. 1952. Mineral Nutrition of the Blueberry as Indicated by Leaf Analysis. M.S. thesis, University of Maryland, College Park.

Poray, R. A. 1958. Burning Blueberry Land. *Me. Ext. Serv. Circ.* 327.

Porter, B. A. 1928. The Apple Maggot. *U.S. Dep. Agr. Tech. Bull.* 66.

Raine, J. 1965. Control of *Dasystoma salicellum*, a New Pest of Blueberries in British Columbia. *Can. J. Plant Sci.* 45:243-245.

Raniere, L. C. 1959. A Note on Blueberry Viruses. *Proc. 27th Ann. Blueberry Open House* 27:20.

Raniere, L. C. 1960. Responses of Cultivated Highbush Blueberry Varieties to the Known Blueberry Viruses. *Proc. 28th Ann. Blueberry Open House* 28:18-20.

Raniere, L. C. 1961. Observations on New or Unusual Diseases of Highbush Blueberry. *Plant Dis. Reptr.* 45:844.

Raniere, L. C. 1964. Necrotic Ringspot Virus of Blueberry, *Vaccinium corymbosum* L., and Its Identity. Ph.D. thesis, Rutgers University, New Brunswick, N.J.

Rayment, A. F. 1965. The Response of Native Stands of Lowbush Blueberry in Newfoundland to Nitrogen, Phosphorus, and Potassium Fertilizers. *Can. J. Plant Sci.* 45:145-152.

Rayner, M. C. 1929. The Biology of Fungus Infection in the Genus *Vaccinium*. *Ann. Bot.* 43:55-70.

Reade, J. M. 1908. Preliminary Notes on Some Species of *Sclerotinia*. *Ann. Mycol.* 6:109-115.

Robbins, W. W. 1931. *The Botany of Crop Plants*. Blakiston's Son & Co., Philadelphia, pp. 517-522.

Roberts, D. A., and Weber, G. F. 1952. Seasonal Relations of Silky Thread Blight, *Rhizoctonia ramicola* W. & R. (abstract). *Phytopathology* 42:287.

Rousi, A. 1963. Hybridization between *Vaccinium uliginosum* and Cultivated Blueberry. *Ann. Agr. Fenniae* 2:12-18.

Royle, D. J., and Hickman, C. J. 1963. *Phytophthora cinnamomi* on Highbush Blueberry. *Plant Dis. Reptr.* 47:266-268.

Rubey, H. 1954. *Supplemental Irrigation for Eastern United States.* Interstate Printers and Publishers, Danville, Ill. Ch. V.

Savage, E. F., and Darrow, G. M. 1942. Growth Response of Blueberries under Clean Cultivation and Various Kinds of Mulch Materials. *Proc. Amer. Soc. Hort. Sci.* 40:338–340.

Savile, D. B. O. 1959. Notes on *Exobasidium. Can. J. Bot.* 37:641–656.

Schaefers, G. A. 1962. Life History of the Blueberry Tip Borer *Hendecaneura shawiana* (Kearfott) (*Lepidoptera: Tortricidae*), a New Pest. *Ann. Entomol. Soc. Amer.* 55:119–123.

Schallock, D. A. 1962. Weed Control in Blueberries. *Proc. 30th Ann. Blueberry Open House* 30:20.

Schallock, D. A. 1965. Weed Control in Blueberries. *Proc. 33rd Ann. Blueberry Open House* 33:5.

Schwartze, C. D., and Myhre, A. S. 1947. Rooting Blueberry Cuttings. *Wash. Agr. Exp. Sta. Bull.* 488.

Schwartze, C. D., and Myhre, A. S. 1948. Fertilizer Responses of Blueberry Hardwood Cuttings. *Proc. Amer. Soc. Hort. Sci.* 51:309–312.

Schwartze, C. D., and Myhre, A. S. 1949. Further Experiments in Fertilizing Blueberry Hardwood Cuttings. *Proc. Amer. Soc. Hort. Sci.* 54:186–188.

Schwartze, C. D., and Myhre, A. S. 1954a. Blueberry Propagation. *Wash. Agr. Exp. Sta. Circ.* 124.

Schwartze, C. D., and Myhre, A. S. 1954b. Growing Blueberries in the Puget Sound Region of Washington. *Wash. Agr. Exp. Sta. Circ.* 245.

Scott, D. H. 1953. The Big Six Varieties of Blueberries. *Trans. Peninsula Hort. Soc.* 43:37–41.

Scott, D. H., and Ink, D. P. 1955. Treatments to Hasten the Emergence of Seedlings of Blueberry and Strawberry. *Proc. Amer. Soc. Hort. Sci.* 66:237–242.

Scott, D. H., and Moore, J. N. 1961. *Status of Blueberry Breeding Program, United States Department of Agriculture and Cooperating Agencies.* U.S. Dep. Agr. Crops Research Div. Mimeo.

Scott, D. H., Moore, J. N., Knight, R. J., Jr., and Hough, L. F. 1960. The Collins Blueberry, a New Variety. *N.J. Agr. Exp. Sta. Bull.* 792.

Seaver, F. J. 1945. Photographs and Descriptions of Cup-Fungi-XXXIX. The Genus *Godronia* and Its Allies. *Mycologia* 37:333–359.

Shaffer, J. D. 1960. Consumer Purchase Patterns for Individual Fresh, Frozen and Canned Fruits and Vegetables. *Mich. St. Univ. Bull.* 8.

Sharpe, R. H. 1954. Horticultural Development of Florida Blueberries. *Proc. Fla. St. Hort. Soc.* 66:188–190.

Sharpe, R. H., and Darrow, G. M. 1960. Breeding Blueberries for the Florida Climate. *Proc. Fla. St. Hort. Soc.* 72:308–311.

Shear, C. L. 1917. Endrot of Cranberries. *J. Agr. Res.* 11:35–42.

Shear, C. L., and Bain, H. F. 1929. Life History and Pathological Aspects of *Godronia cassandrae* Peck (*Fusicoccum putrefaciens* Shear) on Cranberry. *Phytopathology* 19:1017–1024.

Shive, J. W. 1933. Blueberry Nutrition. *N.J. Agr.* 15:6.

Shoemaker, J. S. 1955. *Small Fruit Culture*. 3rd ed. McGraw-Hill Book Company, New York. Ch. V.

Shutak, V. G., and Christopher, E. P. 1952. Sawdust Mulch for Blueberries. *R.I. Agr. Exp. Sta. Bull.* 312.

Shutak, V. G., Christopher, E. P., and McElroy, L. 1949. The Effect of Soil Management on the Yield of Cultivated Blueberries. *Proc. Amer. Soc. Hort. Sci.* 53:253–258.

Shutak, V. G., Hindle, R., Jr., and Christopher, E. P. 1957. Growth Studies of the Cultivated Blueberry. *R.I. Agr. Exp. Sta. Bull.* 339.

Siegler, E. A. 1938. Relations between Crown Gall and pH of the Soil. *Phytopathology* 28:858–859.

Simanton, F. L. 1916. The Terrapin Scale: An Important Insect Enemy of Peach Orchards. *U.S. Dep. Agr. Bull.* 351.

Slate, G. L., and Collison, R. C. 1942. The Blueberry in New York. *N.Y. Agr. Exp. Sta. Circ.* 189.

Smith, K. M. 1957. *A Textbook of Plant Virus Diseases*. 2nd ed. Little, Brown and Company, Boston. 652 pp.

Smith, W. W. 1946. Culture of Lowbush Blueberries. *N.H. Ext. Circ.* 275:7–11.

Smith, W. W., Eggert, R., and Yeager, A. F. 1946. Response of the Lowbush Blueberry to Fertilizer. *Proc. Amer. Soc. Hort. Sci.* 48:263–268.

Snapp, O. I. 1930. Life History and Habits of the Plum Curculio in the Georgia Peach Belt. *U.S. Dep. Agr. Tech. Bull.* 188.

Stace-Smith, R., Wooley, P. H., and Vaughan, E. K. 1953. A New Disease of Cultivated Blueberry (abstract). *Phytopathology* 43:589.

Steere, R. L. 1956. Purification and Properties of Tobacco Ringspot Virus. *Phytopathology* 46:60–69.

Stene, A. E. 1938. Some Observations on the Propagation of the Highbush Blueberry. *Proc. Amer. Soc. Hort. Sci.* 36:158–160.

Stene, A. E. 1939. Some Observations on Blueberry Nutrition Based on Greenhouse Culture. *Proc. Amer. Soc. Hort. Sci.* 36:620–622.

Stene, A. E., and Christopher, E. P. 1941. Some Problems Affecting the Rooting of Hardwood Blueberry Cuttings. *Proc. Amer. Soc. Hort. Sci.* 39:259–261.

Stevens, N. E. 1919. The Development of the Endosperm in *Vaccinium corymbosum*. *Torrey Bot. Club Bull.* 46:465–468.

Stevens, N. E. 1924. Notes on Blueberry and Cranberry Diseases. *Proc. Ann. Conv. Amer. Cranberry Growers' Assoc.* 55:7, 10.

Stevens, N. E. 1926. The False Blossom Situation. *Proc. Ann. Conv. Amer. Cranberry Growers' Assoc.* 57:20–26.

Stretch, A. W. 1959. Experiments on Maintenance of Blueberry Quality in 1958. *Proc. 27th Ann. Blueberry Open House* 27:7–11.

Stretch, A. W. 1960. Keeping Quality of Blueberries, pp. 32–34. *In* J. N. Moore and N. F. Childers (ed.), *Blueberry Research—Fifty Years of Progress*, New Jersey Agricultural Experiment Station, New Brunswick.

Stretch, A. W. 1963. The Present Status of Blueberry Viruses in New Jersey. *Proc. 31st Ann. Blueberry Open House* 31:5.

Struchtemeyer, R. A. 1956. For Larger Yields Irrigate Lowbush Blueberries. *Me. Farm Res.* 4:17–18.

Suit, R. F. 1945. Stunt Disease of Blueberries, p. 40. *In 64th Ann. Rep. N.Y. St. Agr. Exp. Sta.*, Cornell University, Ithaca, N.Y.

Swift, F. C., and Davis, H. D., Jr. 1965. Blueberry Pest Control. *N.J. Agr. Exp. Sta. Leaflet* 347.

Taylor, J. 1958. Stem Canker and Related Blueberry Diseases. *N.C. St. Tech. Bull.* 132.

Tomlinson, W. E., Jr. 1935. Influence of Temperature on Emergence of Blueberry Maggot. *J. Econ. Entomol.* 44:266–267.

Tomlinson, W. E., Jr. 1947. A General Insect Control Experiment in Relation to Stunt Disease. *Proc. 16th Ann. Blueberry Open House* 16:1–2.

Tomlinson, W. E., Jr. 1948. Seven Important Blueberry Insects. *Proc. 17th Ann. Blueberry Open House* 17:10–13.

Tomlinson, W. E., Jr. 1949. Blueberry Insects in 1949. *Proc. 18th Ann. Blueberry Open House* 18:5–8.

Tomlinson, W. E., Jr. 1951. Summer Oil Sprays to Control Blueberry Bud Mite. *J. Econ. Entomol.* 43:727.

Tomlinson, W. E., Jr. 1959. Blueberry Insects. *In Blueberry Growing, Univ. Mass. Pub.* 240:12–13.

Tomlinson, W. E., Jr., Marucci, P. E., and Doehlert, C. A. 1950. Leafhopper Transmission of Blueberry Stunt Disease. *J. Econ. Entomol.* 43:658–662.

Trevett, M. F. 1950. Use of Fertilizer for Maine Blueberries. *Me. Agr. Exp. Sta. Mimeo. Rep.* 8.

Trevett, M. F. 1952. Control of Woody Weeds in Lowbush Blueberry Fields. *Me. Agr. Exp. Sta. Bull.* 499.

Trevett, M. F. 1953. Woody Weed Control in Lowbush Blueberry Fields. *Me. Agr. Exp. Sta. Mimeo. Rep.* 35.

Trevett, M. F. 1955. Some Growth Habits of the Lowbush Blueberry. *Me. Farm Res.* 3:16–18.

Trevett, M. F. 1956. Observations on the Decline and Rehabilitation of Lowbush Blueberry Fields. *Me. Agr. Exp. Sta. Misc. Bull.* 626.

Trevett, M. F. 1961. Controlling Lambkill in Lowbush Blueberries. *Me. Agr. Exp. Sta. Bull.* 600.

Trevett, M. F. 1962. Nutrition and Growth of the Lowbush Blueberry. *Me. Agr. Exp. Sta. Bull.* 605.

Trevett, M. F., and Murphy, H. J. 1958. Chemical Weed Killers, 1958. *Me. Agr. Exp. Sta. Misc. Pub.* 632.

Trevett, M. F. 1965a. Fertilizing Lowbush Blueberries Is Tricky. *Me. Farm Res.* 12:35–42.

Trevett, M. F. 1965b. More about Fertilizing Lowbush Blueberries. *Me. Farm Res.* 13:9–12.

United States Department of Agriculture. 1941. *Climate and Man—Yearbook of Agriculture,* 1941, United States Government Printing Office, Washington, D.C.

United States Department of Agriculture, Agricultural Marketing Service. 1963. Fresh Fruit and Vegetable Unload Totals for 41 Cities. *Rep. No.* 25.

United States Department of Agriculture, Statistical Reporting Service, Crop Reporting Board. 1964a. *Annual Summary Crop Production by States*, United States Government Printing Office, Washington, D.C.

United States Department of Agriculture, Statistical Reporting Service, Crop Reporting Board. 1964b. *Fruits Non-Citrus by States, 1962 and 1963.* Fruits FrNr2-1 (5–64):1–22. United States Government Printing Office, Washington, D.C.

United States Department of Commerce, Bureau of the Census, *United States Census of Agriculture, 1949, 1954,* and *1959.* United States Government Printing Office, Washington, D.C.

United States Department of Commerce, Weather Bureau. 1964. *Climatography of the United States. Decennial Census of United States Climate— Supplement for 1951 through 1960.* United States Government Printing Office, Washington, D.C.

University System of Georgia Agricultural Extension Service. 1957. Fertilizer Recommendations for Georgia. *Ga. Agr. Exp. Sta. Circ.* 371.

Uschdraweit, H. A. 1961. Eine für Deutschland Neve Virose bei *Vaccinium myrtillus*. *Phytopath. Z.* 40:416–419.

Varney, E. H. 1956. Protectant Sprays and Eradicant Ground Treatments for Control of Mummy Berry of Blueberry, 1955. *Proc. 24th Ann. Blueberry Open House* 24:11–16.

Varney, E. H. 1957a. Mosaic and Shoestring, Virus Diseases of Cultivated Blueberry in New Jersey. *Phytopathology* 47:307–309.

Varney, E. H. 1957b. Recently Recognized Virus Diseases of Blueberry in New Jersey. *Proc. 25th Ann. Blueberry Open House* 25:20–22.

Varney, E. H., and Raniere, L. C. 1960. Necrotic Ringspot, a New Virus Disease of Cultivated Blueberry (abstract). *Phytopathology* 50:241.

Vaughan, E. K. 1956. A Strain of *Pseudomonas syringae* Pathogenic on Cultivated Blueberry (abstract). *Phytopathology* 46:640.

Vaughan, E. K., and Boller, C. A. 1954. Experiments on Control of Bacterial Canker of Blueberry. *Plant Dis. Reptr.* 38:867–868.

Ware, L. M. 1930. Propagation Studies with the Southern Blueberry. *Miss. Agr. Exp. Sta. Bull.* 280.

Wasscher, J. 1947. Comparative Anatomy Investigations of Several Species of *Vaccinium*. *Nederland. Dendrol. Vereen. Jaarboek* 15:65–112.

Watt, B. K., and Merrill, A. L. 1964. Composition of Foods. *U.S. Dep. Agr., Agr. Handbk.* 8.

Webber, E. R. 1954. The Cultivation of Blueberries. *World Crops* 6:410–412.

Weeks, E. E., and Nakashima, W. Y. 1964. Washington Blueberry Market Organization and Preferences. *Wash. Agr. Exp. Sta. Bull.* 657.

Weiss, H. B., and Beckwith, C. S. 1945. The Cultivated Blueberry Industry in New Jersey, Including a Report on the Insects of the Cultivated Blueberry. *N.J. State Dep. Agr. Circ.* 356.

White, R. P. 1935. Investigations of *Vaccinium* Diseases, pp. 23–24. In *N.J. Agr. Exp. Sta. Ann. Rep.*, Rutgers University, New Brunswick, N.J.

Whitton, L. 1960. Breeding for Winter Hardiness, pp. 53–54. *In* J. N. Moore and N. F. Childers (ed.), *Blueberry Research—Fifty Years of Progress*, New Jersey Agricultural Experiment Station, New Brunswick.
Wilcox, Marguerite S. 1936. Notes on Blueberry Fungi. *Plant Dis. Reptr.* 20: 106–107.
Wilcox, Marguerite S. 1939. Phomopsis Twig Blight of Blueberry. *Phytopathology* 29:136–142.
Wilcox, Marguerite S. 1940. *Diaporthe vaccinii*, the Ascigerous Stage of *Phomopsis*, Causing a Twig Blight of Blueberry. *Phytopathology* 30:441–443.
Wilcox, R. B. 1940. Blueberry Diseases. *Proc. 9th Ann. Blueberry Open House* 9:9–11.
Wilcox, R. B. 1942. Blueberry Stunt, a Virus Disease. *Plant Dis. Reptr.* 26:211–213.
Wilcox, R. B. 1946. Two Blueberry Diseases Prevalent in New Jersey in 1946. *Proc. 15th Ann. Blueberry Open House* 15:12–13.
Witcher, W., and Clayton, C. N. 1963. Blueberry Stem Blight Caused by *Botryosphaeria dothidea* (*B. Ribis*). *Phytopathology* 53:705–712.
Wood, G. W. 1951. An Annotated List of Lepidopterous Larvae from Commercial Fields in Charlotte County, N.B. *Can. Entomol.* 83:241–244.
Wood, G. W. 1961. The Influence of Honeybee Pollination on Fruit Set of the Lowbush Blueberry. *Can. J. Plant Sci.* 41:332–335.
Wood, G. W. 1962a. Period of Receptivity in Flowers of the Lowbush Blueberry. *Can. J. Bot.* 40:685–686.
Wood, G. W. 1962b. The Blueberry Maggot in the Maritime Provinces. *Can. Dep. Agr. Pub.* 1161.
Wood, G. W. 1965. Evidence of Support of Reduced Fruit Set in Lowbush Blueberry by Pollen Incompatibility. *Can. J. Plant Sci.* 45:601–602.
Wood, G. W., and Barker, W. G. 1964. Preservation of Blueberry Pollen by the Freezedrying Process. *Can. J. Plant Sci.* 44:387–388.
Wood, G. W., Nielson, W. T. A., Maxwell, C. W., and McKiel, J. A. 1954. Life History Studies of *Spaelotis Clandestina* (Harr.) and *Polia Purpurissata* (Grt.) in Lowbush Areas in New Brunswick. *Can. Entomol.* 86:169–173.
Wood, G. W., and Wood, F. A. 1963. Nectar Production and Its Relation to Fruitset in the Lowbush Blueberry. *Can. J. Bot.* 41:1675–1679.
Wood, Mary, and Wright, Carleton. 1954. Blueberries. *N.Y. Agr. Exp. Sta. Ext. Food Mark. Handbk.* 7.
Woodard, O., and Brightwell, W. T. 1952. Planting, Fertilizing and Cultivation of Blueberries. *Ga. Coastal Plain Exp. Sta. Rep.* 35.
Woodbridge, C. G., and Drew, R. H. 1960. A Boron Deficiency Dieback in Highbush Blueberry. *Plant Dis. Reptr.* 44:855–857.
Woods, W. C. 1915. Blueberry Insects in Maine. *Me. Agr. Exp. Sta. Bull.* 244.
Woronin, M. 1888. Über die Sclerotienkrankheit der Vaccinieen-Beeren. *Mem. Acad. Sci. St. Petersburg* 36:1–49.
Worthington, J. T., and Scott, D. H. 1965. Cold Storage Effects on Rooting of Blueberry Cuttings. *Amer. Nurserym.* 121:12, 23.

Wynd, F. L., and Bowden, R. A. 1951. Response of Chlorotic Blueberry Bushes to a Very Insoluble Iron Containing Glossy Frit. *Lloydia* 14:55–57.

Yarbrough, J. A., and Morrow, E. B. 1947. Stone Cells in *Vaccinium*. *Proc. Amer. Soc. Hort. Sci.* 50:224–228.

Young, R. S. 1952. Growth and Development of the Blueberry Fruit (*Vaccinium corymbosum* L. and *V. angustifolium* Ait.). *Proc. Amer. Soc. Hort. Sci.* 59:167–172.

Zentmeyer, G. A., Paulus, A. O., and Burns, R. M. 1962. Avocado Root Rot. *Calif. Agr. Exp. Sta. Circ.* 511.

Zuckerman, B. M. 1957. Field Evaluation of Antibiotics for Control of Crown Gall on the Cultivated Highbush Blueberry. *Plant Dis. Reptr.* 41:674–677.

Zuckerman, B. M. 1959a. Coryneum Canker of Highbush Blueberry (abstract). *Phytopathology* 49:556.

Zuckerman, B. M. 1959b. Fusicoccum Canker of the Highbush Blueberry in Massachusetts. *Plant Dis. Reptr.* 43:803.

Zuckerman, B. M. 1960a. Fungi Collected from Blueberry Stems in Massachusetts. *Plant Dis. Reptr.* 44:416.

Zuckerman, B. M. 1960b. Studies of Two Blueberry Stem Diseases Recently Found in Eastern Massachusetts. *Plant Dis. Reptr.* 44:409–415.

Zuckerman, B. M. 1962. Parasitism and Pathogenesis of the Cultivated Highbush Blueberry by the Stubby Root Nematode. *Phytopathology* 52:1017–1019.

Zuckerman, B. M. 1964. Studies of Two Nematode Species Associated with Roots of the Cultivated Highbush Blueberry. *Plant Dis. Reptr.* 48:170–171.

Zuckerman, B. M., and Bailey, J. S. 1956. A New Gall Disease of Cultivated Blueberry. *Plant Dis. Reptr.* 40:212–216.

Notes on the Contributors

Paul Eck is Associate Professor of Pomology at Rutgers University and project leader for blueberry culture research at the Blueberry-Cranberry Research Center of the New Jersey Agricultural Experiment Station. Dr. Eck did his undergraduate work at Rutgers and holds advanced degrees from the University of Massachusetts and the University of Wisconsin. He conducted research at the Massachusetts Experiment Station in Waltham before joining the Rutgers faculty.

Norman F. Childers, a specialist in fruit nutrition, holds the Maurice A. Blake Chair in Horticulture at Rutgers University. He was educated at the University of Missouri and Cornell University and taught at Ohio State University and served as Assistant Director of the United States Department of Agriculture Experimental Station in Puerto Rico before going to Rutgers. He is the author of *Modern Fruit Science* (3rd edition, 1966) and editor of *Fruit Nutrition: Temperate to Tropical* (2nd edition, 1966).

James N. Moore, who received his Ph.D. from Rutgers University in 1961, was responsible for coordinating New Jersey's blueberry breeding experiments with the federal breeding program from 1957 to 1964. Dr. Moore spent the last three of these years as Research Horticulturist with the USDA at Beltsville working on blueberry breeding, propagation, and pollination. In his current post as Associate Professor in Horticulture at his alma mater, the University of Arkansas at Fayetteville, his duties include the evaluation of blueberry varieties for the upper South. He was a co-developer of the Collins and Darrow varieties.

Walter J. Kender and W. Thomas Brightwell, authors of the chapter on environmental relationships, are authorities on lowbush and rabbiteye blueberry culture, respectively.

Dr. Kender received his graduate training at Rutgers University, where he also became familiar with the culture of the highbush blueberry. Since 1962 he has taught horticulture and acted as project leader for cultural and physiological studies on the lowbush blueberry at the University of

Maine. He was recently co-chairman with Dennis A. Abdalla of a successful Conference of North American Blueberry Workers at Orono, Maine.

Dr. Brightwell has been active in rabbiteye blueberry culture and breeding for over twenty years at the Georgia Coastal Plain Experiment Station at Tifton. He holds degrees from the University of Tennessee, Michigan State University, and Ohio State University. His major efforts have been devoted to improving natural selections of the rabbiteye blueberry, and he is responsible for originating the rabbiteye varieties Callaway, Coastal, Tifblue, Homebell, and Woodard.

George M. Darrow and Donald H. Scott, of the United States Department of Agriculture, who collaborated to prepare the chapter on blueberry varieties, have been the principal project leaders for the highbush blueberry variety breeding program supported by the federal government.

Dr. Darrow received his training at Middlebury College, Cornell University, and Johns Hopkins University and was active in fruit research with the USDA for forty-six years until his retirement in 1957. He is the recipient of many honors, including the Wilder Medal and the Distinguished Service Medal of the USDA. The southern evergreen blueberry *Vaccinium darrowi* and the new highbush blueberry variety Darrow are named for him.

Dr. Scott is Leader of Small Fruit and Grape Investigations at the Plant Industry Station, USDA, Beltsville, Maryland, and project leader and coordinator for the USDA blueberry breeding program. He received his training at North Dakota Agricultural College and the University of Maryland. Over the past twenty years Dr. Scott has collaborated in the origination of new varieties of blueberries, strawberries, blackberries, and grapes.

Charles M. Mainland, who is currently completing the requirements for the Ph.D. at Rutgers University, comes from a blueberry farm in Indiana and began his graduate research on blueberries at Purdue University. In addition to his work on propagation, Mr. Mainland has investigated the effects of growth regulators on blueberry fruit development.

Walter E. Ballinger, of North Carolina State University at Raleigh, is an authority on highbush blueberry production in North Carolina. He grew up on a peach and sweet corn farm near Burlington, New Jersey, and graduated from Rutgers University in 1952. During his graduate years at Michigan State University he began studying the nutritional requirements of the blueberry and has since become a specialist in this field and also in peach nutrition.

Vladimir G. Shutak has been active in highbush blueberry research for nearly twenty years. He holds advanced degrees from the University of Rhode Island and the University of Maryland and has taught horticulture at the University of Rhode Island since 1946. Among his respon-

Notes on the Contributors 357

sibilities are supervision of the experimental fruit farm. He has given particular attention to highbush blueberry fruit development and the influence of mulches on blueberry growth and production. He is the author of over fifty articles and bulletins, including two bulletins on highbush blueberries.

Philip E. Marucci is Research Specialist in Entomology and Extension Specialist in Blueberry and Cranberry Culture at Rutgers University and has been stationed at the Blueberry-Cranberry Research Laboratory, Pemberton, New Jersey, for the past fifteen years. An authority on blueberry insects and their control, he has contributed in an important way to the successful culture of the highbush blueberry in New Jersey. He is frequently called upon to participate in growers' and workers' meetings in blueberry growing regions throughout the country.

Eugene H. Varney and Allan W. Stretch, both plant pathologists, are co-workers at the New Jersey Agricultural Experiment Station.

Dr. Varney was educated at the University of Massachusetts and the University of Wisconsin, and in 1956, after three years as plant pathologist with the USDA, joined the faculty of Rutgers University. He also collaborates officially with Dr. Donald H. Scott's staff in small fruit and grape investigations at the USDA's Plant Industry Station at Beltsville, Maryland. Dr. Varney's work on blueberry and cranberry virus diseases has led to the identification and control of many of the viruses that are currently problems to the industry.

Dr. Stretch, who is with the USDA Agricultural Research Service, Crops Research Division, in charge of blueberry and cranberry disease investigations at the New Jersey Agricultural Experiment Station, received his Ph.D. from Rutgers University in 1962. His research is concerned primarily with blueberry anthracnose disease, cranberry fruit rot, and blueberry viruses, his virus studies dealing particularly with the search for the method of transmission of the red ringspot virus.

Warren C. Stiles and Dennis A. Abdalla are both fruit crop specialists at the University of Maine and the Maine Agricultural Experiment Station.

Dr. Stiles is a graduate of Rutgers University and earned his Ph.D. from Pennsylvania State University in 1958. Before going to Maine in 1963 as Extension Fruit Specialist he taught pomology at Rutgers University, where he also had an opportunity to observe the rapidly expanding highbush industry in New Jersey. He is presently superintendent of the Highmoor Farm Maine Agricultural Experiment Station, Extension Fruit Specialist, and Associate Professor of Pomology.

Mr. Abdalla completed his undergraduate studies at the University of Maryland and received his M.S. degree from Clemson University. Since 1962 he has been the lowbush blueberry specialist at the University of

Maine, where he teaches and conducts blueberry extension programs in production, marketing, and utilization.

Frederick A. Perkins has served as a marketing specialist for the New Jersey Cooperative Extension Service at the Rutgers University College of Agriculture and the New Jersey Agricultural Experiment Station since 1960. He received his university degrees at the University of Maine and before going to New Brunswick was with the Maine Agricultural Extension Service of the University of Maine and the Department of Agricultural Economics of the Maine Experiment Station at Orono. Mr. Perkins has specialized in the development and testing of new consumer packages and master containers for fruits, and has conducted numerous marketing research studies in cooperation with northeastern state universities and the USDA.

John S. Bailey, Associate Professor Emeritus at the University of Massachusetts, has devoted over forty years to research and teaching in pomology. He received his education at Michigan State University and Iowa State University and subsequently joined the Massachusetts Agricultural Experiment Station and the University of Massachusetts, specializing in problems of fruit tree hardiness, fruit drop, blueberry and strawberry culture, peach culture and the inheritance of peach characters. Professor Bailey is the author of numerous articles and bulletins on blueberry culture and has had the prime responsibility for the development of the highbush blueberry industry in Massachusetts.

Index

Aalders, L. E., cited, 20, 45, 84, 193
Abdalla, D. A., cited, 195, 281, 306
abscission, 187; resistance, 57, 63, 73
Aceria vaccinii Keifer, 63, 204, 206, 207, 219, 270–71
acid phosphate, 160
acid soil, *see* soil, pH
Actebia fennica Tausch, 229–30
Actinomycetaceae, 277
Adams variety, 101, 115; breeding experiments with, 51, 52, 66, 95, 107; diseases, 253, 256
advective frosts, 83
aeration, *see* air
Agrobacterium rubi (Hildebrand), Starr and Weiss, 274
Agrobacterium tumefaciens, 274–76
air: disease control and, 121, 122; planting sites and, 83; soil, 81, 86, 90, 134
Alabama, 6, 21; diseases in, 240, 255; rabbiteye of, 5, 22, 83, 108, 109
Alaska, 247, 248
aldrin, 222
Alexander, J. H., 66
Allegheny mountains, 75
Allen variety, 51
allopolyploids, 21, 22
allotetraploids, 15
Alsophila pometaria Harris, 228
Alternaria tenuissima (Sr.) Witts., 261
Altica sylvia May, 232–33
Altica torquata Lec., 228
aluminum sulfate, 158, 159(*tab.*)
amino acids, 149–50, 172
Amling, H. J., cited, 142, 155
Ammate, 169, 170, 265

ammonia, 139, 160
ammonium nitrate, 89, 122, 148, 149, 153, 162
ammonium phosphate, 121, 122
ammonium sulfate, 123, 161, 178, 325; iron chlorosis and, 151; nitrogen supply and, 121, 122, 140, 148, 172; potassium deficiency and, 155; soil pH and, 158
Amphorophora vaccinii, 264
anatomy, 34–44. *See also specific blueberry parts*
Ancylus comptana Froelich, 228
Anderson, J. P., cited, 247, 248
Angola variety, 69, 94, 96; diseases and, 97, 242; Morrow variety and, 101, 107
anions, 148–52
Annapolis Valley, 81
anther, 34
Anthonomus musculus Say, 199, 208, 210
anthracnose, 63, 256–57
Apanteles ornigis Weed, 220
aphids, 264, 271
apple bud moth (*Platynota flavedana* Clem.), 227
Argyrotaenia velutinana Walker, 227
Arkansas, 6, 21, 22
Armillaria root rot (*Armillaria mellea* Bahl), 261
Ashley, E., cited, 90
Ashworth variety, 52, 68
Aspidiotus ancylus Putnam, 204, 206, 207
Atlantic variety, 101; diseases, 52, 239, 241, 244, 248, 250, 267, 277
autopolyploids, 21, 22
autotetraploids, 15

359

bacterial canker (*Pseudomonas syringae* Van Hall), 277–78
bacterial diseases, 274–79
bagworm (*Thyridopteryx ephemeraeformis* Haworth), 228
Bailey, John S., 66, 320–28; cited, 32, 81, 82, 84, 96, 114, 115, 117, 123, 187, 240; on soil management, 132, 133, 140, 142, 150, 151, 152, 153, 155, 156, 158, 161, 163
Bailey, L. H., cited, 14
Bailey, Russell, 66
Bain, H. F., cited, 243
Baker, R. E., cited, 292
Ballinger, Walter E., 132–78; cited, 151, 153, 155, 156, 161, 163
Barnes, E. H., cited, 243, 244
Batchelor, Jackson, cited, 108
Batodendron, 14
bayberry (*Murica colinensis* Miller), 168, 170, 231
Bay of Fundy, 81
Beach, F. H., cited, 325
Beauvaria bassiana, 205
Beckwith, C. S., cited, 22, 86, 133, 134, 160, 183, 185, 291; on insects, 199, 207, 210, 211, 222, 224, 226
bees, 181, 185–86, 187, 210; hive rotation, 194–95
Bell, H. K., cited, 155, 156, 162, 163
Bell, H. P., cited, 31, 37, 38, 84, 193
Bergman, H. F., cited, 250
Berkeley variety, 101, 111, 228; characteristic inheritance and, 58, 68, 97, 98, 107; diseases and, 52, 239, 244, 245, 250, 251, 266, 267, 268
Bessey, O. A., cited, 156
bilberry, 14, 48
birds, 326
black army cutworm (*Actebia fennica* Tausch.), 229–30
blackberries, 302
Black Giant variety, 52, 72, 108, 109, 110, 256
black vine borer (*Brachyrhynus sulcatus* Fabricius), 221–22
black vine weevil, 202
Blasberg, C. H., cited, 162
Blattny, C., 273
Blattny, C., Jr., 273
blood, dried, 160

blossom weevil (*Anthonomus musculus* Say), 199, 208, 210
blueberries (*Vaccinia*): botany of, 13–44; United States industry, 3–13, 302–303. *See also specific varieties; and see specific agricultural and industrial topics, e.g.,* processing
blueberry anthracnose (*Glomerella cingulata* Spaulding and Von Schrenk), 256–57
blueberry bud mite (*Aceria vaccinii* Keifer), 204, 206, 207, 219
blueberry budworm, 226–27
blueberry flea beetle (*Altica sylvia* May and *Altica torquata* Lec.), 228, 232–33
blueberry fruitfly, 202, 204, 215–19
blueberry leaf beetle (*Galerucella vaccinii* Fall), 233
blueberry leafminer (*Gracilaria vacciniella* Ely), 219–20, 228
blueberry maggot, 199, 205, 206, 215–19, 220, 225; lowbush insects and, 232, 233
blueberry midge (*Contarinia vaccinii* Felt.), 228, 235
blueberry sawfly (*Neopareophara litura* Klug), 228, 234
blueberry spittle bug (*Clastoptera saintcyri* Prov.), 228
blueberry stem blight (*Botryosphaeria dothidea* (Moug. ex Fr.) Ces. and deNot.), 249–50
blueberry stem borer (*Oberea myops* Hald.), 199, 206, 224
blueberry stem gall (*Hemadas nubilipennis* Ashm.), 199, 206, 223
blueberry tip borer (*Hendecaneura shawiana* Kearfott), 225
blueberry tip worm (*Contarinia vaccinii* Felt), 228, 235
Bluecrop variety, 58, 61, 68, 104; characteristics, 96, 97, 101, 102; diseases, 239, 250, 257, 266, 267, 268; insects, 206; rooting, 111
Blueray variety, 96, 98, 101, 102, 111; diseases, 52, 239, 250, 257, 266, 268; productivity, 61, 68
Boller, C. A., cited, 127, 128, 140, 153, 247, 262, 278
Bonde, R., cited, 258–60
bone, steamed, 160

Bordeaux mixture, 240, 278
boron, 148–49, 153
Bos, L., cited, 273
Botryosphaeria corticis (cane or stem canker), 69, 97, 236, 240–43, 249, 260
Botryosphaeria dothidea (Moug. ex Fr.) Ces. and deNot., 249–50
Botryotinia fuckeliana (DeBary) Whetzel, 247
botrytis twig and blossom blight (*Botrytis cinerea*), 236, 237, 247–49
Boulanger, L. W., cited, 194
Bowden, R. A., cited, 178
Bowen, H. H., cited, 52, 250
box frames, 115
Boyd, W. M., cited, 264
Brachyrhynus sulcatus Fabricius, 221–22
brake fern, 169
brand labels, 307, 308
Braun, A. J., 278
breeding, 3–4, 10–11, 12, 13, 45–74; chromosome number and, 15; cooperators, 66(*tab.*); disease resistance and, 242; history of, 63–67, 94–95, 108–109; insect resistance, 206; objectives, 57; techniques of, 53–57. *See also* planting; propagation
Brightwell, W. Thomas, 66, 75–93; cited, 22, 49, 52, 53, 72, 73, 108, 110, 130, 131, 175, 188, 198, 202
British Columbia, 24, 75, 79, 227; diseases, 243, 244, 276; lowbush of, 15, 95
Brooks, J. F., cited, 161
Brooks variety, 63, 70, 71, 95; crosses, 103, 106, 107
Brown, Nellie A., 274
budding: propagation method of, 111, 125
bud mite (*Aceria vaccinii*), 63, 204, 206, 207, 219, 270–71
bud-proliferating galls, 276–77
buds, 24, 29, 179, 193; bush characters and, 95, 96–97; cuttings and, 113, 114–15, 120, 122, 130; fertilizers and, 173, 174; frost damage and, 81–82; fungicides and, 240; photoperiod and, 84–85; pruning and, 189, 323
buffalo tree hopper (*Ceresa bubalus* Fabricius), 228
bulking, 288

Bünemann, G., cited, 300
Burchill, J., cited, 193
Burlington variety, 52, 58, 63, 102; bush characters, 96, 97; diseases, 239, 244, 245, 250, 251, 266, 267, 268, 271, 277; insects, 206, 219, 228; scars, 97, 101; storage, 300–301
burning, 5, 89, 169, 249; fertilizing and, 171, 173, 174; insect control and, 204–205, 233; productivity and, 7, 20, 22, 167–68, 195–96
Burns, R. M., cited, 258
Bush, J. T., cited, 108
bushes, 96–97, 100–10 *passim;* size, 51, 52, 53, 57, 70, 71, 95, 128, 140, 179
Butterfield, H. M., cited, 292

Cabot variety, 45, 51, 66, 95, 96, 108; berry characters, 97, 101, 103; cultivation depth and, 134; diseases, 241, 244, 245, 250, 253, 256, 266, 267, 270, 276; insects and, 185, 206, 219, 225; nutrition, 148; rooting, 115, 123
Cain, J. C., cited, 80, 89, 132, 149; on gardening, 322, 323, 325, 326
calcium, 148, 149, 153, 156; lowbush deficiency, 171; rabbiteye deficiency, 175; soil CEC and, 151–52, 155, 163
calcium arsenate, 218, 231, 232, 250
calcium cyanamide, 239
California, 6, 24, 80, 54; harvest, 292–93
Callaway variety, 73, 109, 110
Callistephus chinensis (L.) Nees, 268
callus: formation of, 44
calyx, 31, 34, 187–88
Camp, A. F., cited, 84, 292
Camp, W. H., cited, 14, 15, 21, 24, 32, 45, 47, 75, 88, 94
Canada, 5, 15, 92; diseases, 237, 247, 249, 251, 253, 255, 262, 264, 272; harvest, 280; insects, 218, 226, 227, 229, 230, 231, 232, 233, 234, 235; production, 305; soil management, 172. *See also specific provinces*
Canadian blueberry (*V. myrtilloides* Michaux), 15, 27(*tab.*), 47, 248, 262
cane (stem) canker (*Botryosphaeria corticis*), 57, 69, 97, 236, 240–43, 249, 260. *See also specific varieties*
canning, 3, 305–306, 308; lowbush, 7, 20, 288–89; storage, 291

capping films, 309–10
Captan, 257
carbon dioxide, 90
Carleton, W. M., cited, 136
Carter variety, 95, 104
Cascade Mountains, 79
Catawba variety, 70, 95
cations, 148–52, 153, 163
cells: chlorophyll-containing, 42; epidermal, 34, 37, 38, 42, 44; parenchyma, 37; polyhedral, 42; root formation, 43; stone, 38
Cercosporella sp., 26
chain-spotted geometer (*Cingilia catenaria* Drury), 228, 231
Chamaedaphne calycylata (L.) Moench, 243
Chandler, F. B., cited, 75, 77, 101, 118, 185, 194, 195, 199, 300; on soil management, 138, 156, 167, 168, 171, 172
Charlotte variety, 95
Chatsworth variety, 66, 95, 103, 107
chemical fungus control, *see* fungicides
chemical insect control, *see* insecticides
chemical weed control, *see* herbicides
chemicals, *see specific substances*
Chenopodium amaranticolor L., 268
cherry fruitworm (*Grapholitha packardi* Zell.), 204, 205, 206, 211–15, 226
Childers, Norman F., 3–13
Childs, W. H., cited, 126, 153
Chlamys plicata Fabricius, 228
chlordane, 222, 223
chlorides, 161
chloro IPC, 142
3-chlorophenyl-1,1-dimethylurea (CMU), 142
chloroplasts, 34
chlorosis, 148–49, 150, 151–52, 175, 178. *See also as effect in specific diseases*
Christopher, E. P., cited, 117, 139, 140, 156, 162, 179, 187
chromosome classification, 14–15; sterility and, 45, 70, 73, 74
Cingilia catenaria Drury, 228, 231
ciodrin, 218
Cladosporium, 261
Clara variety, 72, 108, 110
Clark, J. H., cited, 32, 51, 53, 67, 95, 96, 138

classification: bush characters, 96–97; chromosome, 14–15; fruit, 97–99; leaf characters, 94, 95–96. *See also grade standards*
Clastoptera saint-cyri Prov., 228
clay, 86, 138, 144
Clayton, C. N., cited, 242, 249, 258
clean cultivation, 132–36, 138, 139, 140, 161, 167, 175; insect control and, 205, 226
climate, 75–85; highbush and, 11, 13; hybridization and, 4, 48, 52, 73; insects and, 204; planting and, 127; pollination and, 186–87; rest requirement and, 13, 22, 53, 57, 69–70, 114; softwood cutting and, 122. *See also specific climate conditions, i.e.,* cold; drought; heat; light; rainfall; snow
clones, 128, 193, 194, 242
Coastal variety, 73, 109, 110
cold, 75, 78; blueberry maggot and, 215; hardiness and, 48, 68, 80–83, 96; pollination and, 186–87, 194; rest requirement and, 52, 79–80, 96–97, 109, 114
cold frame, 116, 117, 131
cold storage, 290–91
Colgrove, M. S., Jr., cited, 163
Collins variety, 70, 103, 111, 191; diseases, 52, 250, 266, 268; harvesting, 95, 97
Collison, R. C., cited, 138, 161
color, 25–27(*tab.*), 31, 38, 307; berry development, 187–88; breeding for, 49, 51, 52, 55, 57, 58, 69, 70, 71, 72, 97; light intensity and, 85; shoestring disease and, 273; variety rating, 100–10 *passim*
common brake (*Pteris sp.* Linnaeus), 168
Concord variety, 97, 103, 111; cuttings, 114, 125; diseases, 241, 250, 253, 256, 257, 267; insects, 206, 207, 225
Connecticut, 216, 267
Connors, I. L., 243, 244
Conotrachelus nenuphar, 204, 205, 206, 208–10, 233
consumption, 306–307, 315–17
containers, 309–10
Contarinia vaccinii Felt., 228, 235
Cooke, W. B., cited, 250, 261
cooling process, 312–13
copper, 155, 174, 240, 278

copper sulfate, monohydrated, 250
corn rootworm (*Diabrotica duodecimpunctata* Fabricius), 228
corolla, *see* flower
cortex, 44
Coryneum canker (*Coryneum microstictum* Berk. and Br.), 245
costs: burning, 196; freeze-drying, 290; hand packing, 312; harvesting, 97, 281, 291, 298; mulching, 140, 142; spraying, 206; weed control, 169, 171
cottony maple scale (*Pulvinaria innumerabilis* Rathv.), 207
cover crops, 136–37, 138, 139, 167
Coville, F. V., cited, 32, 70, 73, 86, 90; on disease control, 274; on harvesting, 291; hybridization experiments, 10–11, 45, 47, 63–67, 68, 95; on plant development, 181, 184, 185; on soil management, 132, 158, 160
Coville variety, 68, 97, 104; diseases, 239, 250, 251, 256, 257, 266, 268, 277; flavor, 58, 98; insects, 228; pollination, 181, 183, 186, 187; ripening period, 63, 101; rooting, 111
Crabbe selections, 242
Crabbe-4 variety, 101, 104, 107, 108
cracking: resistance to, 97, 100–10 *passim*
Craig, D. L., cited, 84
cranberry, 11, 14, 92, 302; diseases, 243, 245–46; insects, 210, 222, 226, 231; pollen of, 181
cranberry fruitworm (*Mineola vaccinii* Riley), 204, 205, 206, 210–11, 226
cranberry rootworm (*Rhabdopterus picipes* Oliv.), 199, 204, 222–23
Crater Lake, 6
Creelman, D. W., 243, 244
Croatan variety, 69, 94, 96, 97, 99, 104, 242; packaging, 310, 311(*tab.*)
cross-pollination, 11, 183–85, 322
Crowley, D. J., cited, 24, 160, 236, 247, 248
crown gall (*Agrobacterium tumefaciens*), 274–76
crown girdler (*Cryptorhynchus obliquus* Say), 221
Crumenula urceolus (Fries) de Notaris, 243
Cryptorhynchus obliquus Say, 221
Cucumis sativus L., 268

cultivation, *see* clean cultivation
Curran, C. H., cited, 215
currant fruit weevil (*Pseudenthonomus validus* Dietz), 228, 233
Cuscuta campestris Yuncker, 264
cuttings, 11, 166; hardwood, 111–22; pruning, 189–90; softwood, 111, 122–25
cutworms, 205, 229–31, 235
Cyanococcus, 14
cygon, 218

Darrow, George M., 22, 32, 45, 94–110; on diseases, 247, 248; on environment, 79, 80, 81, 82, 84; on hybridization, 47, 49, 51, 52, 53, 54, 55, 67, 68, 69, 70, 71, 72, 73, 95; on plant development, 183, 187, 198; on soil management, 137, 138, 162, 175, 178
Darrow, William, 66
Darrow variety, 68, 97, 101, 104, 267
Dasystoma salicellum Hbn., 227
Datana worm, 225
Datura stramonium L., 260, 268
Davis, H. D., Jr., 202
DDT, 205, 207, 209, 215, 221, 223, 225, 226, 227; lowbush insects and, 230, 231, 232, 233, 234
deerberry, 14, 47
dehiscence, 34, 35
dehydration, 290
Delaware, 142
Demaree, J. B., 69, 95; on plant diseases, 240, 241, 247, 248, 252–53, 254, 255, 256, 257, 260, 261, 262, 274, 276–77
Dermen, H., cited, 49, 73, 74
dewberries, 302
Dewey, D. H., cited, 300
Dexon, 258
Diabrotica duodecimpunctata Fabricius, 228
diazinon, 218
dibromochloropropane, 279
dichlone, 240
dichloropropanes, 279
dieback, 153, 268
dieldrin, 208, 210, 232
diesel oil, 142
dinitro-o-cresol, 239
dinitro-o-phenol, 239
Dinitro Weed Killer, 142

diploids, 15, 20–21, 22; hybridization and, 45, 47(*tab.*), 49, 63, 73–74, 193–94; leaf character and, 31–32

disease: bush character and, 96, 97; hybridization and, 4, 49, 52, 57, 63, 69, 97; insects and, 214–15, 219; propagation methods and, 121; pruning and, 189, 190, 323. *See also specific diseases*

distribution, 15–24

Dithane M-45, 257

diuron, 142

Dixi variety, 38, 63, 97, 104; diseases, 239, 241, 250, 253, 256, 258, 267, 276; fruit size, 101, 179

Doak, K. D., cited, 32

dodder, 142, 263–64

Doehlert, C. A., cited, 89, 90, 188, 321, 322, 323, 325, 326; on diseases, 262, 264, 265; on harvesting, 291; on insects, 208, 214; on propagation, 113, 114, 115, 117, 118, 120; on soil management, 132, 133, 134, 136, 148, 152, 153, 155, 160, 162, 163, 166, 167

dogwood borer (*Thamnosphecia scitula* Harris), 227–28

Doran, W. L., cited, 118, 123

Dothichiza caroliniana, 97, 253

Dothiorella ribis Grov. and Dugg., 261

double spot (*Dothichiza caroliniana*), 97, 253

drainage: ditches, 144, 146; roots and, 153, 258

Drake, M., cited, 152, 155

Drew, Mrs. A. M., 95

Drew, R. H., cited, 153

Driggers, B. F., 223

drought resistance, 78, 84, 90–92; berry characters and, 101; breeding for, 49, 57, 71; bush characters and, 96; clean cultivation and, 137; fertilizers and, 166; irrigation and, 143, 144; pollination and, 187; rabbiteye, 13, 22, 57, 71, 73

drying, 3, 12, 322

dryland blueberry, 6, 21, 49

Dunfee variety, 66

dusting, 202, 207, 218, 239, 250–51. *See also* spraying, *and see specific dusts*

dylox, 218

Dyrene, 253, 257

Earliblue variety, 96, 103, 104, 107; breeding experiments, 45, 58, 63, 70; diseases, 52, 239, 244, 245, 250, 258, 266, 268; harvesting, 95, 97; pollination, 181, 183, 186; pruning, 191; rooting, 111

East, The, 214, 276. *See also specific states*

Eaton, E. L., cited, 15, 81, 95, 125, 196; on soil management, 137, 162, 168, 172

Eberhardt, Joseph, 95

Eck, Paul, 3–13, 155; on botany, 14–44

ecology, 15–24, 75–93; insects and, 202, 204–205, 229. *See also specific ecological factors, e.g.,* water

economics, 302–19. *See also specific economic factors, e.g.,* costs

edaphic conditions, 75, 85–93

Edmonston, M. N., 307

Eggert, F. P., cited, 193, 195

Eggert, R., cited, 128, 130, 172

Eglitis, M., 247, 248

Elizabeth variety, 110*n*

Elliott, Arthur, 66

emasculation, 53

Emmett, H. E. G., cited, 90

environment, 75–93. *See also* ecology; *and see specific environmental factors, e.g.,* climate

epidermis, 34, 36, 37, 38, 42, 44

Ericaceae, 14

Ethel variety, 52, 72, 108, 110

ethylene dibromide, 221, 279

ethylene dichloride, 221

European lecanium (*Lecanium corni* Bouché), 207

Euvaccinium, 95, 277

Evelyn variety, 95, 277

Everett, H. F. M., cited, 243

evergreen blueberry, 6, 31, 49, 255

Everson, J. N., cited, 151

Evinger, E. L., cited, 138

Exobasidium vaccinii Wor., 251–52

fall cankerworm (*Alsophila pometaria* Harris), 228

fall webworm (*Hyphantria cunea* Drury), 224–25

farkleberry, 14

Fellers, C. R., cited, 156

Ferbam, 122, 240, 249, 251, 253, 256, 257
Fernald, M. L., cited, 14
ferric citrate crystals, 151
ferric oxide, 178
ferrous sulfate, 151, 178
fertilizers: amounts, 140, 142, 163–65, 166, 167, 321, 325; application methods, 133, 140, 163, 166–67, 325; cuttings and, 120, 121–22, 123; highbush practices, 138, 156–67; iron chelates in, 151; lowbush, 7, 165, 167, 171–74, 249; magnesium deficiency and, 152; manure, 158–60; rabbiteye, 165, 178; ratios, 162–63; seedlings and, 126; seed planting and, 128; soil reactions, 156–58. *See also specific substances*
Filmer, R. S., cited, 181, 183, 184, 185, 186, 187
fire hazards, 140, 196, 225, 324
firmness, 51, 57, 58, 97, 307
Fisher, H. E., cited, 32, 45
fish meal, 172
flavor, 3, 6, 7, 31; breeding for, 49, 52, 57, 58, 97–98; rabbiteye, 13; stone cells and, 38; temperature and, 84; variety rating, 100–10 *passim*
Flint, W. D., cited, 229
Florida, 22, 49, 57, 292; climate, 48, 52, 73, 79, 80; diseases, 240, 252, 255, 261, 264; highbush, 20–21, 22, 69; rabbiteye, 5, 11, 22, 72, 75, 84, 86, 108, 109, 178; soil, 86, 178
flower, 24, 25–27(*tab.*), 29–31; anatomy, 34–37, 181–83; leaf buds and, 113; mummy berry control and, 239, 240; pruning and, 179; removal in planting, 127; weather influences, 81, 82, 186–87. *See also* buds; pollination
forcing, 123
forest tent caterpillar (*Malacasoma disstria* Hbn.), 228
Fox, J. A., cited, 242
frames, 115–18, 123, 125, 130–31; planting in, 126–27; ventilation of, 121
Franklin, H. J., cited, 114, 115, 117, 231; on soil management, 132, 133, 140, 156, 163
Frankliniella vaccinii Morgan, 219, 231
Fraser variety, 95

freeze-drying, 289–90
freezing, 3, 305, 306; lowbush, 7, 284–88
French, A. P., cited, 32, 96
French, B. C., cited, 306–307
frost, 75, 77(*tab.*), 78, 81–83, 320; peat soil and, 88, 92
fruit, 25–27(*tab.*), 31; berry growth, 187–88; classification by characters, 97–99; development, 179–98; maturation, 84 (*See also* ripening); mineral composition, 155–56, 157(*tab.*), 172; nitrogen supplements, 162. *See also specific fruit characteristics, e.g.,* size (of fruit)
fruitworms, 202, 220. *See also specific worms*
Fulton, B. B., cited, 210–11, 239, 251, 261, 273
fumigants, 221, 279
fungicides, 7, 121, 236, 239–40, 245, 246–47, 250, 252; bacterial diseases and, 278; legal restrictions on, 253, 257
fungus diseases, 69, 73, 236, 237–61; gall formation and, 274; insect control and, 205; nematodes and, 278; pruning and, 189; root system and, 32–33; scar and, 58; striking of cuttings and, 118; ventilation and, 121. *See also specific fungi*
Fusicoccum putrefaciens, 243–45

G-3 variety, 267
G-72 variety, 267
G-80 variety, 69, 267
G-90 variety, 69
Galerucella vaccinii Fall, 233
Galletta, G, J., 66; cited, 67, 95
Galletta, S. A., 66
galls, 274–78
Garden Blue variety, 73, 109, 110
gardens, 13, 110, 320–28
Gaston, H. P., cited, 296
Gaylusaccia, 4, 14, 204
Gaylusaccia baccata (Wang.) K. Koch, 277
Gelechia trialbamaculella Cham., 226
Gem variety, 95
Georgia, 6, 79; breeding objectives in, 57, 72; diseases, 237, 240, 255; highbush, 21, 80; rabbiteye, 5, 11, 22, 75,

83, 108, 109, 110, 130–31, 178; soil, 88, 138, 142, 178
Giffin, Elspeth C., cited, 37, 268
Gilbert, F. A., cited, 67, 96, 100
Gilbert, S. G., cited, 32
glass frit, 178
Gloeocercospora leaf spot (*Gloeocercospora inconspicua*), 260
Gloeosporium fructigenum, 256, 258
Gloeosporium minus, 260
Glomerella cingulata Spaulding and von Schrenk, 256–57
Godronia cassandrae Peck, 243
Goheen, A. C., 240, 248, 262, 278
Gorham, R. P., 231
Gracilaria vacciniella Ely, 219–20, 228
grade standards, 307–308, 314–15; hand picking and, 296, 310
grafting, 264, 270, 273
Grapholitha packardi Zell., 204, 205, 206, 211–15, 226
Graptolitha bethunei Grote and Robinson, 228
Gray, A., cited, 14
Gray, Elizabeth G., cited, 243
Great Lakes, climate of, 78
Greenfield variety, 70, 95
green fruitworm (*Graptolitha bethunei* Grote and Robinson), 228
green peach aphis (*Myzus persicae* Sulzer), 228
Griggs, W. H., cited, 139
Grover variety, 51, 66, 95, 106; diseases, 253, 256
grower prices, 318
growing season, 75, 77, 78, 79, 81
growth habit, 25–27(*tab.*), 51, 70; fertilizers and, 171; fruiting, 179, 193, 198; highbush, 29, 133, 179; nutrients and, 161; root system and, 132, 133, 140
guthion, 209, 218
gypsy moth (*Porthetria dispar* Linné), 228

Haasis, F. A., cited, 258
Hagler, T. B., cited, 175
Hagood, Rev. H. H., 108
Hagood variety, 130
Hall, I. V., cited, 15, 20, 33, 44, 45, 84, 85, 92, 93, 193, 271–72, 273
hand packing, 309, 310–12

hand picking, 5, 292–96, 298; packaging and, 308, 310
hand raking, 280–81
Harbout variety, 95
hardback (*Spires latifolia* Borlch), 168
hardiness: breeding for, 48, 49, 52, 57, 68, 70, 71; cold and, 80–83, 96; variety rating, 100–10 *passim*
Harding variety, 51, 52, 66, 95, 108; diseases, 250, 253, 256; insects, 206, 219; leaves, 42
Harmer, P. M., cited, 158
Harrison, Gale, 66
harvesting, 308–309, 326; botrytis and, 248; breeding techniques and, 53; costs, 97, 281, 291, 298; highbush, 5, 291–300; lowbush, 7, 12, 280–84; soil management and, 137, 139; wild blueberries, 3, 5, 6
Haut, I. C., 66
hay, 138, 139, 140, 167; burning cover, 196
Heald, F. D., 247
heat, 48, 75, 83–84, 93; cherry fruitworm and, 211; lowbush and, 15, 194; Michigan, 78; pollination and, 187; rabbiteye and, 13, 22, 57, 71, 73; red ringspot and, 267
Heddon, S., cited, 296
Helminthosporium inaequale Shear, 261
Hemadas nubilipennis Ashm, 199, 206, 223
Hemerocampa leucostigma Smith and Abbott, 228, 234
Hemicycliophora sp., 278
Hendecaneura shawiana Kearfott, 225
Herbert variety, 104, 111; berry characters, 58, 97–98, 101; diseases, 52, 239, 250, 266, 268; productivity, 61; ripening period, 63
herbicides, 7, 142–43, 168, 169–71, 195. See also specific herbicides
heteroploids, 45, 47(*tab.*), 53
hexaploids, 15, 22, 31; hybridization and, 45, 47(*tab.*), 49
Hickman, C. J., cited, 258
highbush blueberry, 24, 29; berry growth, 187–88; development, 179–93; diploid, 15, 20–21; diseases, 242, 250, 253, 254, 255, 256, 270; ecology, 20–21, 22, 72, 75, 77, 79–80, 81, 84, 85,

86; fertilizer practices, 156–67; floral characteristics, 181–83; fruiting habit, 179; harvesting, 5, 291–300; hybridization, 3, 5, 10–11, 12–13, 45, 49, 53, 57, 61, 63–69, 70–74, 94; insects and, 199, 204–29; leaves, 31–32; nutrition, 148–56; ornamentals, 328; pollination, 179–87; propagation methods, 111–28; pruning, 188–93; root system, 32–33; scar and, 58; soil management systems, 132–47; storage, 300–301; variety characteristics. See also specific varieties

Highlands, M. E., cited, 156, 288

Hilborn, M. T., cited, 247, 248, 249, 251, 255, 256, 258–60

Hill, R. G., Jr., cited, 142, 151

Hindle, R., Jr., cited, 85, 156, 179, 187

Hitz, C. W., cited, 128, 130, 142

Hodges, M. A., cited, 156

hoeing, 239, 296

Hoerner, J. L., 212

Holley, R. W., cited, 149

Homebell variety, 73, 81, 109, 110

homoploids, 45, 47

Honey, E. E., cited, 236

hops, 324

Hough, L. F., 66; cited, 67, 95

Houser, J. R., cited, 234

Howitt, A. J., cited, 227, 228

Hruschka, H. W., cited, 300, 309, 310

huckleberry, 4, 14, 31, 92, 204, 216

Huguelet, J. E., cited, 251, 268

Huntington, H. G., quoted, 202

hurtleberries, 4

Hutchinson, M. T., cited, 210–11, 215, 225, 262; on diseases, 263, 264, 265, 266, 267, 270, 271, 273, 278

hybridization, 4, 45–51, 63–67; highbush, 3, 10–13, 20–21, 45, 49, 53, 57, 61, 63–69; highbush × lowbush, 70–71, 94–95; inheritance and, 51–53; lowbush, 21, 45, 49, 61; pollination methods and, 193–94, 195; rabbiteye, 11, 13, 49, 52, 57, 58, 61; rabbiteye × highbush crosses, 73–74

Hyland, Fay, cited, 251

Hyphantria cunea Drury, 224–25

Iaeniothrips vaccinophilus Hood, 231
Idiodonus cruentatus Panz, 273

Illinois, 257, 267, 268; market, 316, 317, 319

imports, 305

Indiana, 5, 15

Indians, 3, 7, 93

Individually Quick Frozen (IQF) units, 284–88

indolebutyric acid, 123, 125

indolepropionic acid, 123, 125

industry, 3–13. See also specific industrial aspects, e.g., marketing

inflorescence, 24–29, 53. See also buds; flowers

inheritance, 51–53. See also breeding

Ink, D. P., cited, 54

insecticides, 7, 202, 227; bee populations and, 185; ladybirds and, 205; residual, 218; stunt virus and, 264–65

insects, 199–235; disease vectors, 237, 264, 270–71, 273; pollination and, 31, 181, 185–86, 187, 193–95, 208, 210. See also specific insects

iron, 89, 148, 149, 155, 156, 273; glass frit and, 178

iron chelates, 151

iron chlorosis, 149, 150, 151

irrigation, 85, 143–47, 167

Isidro, Dennis S., cited, 306

Ivanhoe variety, 96, 98, 101, 104; diseases, 239, 250; rooting, 111

Janifer, C. S., Jr., cited, 83

Japanese beetle (*Popillia japonica* Newman), 223

Jenkins, W. R., cited, 279

Jersey variety, 106; bush characters, 96; crosses, 51, 58, 63, 101, 102, 104, 110; diseases, 52, 239, 241, 244, 245, 250, 253, 256, 257, 258, 261, 266, 267, 268, 271, 276, 277; insects and, 206, 208, 228; pollination and, 186, 187; rooting, 111, 123, 125; soil management, 153

Johnson, F., cited, 247, 248

Johnston, Stanley, 66; on breeding, 51, 52, 53, 70, 94–95; on environment, 81, 86, 89; on harvesting, 291; on home gardening, 322; on plant development, 184, 185, 188; on plant disease, 240; on propagation, 114, 115, 116, 118, 120, 121, 122, 125, 126, 130; on soil management, 137, 158, 160, 162

Judkins, W. P., cited, 147, 321, 325
June variety, 45, 51, 58, 70, 106; bush characters, 97; crosses, 95, 102, 108; diseases, 239, 241, 253, 256, 271, 278; fruit size, 101; insects, 206; pollination, 183; rooting, 111

Kalmia angustifolia L., 168, 169, 231
Kampe, D., cited, 136
Karschia lignyota (Fr.) Sacc., 261
Katharine variety, 45, 67, 95, 107, 108, 110, 250
Keifer, H. H., 219
Kelley, J. L., cited, 114, 115, 117, 133, 140, 156, 163
Kender, Walter J., 75–93, 195
Kengrape variety, 95, 125
Kenlate variety, 95
Kentucky, 21
Kenworthy, A. L., cited, 155, 156, 162, 163
Keweenaw variety, 94
Kinsman, G. B., cited, 186, 187, 194; on insects, 229, 231, 232, 234
Knight, R. J., Jr., cited, 67
Knox, E. M., cited, 34
korlan, 218
Kramer, A., cited, 138, 148, 151, 163, 175
Krohn, J., cited, 93
Kunkel, L. O., cited, 264
Kushman, L. J., cited, 161, 300, 309, 310

labor: fertilizer application, 174; harvesting, 280–81, 291, 292–93, 296, 298, 300, 310; packaging, 309, 310; pruning, 190, 191, 193; sod culture and, 137
Labrador, 15
ladybird beetles, 205
Lagerstedt, Harry, 66
Lake Michigan, 78, 81
Lake Superior, 78
lambkill (*Kalmia angustifolia* L.), 168, 169, 231
Langille, W. M., cited, 71
Larsen, R. P., cited, 162, 163
Lathrop, F. H., 215, 216, 218
Latimer, L. P., cited, 128
leaf, 25–27(*tab.*), 31–32; abnormal shapes, 273; anatomy, 42; buds, 113; chlorosis, 148–49, 150, 151–52, 175, 178, 267–68; mineral composition, 153–55, 174; scorch, 84, 148, 175; variety classification by, 94, 95–96
leafhopper, 214–15, 264, 273
leafminer, 219–21
leafroller, 202, 227
leaf rust (*Pucciniastrum myrtilli* (Schum.) Arth.), 97, 255–56
leaf tiers, 235
Lecanium corni Bouché, 207
Lecanium nigrofasciatum Pergande, 207
Lee, W. R., 193
Lepidosaphes ulmi L., 207
Leven, J. H., 296
Liebster, G., cited, 162
light, 84–85, 167
lime, 151, 152, 153, 158; hydrated, 172, 218, 250
lime sulfur, 240
List, G. M., cited, 212
Lister, R. M., cited, 268, 270
Lockhart, C. L., cited, 171, 240, 249, 251–52, 255, 258, 271–72, 278
Longley, A. E., cited, 15, 73
Longyear, B. O., cited, 236
Louisiana, 6, 22, 80
lowbush blueberry, 5, 7, 12, 20, 24; development, 193–96; diploid, 15, 21; diseases, 249, 250, 251, 253, 255, 256, 258–60; ecology, 21, 22, 75, 77, 81, 84–85, 86, 92–93; fertilizer practices, 171–74; fruiting habit, 193; harvesting, 7, 12, 280–84; hybridization, 21, 45, 49, 61, 70, 94; insects, 199, 217, 219, 229–35; leaves, 31–32; ornamentals, 328; pollination, 193–95; processing, 284–90; propagation methods, 128–30; pruning, 195–96; roots, 32–33, 44; soil management, 167–71; storage, 290–91
Ludwig, R. A., cited, 84
Lygus pratensis Linné, 228

McAllister, L. C., Jr., cited, 215
McCrum, R. C., cited, 262
McElroy, L., cited, 139
machinery: bulking, 288; burning, 196; destemming, 289; grading, 296; harvest, 12, 57, 63, 227, 281–84, 291, 296–300, 308–309; hoeing, 205; IQF units, 284–88; packaging, 309, 310–12; puff drying, 290; winnowing, 281, 300, 309, 312. *See also* tools

McKeen, W. E., cited, 243, 244, 245
Macrodactylus subspinosus Fabricius, 228
Macrophoma sp., 261
magnesium, 139, 156, 175, 273–74; fertilizers and, 161, 163; leaf chlorosis and, 148–49, 152–53
magnesium sulfate (Epsom salts), 151, 152–53
Mahlstede, J. P., cited, 33, 43
Mahn, F., cited, 93
Maine: bees in, 185, 194; climate, 48, 81, 82; diseases in, 243, 247, 249, 250, 251, 253, 254, 255, 261; harvesting in, 280; highbush, 20–21, 68, 75, 80, 81; insect control, 199, 210, 216, 217, 218, 219, 226, 229, 230, 231, 232, 233, 234, 235; lowbush, 6, 10, 20, 75, 77, 86, 92, 173, 195, 217, 219, 226, 229; processing in, 288, 306; production, 303, 304, 305; soil management, 86, 92, 173; weed control in, 169, 171
Mainland, Charles M., 111–31
Malacasoma disstria Hbn, 228
malathion, 215, 218, 220, 223, 225, 227, 232
maneb, 240, 257
manganese, 148, 149, 155–56, 172
manganese sulfate, 151
manures, 158–60, 321
Maramorosch, K., cited, 264
marketing, 6, 7, 10, 11, 97, 302–19; areas, 3, 315–17; insects and, 210, 211, 216, 218; organizations, 307, 308, 314–15; outlets, 315, 316; season, 4, 98–99, 109–10, 291–92, 306, 315; standards, 58, 307–308, 309
Markin, Florence L., cited, 247, 254
Marucci, Philip E., cited, 262, 264; on development, 179–98; on insect control, 199–235
Maryland, 49, 63, 71, 95, 99, 104; diseases, 255, 260, 261; production, 303
Mason, I. C., cited, 118, 138, 167, 168, 171, 172; on development, 185, 194, 195, 199
Massachusetts, 77, 92, 142, 280; diseases in, 240, 243, 244, 245, 246, 261, 265, 276; highbush in, 5, 63, 132, 133; insects, 210, 212, 217, 218, 228, 231, 233; production, 303, 304; propagation, 114, 115, 117, 118, 123; soil management in, 132, 133, 150, 161
Matthews, J. R., cited, 34
maturation, *see* ripening
Maxwell, C. W. B., cited, 229, 230, 231, 232, 233, 235
Meader, E. M., 66, 70; cited, 123, 183, 262
meadow sweet (*Spires latifolia* Borlch), 168
meiosis, 73
Melanospora destruens Shear, 261
Meloidogyne incognita Kofoid and White, 278
Menditoo variety, 73, 109, 110
mercury, 240
Merriam, O. A., cited, 156
Merrill, A. L., cited, 156, 158, 161
Merrill, T. A., cited, 53, 181, 183, 184, 185, 187
mesophyll, 34, 36, 37, 42
Metcalf, C. L., cited, 229
mice, 140
Michigan, 78; diseases in, 237, 239, 240, 244, 245, 251, 261, 262, 265, 267, 268, 271, 273; harvesting, 291, 292, 296; highbush in, 5, 21, 68–69, 75, 80, 81, 96, 99, 106, 111, 184, 185; insects, 210, 214, 219, 226, 227; marketing, 314, 315, 319; processing in, 288, 306; production, 303, 304, 305; propagation methods in, 114, 115, 117, 118, 120, 121, 122, 123, 130; soil management in, 86, 136, 137, 143, 151–52, 153, 155, 158, 160, 161, 162, 163
Michigan–1 variety, 61, 63
Michigan–19H variety, 61
Microsphaera penicillata var. *vaccinii* (Schw.) Cooke, 52, 236, 250–51, 265
Midwest, The, 3. *See also* specific place-names
Mikkelsen, D. S., cited, 152, 155
mildew, 63, 97. *See also* specific varieties
milkweed bug (*Oncopeltus fasciatus* Dallas), 228
Mineola vaccinii Riley, 204, 205, 206, 210–11, 226
Minnesota, 75
Minton, N. A., cited, 175
Mississippi, 5, 237, 240

Missouri, 21
mist, 123, 131
mites, 57, 63, 202, 219. *See also specific varieties*
moisture, *see* water
Monilinia vaccinii-corymbosi, 261
Monuron (CMU), 142
Moore, James N., 45–74; cited, 250, 265, 267
morphology, 24–33. *See also* anatomy
Morrow, E. B., cited, 49, 51, 53, 55, 67, 71, 72, 73, 81, 95, 100, 241
Morrow variety, 101, 107, 242
mosaic, 270–71. *See also specific mosaics*
mounding, 90, 93, 111, 125; in propagation structures, 118, 120; soil aeration and, 134
Mount Adams, 6
mountain blueberry, 6, 21–22, 24, 29
Mount Hood, 6
Mount Rainier, 6
mowing, 168–69, 170, 195
Mowry, H., cited, 84, 88, 292
muck soil, 143, 146, 155, 158, 161
mulches, 13, 86, 90; botrytis and, 249; in gardening, 324–25; highbush, 137–42; insect control and, 226, 234; lowbush, 167; potassium and, 155; rabbiteye, 175
mummy berry (*Monilinia vaccinii-corymbosi* (Reade) and *Sclerotinia vaccinii* Wor.), 237–40; botrytis and, 236, 249; resistance to, 63, 69, 97; sod culture and, 133, 137
Munsell Purple-Blue Chart (Morrow, Darrow and Rigney), 55, 101
muriate of potash, 161, 172
Murica carolinensis Mill, 168, 170, 231
Murphy, H. J., cited, 143
Murphy variety, 69, 94, 99, 107, 242; bush characters, 96, 97
mycorrhizal fungus, 32–33
Myers variety, 52, 72, 108, 109, 110, 256
Myhre, A. S., cited, 84, 88, 92, 95, 111, 142, 221, 324; on propagation, 114, 115, 117, 118, 120, 121, 122, 123, 125, 126, 128
Myrica asplenifolia L., 168–69, 170, 231
Myrica gale Linnaeus, 168
Myzus persicae Sulzer, 228

N51G variety, 277
NC-245 variety, 69
NC-683 variety, 69
Nakashima, W. Y., cited, 318
Neary, M. E., cited, 215
necrotic ringspot, 267–70, 279
Neilson, W., cited, 230, 231, 232, 233, 235
Nelson, E. K., cited, 156, 245, 251, 268
nematodes, 268, 278–79
Neopareophora litura Klug, 228, 234
New Brunswick, Canada, 193, 194, 195, 229, 243; insects, 230, 232, 235
Newcomber, E. H., cited, 15
New England, 68–69, 75–77, 142. *See also specific states*
New Hampshire, 63, 95, 193, 262, 280, 303
New Jersey, 71, 83; diseases, 237, 239, 240, 246, 247, 248, 249, 255, 256, 257–58, 260, 261, 262, 264, 265, 267, 268, 270, 271, 273, 278; harvesting in, 291, 296; highbush in, 5, 11, 21, 66, 68–69, 86, 95, 96, 99, 106, 107, 110n, 111; insect control, 199, 202, 204, 205, 207, 208, 210, 211, 212, 215, 217, 218, 219, 220, 221, 223, 225–26, 227, 235; irrigation, 146; marketing in, 307, 308, 312, 314, 315, 318, 319; pollination in, 181, 183, 185, 186; processing in, 288, 306; production, 303, 304, 305; propagation methods, 114, 115, 117, 118, 120, 127; soil management, 77–78, 86, 92, 136, 137, 142, 152, 160, 161, 162, 163; weed control, 142
New York State, 15, 21, 161, 225; highbush in, 5, 21, 68; plant diseases, 237, 257, 261, 262, 267
Nickels, C. B., 216
Nicotiana rustica L., 268
Nicotiana tabacum L. var. Samsun and White Burley, 268
nitrates, 139, 148, 149
nitrogen, 148, 149, 153, 155, 156; fertilizer combinations, 160, 161–62, 325; hardwood cuttings and, 121–22; lowbush application, 172–74; manure, 158; metabolism, 33, 88, 89, 148, 149; rabbiteye and, 175

Nocardia minima (Jensen) Waksman and Henrici, 277
Nocardia vaccinii, 276
North, The, 3, 5, 57, 96, 101, 127; diseases in, 236, 237, 240, 243, 255, 258. *See also specific states*
North Carolina, 21, 22, 71, 82, 303; breeding objectives, 57, 98–99; climate, 78–79; diseases, 240, 241, 242, 249, 252, 253, 255, 256, 257, 258, 260, 262, 265, 270, 278; harvesting in, 291, 296; highbush in, 5, 69, 75, 80(*tab.*), 96, 98, 99, 104, 107, 108, 184; insect control, 202, 204, 210, 215, 219, 223; irrigation, 146; marketing, 308, 309, 310, 314, 315, 319; production, 304, 305, 306; propagation methods, 114, 117, 118, 127; rabbiteye of, 5, 11, 109; soil management in, 136, 137, 142, 161, 162, 163
Nova Scotia, Canada, 168, 243, 244; highbush, 21, 75, 81, 95; lowbush, 193, 194, 229, 230, 231, 232, 234, 235
nutrients: cation concentrations, 163; deficiencies, 148–56, 171, 175, 321; irrigation and, 147; mineral, 148–56, 321, 325; organic, 88–89, 92, 126, 132, 133, 137; root development and, 133. *See also specific substances*

oak leaf mulches, 138–39
oats, 136
Oberea myops Hald., 199, 206, 224
Oecanthus niveus DeGreer, 228
Ohio, 5, 104, 142, 234
Olympia variety, 95
Oncopeltus fasciatus Dallas, 228
Ontario, Canada, 21
Opius ferrugineus Gahan, 218–19
Opius melleus Gahan, 218–19
Oregon, 6, 75, 79, 303; diseases, 247, 265, 277; harvesting in, 291
ornamentals, 320, 328
O'Rourke, F. L., cited, 113, 114, 125, 129
ovary, 14, 31, 35–37, 187; frost damage, 81, 82
Owens variety, 256
Oxycoccus, 14
oxygen, 90

oyster shell scale (*Lepidosaphes ulmi* L.), 207
Ozark Mountains, 21

Pacific Northwest, *see* West, The; *and see specific placenames*
Pacific variety, 95
packaging, 307–12; canned, 288–89; frozen, 288; storage, 290–91
Paleacrita vernata Peck, 228
Pandemis moth (*Pandemis limitata* Robinson), 228
paradichlorobenzene (PDB), 221
parathion, 208, 209, 215, 220, 226, 232
parenchyma, 42, 43
Paria cannella Fabricius, 228
Paulus, A. O., cited, 258
peat, 92, 93, 151; botrytis and, 248–49; mulches, 138, 139, 167, 324; nitrogen and, 33, 88; planting in, 126, 128, 160, 321; as rooting medium, 115, 118, 120, 122, 123, 126, 132, 148; soil pH and, 158, 321
Pecan variety, 80
Pelletier, E. N., cited, 247, 248, 249
Pemberton variety, 48, 96, 107, 179; diseases, 239, 241, 244, 245, 250, 258, 266, 267, 268; insects, 228; pruning, 191; rooting, 111; scar, 97, 101; soil management, 153
Pennsylvania, 5, 15, 21, 254, 262
pentaploids, 73–74
periwinkle (*Vinca rosea* L.), 263–64
Perkins, Frederick A., 302–19, 309
Perlite, 123, 131
Perlmutter, F., cited, 84
Peronea minuta Robinson, 228
Persing, D. P., cited, 318
Pestalozzia quepini vaccinii Shear, 261
Peterson, W. H., cited, 156
Petunia hybrida Vilm. var. Fire Chief, 268
Pezizella lythri-(Desm.) Shear and Dodge, 261
Phaseolus vulgaris L. var. Pinto, 268
phenyl mercury, 245; acetate, 257
Phipps, C. R., cited, 194; on insects, 199, 226, 229, 230, 231, 232, 233, 234
phloem, 34, 42, 43
Phomopsis, gall formation and, 274

Phomopsis twig and cane blight, 245–47
phosphate rock, 160
phosphoric acid, 172
phosphorus, 122, 139, 148, 155, 156, 162; lowbush productivity and, 172, 173; rabbiteye, 175
Phyllophaga sp., 223
Phyllostictina leaf and berry spot (*Phyllostictina vaccinii*), 260
Phytophthora cinnamoni Rands, 258
Phytophtora root rot, 257–58
Pickett, A. D., 215, 231, 232, 233
Pieris mariana L., 225
pine needle mulch, 324
Pioneer variety: characters, 51, 95, 96, 97, 107, 179; crosses, 45, 52, 58, 63, 66, 101, 102, 104; diseases, 241, 244, 245, 250, 251, 253, 256, 266, 276, 277; insects, 206, 219, 225; pollination, 185; rooting, 111, 115, 125; soil management and, 139, 153
pistil, 31, 35, 181, 185, 195; frost damage, 81, 82
placenta, 36, 37
planting, 126–28, 131; fertilization and, 167, 321; pruning and, 190–91; sites, 83, 88, 92–93, 320, 321–22; solid and mixed block, 183–85, 206; spacing, 323; transplanting, 54–55, 130, 322. *See also* breeding; propagation
plastic mulch, 324–25
Platynota flavedana Clem., 227
Pleospora obtusa (Fckl.) Hoshn., 261
plum curculio (*Conotrachelus nenuphar*), 204, 205, 206, 208–10, 233
Polia purpurissata Grt., 230, 231
pollen, 35, 181, 183, 187, 193–94
pollination, 179–87, 193–95, 208, 210, 227; breeding techniques, 53; cross-pollination *vs.* self-pollination, 183–85, 322; flower morphology and, 31, 181, 183
polyploids, 15
Popenoe, J., cited, 152
Popillia japonica Newman, 223
Porter, B. A., 216
Porthetria dispar Linné, 228
potassium, 122, 139, 148, 149, 153, 156, 162; fertilizer amounts and, 163; lowbush productivity and, 172, 173; muck soils and, 155, 161; rabbiteye, 175

potassium chloride, 160, 161, 178
potassium indolebutyrate, 123
potassium sulfate, 160, 172
powdery mildew (*Microsphaera penicillata* var. *vaccinii* (Schw.) Cooke), 52, 236, 250–51, 265
precipitation, 75, 76(*tab.*), 78–79, 173, 186–87, 249; irrigation and, 85, 143, 144, 146; snow, 81, 95
Premerge, 143
prices, 316, 318–19
Pristiphora idiota Nort., 234
processing, 3, 5, 12, 284–90, 300; method percentages, 305–306
productivity, 3, 96, 100–10 *passim*, 137; aeration and, 89; average yields, 291, 326; fertilizers and, 156–67, 172; frost damage and, 82; inheritance and, 51, 57, 58, 61; insect damage, 199; irrigation and, 143; light and, 167; mulches and, 138, 139, 140; national production value, 302–303; nutrition and, 148, 153, 156; planting and, 127–28; pollination and, 183–85, 186, 187, 193–94; pruning and, 179, 188, 190, 191, 195–96, 323; sales and, 314–17; soil moisture, 90; stem length and, 171
propagation methods, 11, 12, 15; budding and, 125; cuttings, hardwood, 111–22, 129–30; cuttings, softwood, 122–25, 130; lowbush, 128–30; media, 115, 116, 117, 118, 121, 122, 123, 130–31; mounding, 90, 93, 111, 118, 120, 125, 134; planting, 126–28, 131; rabbiteye, 130–31; seed, 126, 128; striking methods, 118, 120; structures, 115–18, 121, 123, 125, 126–27, 130–31
pruning: berry characters and, 101; diseases and, 246, 249, 276, 323; facility, 96, 97, 139; highbush, 179, 188–93; insect control and, 206, 207, 223; light and, 85; lowbush, 195–96; rabbiteye, 198
Pseudomonas syringae Van Hall., 277–78
Psyllids (*Psylla* spp.), 228
Pteris sp. Linnaeus, 168
Pucciniastrum goeppertianum (Kühn) Kleb., 251, 254–55
Pucciniastrum myrtilli (Schum) Arth., 97, 255–56
puff-drying, 290

Puget Sound, 6, 24, 84, 121
Putnam scale (*Aspidiotus ancylus* Putnam), 204, 206, 207

quack grass, 139
Québec, Canada, 243, 280

rabbiteye blueberry, 5–6; development, 198; diseases, 243; ecology, 22, 71–72, 75, 79, 80, 82–83, 84, 86, 88; fertilizer practices, 178; harvesting, 291–300; hybridization, 11, 13, 49, 52, 57, 58, 61, 70, 71–74, 94; insects, 202; propagation methods, 130–31; soil management for, 175; storage, 300–301; variety characteristics, 108–10. *See also specific varieties*
Race, S. R., cited, 278
raceme, 24
radiation frosts, 83
Raine, J., cited, 227
rainfall: botrytis and, 249; irrigation and, 143, 144, 146; nitrogen and, 173; pollination and, 186–87
raking, 280–81
Ramularia effusa Pk., 261
Ramularia vaccinii Pk., 261
Rancocas variety, 48, 70, 107; characters, 96, 97; cultivation depth, 134; diseases, 239, 241, 244, 250, 253, 256, 257, 263, 266, 268, 271, 278; hybridization and, 95, 104; insects, 206, 210, 219, 225; leaves, 42, 94; magnesium, 152; pollination, 183; rooting, 111
Raniere, L. C., cited, 249, 257, 258, 265, 266, 267–68, 270
raspberry, 302
raspberry ringspot, 270
rays, 43
Reade, J. M., cited, 236
red-banded leafroller (*Argyrotaenia velutinana* Walker), 227
red leaf disease (*Exobasidium vaccinii* Wor.), 251–52
red ringspot disease (RRS), 265–67
Redskin variety, 70, 95
red spider (*Tetranychus telarius* Linné), 228
red-striped fireworm (*Gelechia trialbamaculella* Cham.), 226
Reed, J. P., cited, 278

rest requirement, 13, 22, 109; breeding and, 53, 57, 69–70, 72, 73; bud hardiness and, 96–97; frost incidence and, 79–80; photoperiod, 84–85
retail prices, 319
Rhabdopterus picipes Oliv., 199, 204, 222–23
Rhabdospora oxycocci Shear, 261
Rhagoletis mendax, see blueberry maggot
Rhagoletis pomonella Walsh, 215
Rhizoctonia ramicola, W. and R., 261
rhizome, 33, 44, 93, 111; buds, 193; nitrogen and, 173, 174; polarity, 128–29
Rhode Island, 117, 121, 139, 162
Rhyncagrotis anchocelioides Grn., 226
Rhytisma vaccinii (Schw.), 261
Rich, A. E., cited, 262
Rigney, J. A., cited, 55
ripening: breeding for speed of, 49, 51, 52, 57, 61, 63, 68, 69, 70, 71, 73; light and, 97; markets and, 98–99, 109–10; pollination and, 183, 184; temperature and, 84; variety rating, 85, 100–10 *passim*
Robbins, W. W., cited, 24
Roberts, A. N., cited, 163
Roberts, D. A., cited, 261
roguing, 121, 243, 264–65, 267, 273
Rollins, H. A., cited, 139
root-gall disease, 276
rooting, *see* propagation
root-promoting substances, 114, 115, 118, 123–25, 128–29, 148
root rot (*Armillaria mellea* Bahl), 261
roots, 32–33, 43–44; fertilizing and, 166; heat and, 83–84; mounding and, 90; mulches and, 138, 140; nematodes and, 278; organic matter and, 89, 132; rabbiteye, 175; rhizomes and, 128–29; soil properties and, 132, 133, 153
Rose, Sayre, 66
rose chafer (*Macrodactylus subspinosus* Fabricius), 228
rotenone, 218
rots, 97, 122. *See also specific diseases*
Rousi, A., cited, 48
Royle, D. J., 258
Rubel variety, 38, 96, 107; crosses, 51, 52, 66, 95, 102, 103, 108; diseases, 239, 241, 244, 245, 250, 256, 267, 276, 277; magnesium, 155; pollination, 183, 184,

210; rooting, 111, 115, 123, 125; soil management for, 137, 158, 161
Russell variety, crosses, 51, 63, 70, 71, 95, 106, 107

salt water, 146
Sam variety, 66, 95, 107, 253, 256
sand: irrigation and, 143, 147; mineral fertilizers and, 160, 161; mulches, 137, 138; nutrients, 148, 149, 152–53, 158, 321; organic matter in, 86, 89, 137; plowing, 93; as rooting medium, 118, 120, 122, 126
Sapp, M. A., cited, 72
Sapp, W. B., cited, 108
Satilla River, 22
Savage, E. F., cited, 138
Saville, D. B. O., cited, 244, 251
sawdust: mulches, 138–39, 140, 142, 167, 249, 324; propagation beds and, 117, 321
scale insects, 206–208
Scammell, H. B., 66
Scammell variety, 97, 107, 110, 111; diseases, 239, 241, 277; insects, 206, 219
Scaphytopius magdalensis Prov., 204, 205, 264
Scaphytopius verecundus Van Duz., 215, 264
scar, 100–10 *passim;* breeding and, 49, 51, 55, 57, 58, 73, 97
scarab beetle (*Serica cucullata* Dawson), 235
Scarabeid root grubs, 223
Schaefers, G. A., cited, 225
Schallock, D. A., cited, 142
Schrader, A. L., cited, 138, 148, 151, 163, 175
Schwartze, C. D., 66; cited, 84, 88, 92, 95, 111, 142, 221, 324; on propagation, 114, 115, 117, 118, 120, 121, 122, 123, 125, 126, 128
Sclerotinia sclerotiorum (Lib.) D By., 261
Sclerotinia vaccinii Wor., 237–40
Scott, Donald H., 94–110; cited, 49, 52, 53, 54, 57, 67, 68, 71, 72, 73, 74, 95, 100, 250
Seaver, F. J., cited, 243

seed, 3, 38, 111, 126, 128; harvest for breeding, 53–54; huckleberry, 4, 14, 31
Semesan, 115
sepals, 34, 36
septa, 42
Septoria albopunctata Cke., 252–53, 258
Septoria difformis Cke., 261
Septoria leaf spot (*Septoria albopunctata* Cke.), 252–53, 258
Septoria vaccinii E. and E., 261
Serica cucullata Dawson, 235
70-30 gypsum-cryolite dust, 233
shade, 15, 167, 320
Shaffer, J. D., cited, 307
Sharpe, R. H., 66
Sharpe, Ralph, cited, 52, 69, 70, 72
sharp-nosed leafhopper (*Scaphytopius magdalensis* Prov.), 204, 205, 264
Shaw, Carlton, 66
Shear, C. L., cited, 243
shipping, *see* transportation
Shirley variety, 95, 96
Shive, J. W., cited, 148
Shoemaker, J. S., cited, 137, 140, 293
shoestring disease, 271–73
Shutak, Vladimir G., 179–98; cited, 85, 139, 140, 156, 162, 179, 187
Simanton, F. L., cited, 207
simazine, 142
Sinox, 143
size (of fruit): breeding for, 49, 51, 52, 57, 58, 72, 73, 95, 97; market season and, 292; measurement, 55, 308; mulches, 140; pollination and, 183, 184, 186, 187, 193–94; pruning and, 188, 191, 323; variety rating, 100–10 *passim*
Slate, G. L., cited, 80, 132, 138, 161, 322, 323, 325, 326
Smith, K. M., cited, 268
Smith, N. R., cited, 274, 276–77
Smith, W. W., cited, 70, 128, 167, 168, 172
Snapp, O. I., 208
snow, 81, 95
snowy tree cricket (*Oecanthus niveus* DeGreer), 228
sod culture, 137, 223
sodium nitrate, 122, 151, 160, 325

Index

soil, 320–21; aeration, 89–90; CEC, 148–52, 163; cranberry rootworm and, 222; erosion, 138, 146, 147, 196; fertilizer reactions, 156–58; highbush and, 3, 10, 11, 13, 21, 22, 85, 86, 90, 132–47, 158, 163; lowbush, 15, 21, 22, 86, 92–93, 167–71; moisture, 90–92, 93, 137, 138, 139, 140, 143–47, 153; nematodes, 279; organic matter, 88–89, 92, 126, 144, 150, 156, 158, 160–61, 324; peaty, 33, 88, 248–49; pH of, 3, 10, 22, 63, 90, 137, 138, 139, 151, 158, 159(tab.), 162–63, 267, 274, 321, 325; plant size and, 128; for rabbiteye, 6, 13, 22, 86, 175; structure, 86–88, 89, 90. *See also* muck soils; sand
solar frame, 115–16
Sooy variety, 66, 95, 107
South, The, 3, 49, 57, 72, 73; diseases in, 236, 240, 241, 243, 255. *See also specific states*
South Carolina, 5, 22, 88
southern gooseberry, 14
Spaelotis clandestina Harr., 230
span worms, 235
Sparganothis fruitworm (*Sparganothis sulfureana* Clemens), 226
sparkleberry, 14
Spayd, George, 66
Sphaeropsis malorum Pk., 261
sphagnum moss, 54, 114, 115, 126, 128
Spires latifolia Borlch, 168
spraying: bait spraying, 218; bee populations and, 185; boron, 153; ferrous sulfate, 178; fungicide, 240; insecticide, 202, 205, 206, 207, 215, 225; oil, 207–208, 219; pollination and, 185–86, 187, 208, 210; pruning and, 206; for stunt disease, 215; 2,4-D, 169–70; weed killers, 255. *See also specific products*
spring cankerworm (*Palaecrita vernata* Peck), 228
Sproston, T., cited, 240
Stace-Smith, R., cited, 277
staggerbush (*Pieris mariana* L.), 225
stamens, 31, 34, 181, 187
Stanley variety, 51, 97, 108; crosses, 58, 101, 102, 103, 104, 107, 108; diseases, 52, 239, 241, 244, 250, 256, 266, 267, 268, 270; insects, 228; pollination, 183, 186, 187; rooting, 111, 123; soil management, 153
Steere, R. L., cited, 270
stele, 44
stem: anatomy, 42–43, 44; productivity and length of, 171, 173, 174; soil organic matter and, 89
stem and leaf fleck (*Gloeosporium minus*), 260
stem borers, 202
stem canker (*Botryosphaeria corticis*), 69, 97, 236, 240–43, 249, 260
Stene, A. E., cited, 117, 121
sterility, 45, 70, 73, 110; pollination methods and, 193–94
Stevens, N. E., cited, 38, 245, 246, 247, 262
stigma, 36, 181, 187
Stiles, Warren C., 320–28; on harvesting, processing, and storage, 280–301
stomates, 34, 38, 42
storage, 288, 290–91, 300–301; botrytis and, 248; capping films and, 309–10; freeze drying and, 290; rhizome, 129; seed, 54; whips, 114
Strasseria oxycocci Shear, 261
strawberries, 221, 302, 324
strawberry leafroller (*Ancylus comptana* Froelich), 228
strawberry rootworm (*Paria cannella* Fabricius), 228
straw mulches, 138–39, 140, 324
Stretch, Allan W., cited, 256, 265, 267, 312; on diseases, 236–79
Struchtemeyer, R. A., cited, 167
stunt disease, 97, 214–15, 236, 262–65
style, 36, 37, 183, 187
Sudan grass, 136, 138
Suit, R. F., cited, 264
sulfur, 148, 149, 158, 159(tab.), 172, 175; mulches and, 321–22; powdery mildew and, 251; wettable, 240
Sul-Po-Mag, 161
superphosphate, 160–61
supplies, *see* tools and supplies
Suwannee River, 22
sweet fern (*Myrica asplenifolia* L.), 168–69, 170, 231
sweet gale (*Myrica gale* Linnaeus), 168
Swift, F. C., cited, 202

tankage, 121
tannic acid, 158
tarnished plant bug (*Lygus pratensis* Linné), 228
taxonomy, *see* classification
Taylor, J., cited, 241, 249–50, 260
temperature: cooling process, 312–13; soil, 138, 139, 140; storage, 290, 291, 300–301, 326; transport, 313. *See also* climate; cold; heat
Tennessee, 21, 22, 81
tensiometers, 144
termites, 223
terrapin scale (*Lecanium nigrofasciatum* Pergande), 207
Tetranychus telarius Linné, 228
tetraploids, 15, 20, 24, 31, 32; hybridization and, 45, 47(*tab.*), 49, 57, 63, 70, 73, 193–94
Tetylenchus christiei, 278
Tetylenchus joctus Thorne, 278
Texas, 22
Thamnosphecia scitula Harris, 227–28
Thelephora (*Thelephora terrestris*), 260–61
Thiodan, 219
1316-A variety, 183, 186, 187
thrips (*Frankliniella vaccinii* Morgan), 219, 231
Thyridopteryx ephemeraeformis Haworth, 228
Tifblue variety, 73, 81, 109, 110
tipping, 189, 190, 191
tobacco rattle viruses, 270
tobacco ringspot, 268, 270, 279
tomato blackring, 270
tomato ringspot, 270
Tomlinson, W. E., Jr., cited, 95, 208, 214, 218, 219, 222–23, 233; on diseases, 247, 248, 262, 264
tools and supplies: for cutting, 114–15; frame materials, 115, 116–17; gardening, 321; hand picking, 293, 296; pruning, 189, 190, 191, 193, 276; raking, 281; screens, 126; soil moisture measurement, 144, 146; spray, 170; tillage, 134–36. *See also* machinery
trademarks, 307, 308
transplanting, 54–55, 130, 322

transportation, 3, 309; berry firmness and, 58; berry scar and, 97; frozen berries, 288; fresh berries, 313
trees, 92, 93, 168, 169, 170; gardening, 320; witches'-broom and, 254–55
Trevett, M. F., cited, 33, 89, 143, 167, 168, 169, 170, 173, 174, 193
Trialeurodes sp., 228
Trichodorus sp., 278
Twearley variety, 95
Tweedie, H. C., 243, 244
twig blight, 245–49
2,4,5-T, 169, 170, 252
2,4-D, 169–70, 252, 265

United States: blueberry consumption in, 306–307; blueberry industry in, 3–13; federal marketing standards, 307–308; small fruits production value in, 302, 304–305. *See also* specific states and placenames
United States Census of Agriculture for 1949, 1954, and 1959, 302, 303, 304
United States Department of Agriculture, 307; cooperative breeding program, 65(*tab.*), 66(*tab.*), 68–70, 72, 95, 108–109
United States Department of Agriculture, Agricultural Marketing Service, 315
United States Department of Agriculture, Bureau of Plant Industry: breeding, 4, 10–11, 63–67; size classifications, 308
United States Weather Bureau, 144
Uschdraweit, H. A., cited, 273
US-1 variety, 63

Vaccinioideae, 14
Vaccinium, 4–5, 14–24, 204; morphology, 25–27(*tab.*)
Vaccinium alto-montanum Ashe, 15, 21, 22, 24, 26(*tab.*), 38, 49, 263
V. amoenum Aiton, 15, 22, 25(*tab.*), 32, 263
V. angustifolium Aiton (*V. lamarckii* Camp), 5, 15, 20, 24, 25(*tab.*), 26(*tab.*), 38, 49, 125; diseases and, 248, 261, 263; growth habit, 29; hybridization, 45, 94; hybrids, 63, 68, 70; leaves, 31, 32; pollination, 193–94, 195; shoestring disease, 272
V. arboreum Marsh, 252

V. arkansanum Ashe, 22, 24, 25(*tab.*)
V. ashei Reade, 5–6, 15, 22, 25(*tab.*), 31, 32, 38, 81, 108, 125; heat resistance, 71–72, 73; hybridization and, 45, 49, 51, 58, 61, 63, 70
V. atrococcum Heller, 15, 20–21, 24, 25(*tab.*), 63, 113, 125, 261, 262
V. australe Small, 5, 15, 21, 24, 25(*tab.*), 32, 38, 94, 261; hybridization and, 45, 52, 69, 72, 73, 94; root-inducing substances, 125
V. australe × *V. lamarckii*, 278
V. brittonii Porter, 5, 15, 20, 21, 24, 25(*tab.*), 32
V. caesariense Mackenzie, 15, 20–21, 24, 25(*tab.*), 31, 32, 263
V. canadense, 15
V. constablaei Gray, 15, 22, 25(*tab.*)
V. corymbosum Lamarck, 15, 262
V. corymbosum Linnaeus, 5, 21, 24, 26(*tab.*), 38, 94; flowers, 34–37; hybrids, 45, 52, 63, 68, 73, 94; leaves, 31, 32, 42; stem, 42–43
V. darrowi Camp, 15, 22, 24 26(*tab.*), 31, 32, 48, 263; hybrids, 49, 52, 69, 70
V. elliottii Chapman, 15, 22, 24, 26(*tab.*), 263
V. fuscatum Aiton, 15, 24, 26(*tab.*), 31, 32
V. hirsutum Buckley, 15, 21, 26(*tab.*), 32, 42
V. irrigatum Aiton, 261
V. lamarckii Camp, 20, 21, 24; hybrids, 45, 48, 51, 70. See *Vaccinium augustifolium* Aiton
V. marianum Watson, 15, 21, 24, 26(*tab.*)
V. melanocarpum Mohr, 47
V. membranaceum Douglas, 6, 26(*tab.*), 29
V. myrsinites Lamarck, 15, 22, 27(*tab.*), 31, 32, 48, 49; hybrids, 52, 53, 69, 73
V. myrtilloides Michaux, 15, 24, 27(*tab.*), 32, 47; botrytis and, 248; pollination, 193–94, 195; stunt, 262
V. myrtillus Linnaeus, 273
V. ovatum Pursh, 6, 24, 27(*tab.*), 31
V. pallidum Aiton, 6, 15, 21, 24, 27(*tab.*), 125
V. pennsylvanicum, 261
V. simulatum Small, 15, 21, 22, 24, 27(*tab.*), 32

V. stamineum Linnaeus, 47, 262
V. tenellum Aiton, 15, 22, 27(*tab.*), 32, 38, 49, 263
V. uliginosum Linnaeus, 48
V. vacillans Torrey, 15, 21, 24, 27(*tab.*), 129, 138, 261, 262
V. virgatum Aiton, 15, 22, 27(*tab.*), 32
variegation, 270, 273
Varney, Eugene H., 236–79
Vaughan, E. K., cited, 277, 278
Veenstra, H. A., cited, 251
Vermont, 162
VHPF, 121
Vigna sinensis (Torner) Savi, 268
Vinca rosea, L., 263–64
Virginia, 21, 22, 262
virus diseases, 236, 261–74; insect vectors, 219; propagation methods and, 113; resistance, 63, 97
vitamin C, 156

Walker, W. M., cited, 108
Walker variety, 52, 108
Wareham variety, 51, 94, 97, 101, 104, 108; diseases, 250, 276; rooting, 123
warehouse storage, 291
Washington State, 6, 79, 84, 318; diseases, 247, 248, 254, 265, 276, 277; harvesting in, 291; highbush in, 5, 75, 95, 111; insects, 202, 221; production, 303, 304, 305, 306; propagation methods, 114, 115, 117–18, 120, 121, 122, 123, 126, 128; soil, 92, 153, 160; weed control, 142
Washington variety, 95
wasp parasite (*Trichogramma minutum* Riley), 205
Wasscher, J., cited, 34, 42
water, 3, 33, 85, 90, 92; heat and, 84; mists, 123, 131; mulches and, 137; propagation methods and, 118, 120, 121, 123, 125, 322; soil drainage, 86, 88, 89, 90–92, 93, 116; soil organic matter and, 88–89; sources, 146–47. See also irrigation; precipitation
Watson, D. P., cited, 33, 43, 300
Watt, B. K., cited, 156
weather, *see* climate
Webber, E. R., cited, 122
Weber, G. F., cited, 261

weeds, 7, 93, 168–71, 175; chemical control, 142–43, 168, 169–71; insects and, 210, 211, 231; mulches and, 137–38, 140; production and, 85, 132–33, 172; tillage and, 134. *See also specific plants*
Weeks, E. E., cited, 318
Weiss, H. B., cited, 207, 222, 226
West, The, 3, 24, 79, 214, 215; diseases, 237, 247, 262. *See also specific states*
West Virginia, 6, 15, 21, 75, 262
Weymouth variety, 97, 186, 187; crosses, 45, 63, 70, 94, 101, 103, 104, 107, 108; diseases, 239, 240, 241, 250, 253, 256, 258, 266, 267, 268, 277; insects, 206, 219, 228
whinberries, 4
whips, 113–14
White, Elizabeth C., 10–11, 66, 95
White, R. G., cited, 196
White, R. P., cited, 256
white fly (*Trialeurodes* sp.), 228
white-marked tussock moth (*Hemerocampa leucostigma* Smith and Abbott), 228, 234
white root grubs, 223
Whitton, L., 52, 71, 195
wholesale prices, 318–19
whortleberries, 4, 14
Wilcox, Marguerite S., cited, 240, 241, 245–46, 252–53, 254, 255, 256, 257, 260, 261
Wilcox, R. B., cited, 22, 246, 262, 264
wild roe (*Rosa blanda* Aiton), 168
wild spirea (*Spires latifolia* Borlch), 168
Windom, Tom, 95
winnowing, 281, 289, 300, 312
Witcher, W., 249
witches'-broom (*Pucciniastrum goeppertianum* (Kühn) Kleb.), 251, 254–55
witches'-broom rust, 52, 63

witches'-broom virus, 273
W-marked cutworm (*Spaelotis clandestina* Harr.), 230
Wolcott variety, 69, 94, 108, 184; bush characters, 96, 97; chlorides and, 161; diseases, 242, 258; nitrogen and, 162; ripening period, 99
Wood, G. W., cited, 193, 194, 195; on insects, 218, 230, 231, 232, 233, 235
Wood, Mary, cited, 326
Woodard, O. J., cited, 49, 52, 53, 67, 72, 73, 81
Woodard variety, 73, 81, 109, 110
wood ashes, 172
Woodbridge, C. G., cited, 153
Woods, W. C., cited, 229, 233
Wooley, P. H., cited, 277
Woronin, M., cited, 236
wrapping films, 309–10
Wright, Carleton, cited, 326
Wynd, F. L., cited, 178

Xiphinema americanum Cobb, 268, 279
xylem, 42, 43

Yarbrough, J. A., cited, 38
Yeager, A. F., cited, 70, 172
yellow-headed fireworm (*Peronea minuta* Robinson), 228
Yellow River, 22
yield, *see* productivity
Young, R. S., cited, 143, 187

Zentmeyer, G. A., cited, 258, 261
zinc, 155, 174, 240
zinc sulfate, 151
zineb, 240, 249
ziram, 240, 257
Zuckerman, B. M., cited, 244, 245, 276, 278